Molecular Biology Intelligence Unit

Shh and Gli Signalling and Development

Carolyn E. Fisher, B.Sc. Hons., Ph.D.
Immunobiology Group
MRC/UoE Centre for Inflammation Research
The Queen's Medical Research Institute
Edinburgh, U.K.

Sarah E.M. Howie, B.Sc. Hons., Ph.D.
Immunobiology Group
MRC/UoE Centre for Inflammation Research
The Queen's Medical Research Institute
Edinburgh, U.K.

Landes Bioscience / Eurekah.com
Georgetown, Texas
U.S.A.

Springer Science+Business Media
New York, New York
U.S.A.

SHH AND GLI SIGNALLING AND DEVELOPMENT

Molecular Biology Intelligence Unit

Landes Bioscience / Eurekah.com
Springer Science+Business Media, LLC

ISBN: 0-387-39956-9   Printed on acid-free paper.

Copyright ©2006 Landes Bioscience and Springer Science+Business Media, LLC

All rights reserved. This work may not be translated or copied in whole or in part without the written permission of the publisher, except for brief excerpts in connection with reviews or scholarly analysis. Use in connection with any form of information storage and retrieval, electronic adaptation, computer software, or by similar or dissimilar methodology now known or hereafter developed is forbidden.
The use in the publication of trade names, trademarks, service marks and similar terms even if they are not identified as such, is not to be taken as an expression of opinion as to whether or not they are subject to proprietary rights.
While the authors, editors and publisher believe that drug selection and dosage and the specifications and usage of equipment and devices, as set forth in this book, are in accord with current recommendations and practice at the time of publication, they make no warranty, expressed or implied, with respect to material described in this book. In view of the ongoing research, equipment development, changes in governmental regulations and the rapid accumulation of information relating to the biomedical sciences, the reader is urged to carefully review and evaluate the information provided herein.

Springer Science+Business Media, LLC, 233 Spring Street, New York, New York 10013, U.S.A.
http://www.springer.com

Please address all inquiries to the Publishers:
Landes Bioscience / Eurekah.com, 810 South Church Street, Georgetown, Texas 78626, U.S.A.
Phone: 512/ 863 7762; FAX: 512/ 863 0081
http://www.eurekah.com
http://www.landesbioscience.com

Printed in the United States of America.

9 8 7 6 5 4 3 2 1

Library of Congress Cataloging-in-Publication Data

Shh and Gli signalling and development / [edited by] Carolyn E. Fisher,
    Sarah E.M. Howie.
      p. ; cm. -- (Molecular biology intelligence unit)
    Includes bibliographical references and index.
    ISBN 0-387-39956-9 (alk. paper)
    1. Morphogenesis--Molecular aspects. 2. Transcription factors.
  3. Cellular signal transduction. 4. Developmental genetics.   I. Fisher,
  Carolyn E. II. Howie, Sarah E. M. III. Series: Molecular biology
  intelligence unit (Unnumbered)
    [DNLM: 1. Embryonic Induction--physiology. 2. Organogenesis
  --physiology. 3. Signal Transduction--physiology. 4. Transcription
  Factors--genetics. QU 375 S554 2006]
  QH491.S47 2006
  571'.833--dc22

2006025296

CONTENTS

Preface xi

1. **Introduction** 1
   Carolyn E. Fisher and Sarah E.M. Howie
   The Concept of Developmental Biology... 1
   Introduction to Morphogens: Shh 2
   The Hh Pathway in *Drosophila* 2
   The Shh Pathway in Vertebrates 4
   Gli Transcription Factors 5
   Roles for Shh in Vertebrates 6
   Clinical Aspects 8
   Final Thoughts 8

2. **Sonic Hedgehog Signalling in Dorsal Midline
   and Neural Development** 12
   Silvia L. López and Andrés E. Carrasco
   The Hedgehog Pathway 12
   Shh Signalling and Floor Plate Formation 13
   Shh in Neural Development 17
   Closing the Idea 19

3. **Role of Hedgehog and Gli Signalling in Telencephalic
   Development** 23
   Paulette A. Zaki, Ben Martynoga and David J. Price
   The Hedgehog Signalling Pathway 23
   Overview of Telencephalic Development 24
   Role of Shh in Telencephalic Dorsoventral Patterning 25
   Role of Shh in Cell Death and Proliferation 28
   Role of Shh in Cell Type Specification 29
   Telencephalic Phenotypes of Gli Mutants 29
   Role of Gli3 in Cell Death 30
   Loss of Gli3 Partially Rescues *Shh*[-/-] Telencephalic Phenotypes 30

4. **Role of Shh and Gli Signalling in Oligodendroglial Development** 36
   Min Tan, Yingchuan Qi and Mengsheng Qiu
   Early Oligodendrocyte Precursors Originate from the *Olig1/2+*
     Ventral Neuroepithelium and Share the Same Lineage
     with Motor Neurons 36
   Ventral Oligodendrogenesis Is Induced by Sonic
     Hedgehog Signalling 37
   A Shh-Independent Pathway for Oligodendrogenesis
     in the Developing Spinal Cord 37
   Differential Roles of *Gli* Genes in Ventral Oligodendrogenesis 39

5. **The Role of Sonic Hedgehog Signalling in Craniofacial Development** 44
   Dwight Cordero, Minal Tapadia and Jill A. Helms
   Overview of the Anatomy of Craniofacial Development 44
   Sonic Hedgehog in Development of the Upper Face 46
   Sonic Hedgehog in the Development of Lower Facial Structures 51
   Human Craniofacial Disorders 52
   Importance and Future Directions 54

6. **Multiple Roles for Hedgehog Signalling in Zebrafish Eye Development** 58
   Deborah L. Stenkamp
   Hedgehog Signalling and Photoreceptor Differentiation 59
   Hedgehog Signalling and Ganglion Cell Differentiation 62
   Hedgehog Signalling and Retinal Neurogenesis 63

7. **Sonic Hedgehog Signalling during Tooth Morphogenesis** 69
   Martyn T. Cobourne, Isabelle Miletich and Paul T. Sharpe
   Early Generation of the Tooth 70
   The Shh Pathway Is Active in the Developing Tooth 70
   Long and Short Range Shh Signalling in the Tooth 72
   Shh Interacts with Multiple Gene Families in the First Branchial Arch 72
   The Functional Significance of Shh during Development of the Tooth 75

8. **Limb Pattern Formation: Upstream and Downstream of Shh Signalling** 79
   Aimée Zuniga and Antonella Galli
   Limb Bud Initiation 82
   Early Limb Bud Polarisation and Establishment of the ZPA 82
   Molecular Mechanisms of the Shh Response 84
   Maintenance of the ZPA by Signal Relay 88
   Digit Patterning by Shh: The End of the Spatial Gradient Model? 88

9. **Sonic Hedgehog Signalling in the Developing and Regenerating Fins of Zebrafish** 93
   Fabien Avaron, Amanda Smith and Marie-Andrée Akimenko
   The Zebrafish Hedgehog Genes 93
   Overview of the Zebrafish Pectoral Fin Bud Development 93
   Shh and Twhh Expression during Fin Bud Development 95
   Shh, Retinoic Acid Regulation and the ZPA 96
   Mutants of the Hh Pathway and Fin Bud Development 96
   Fin Ray Morphogenesis and Regeneration 99
   The Hedgehog Pathway and Fin Ray Patterning: Role of the Epithelial-Mesenchymal Interactions 101

10. **Hedgehog Signalling in T Lymphocyte Development** 107
   Susan Outram, Ariadne L. Hager-Theodorides and Tessa Crompton
   Introduction to T Cell Development 107
   Effect of Hh Signalling on Thymocyte Development in the Mouse ... 109

11. **Hedgehog Signalling in Prostate Morphogenesis** 116
   Marilyn L.G. Lamm and Wade Bushman
   Prostate Morphogenesis 116
   Mesenchymal-Epithelial Signalling in Prostate Morphogenesis:
     Role of Androgens 117
   Epithelial-Mesenchymal Signalling in Prostate Morphogenesis:
     Sonic Hedgehog-Gli Pathway 117
   Shh Signalling during the Budding Phase of Prostate
     Morphogenesis 120
   Shh Signalling in Prostatic Ductal Branching 121
   Shh Signalling during Ductal Outgrowth and Differentiation 121

12. **Sonic Hedgehog Signalling in Visceral Organ Development** 125
   Huimin Zhang, Ying Litingtung and Chin Chiang
   Esophagus 126
   Lung 127
   Stomach 129
   Pancreas 131
   Intestine 132
   Kidney 133

13. **Shh/Gli Signalling during Murine Lung Development** 137
   Martin Rutter and Martin Post

14. **New Perspectives in Shh Signalling?** 147
   Carolyn E. Fisher
   Megalin 147
   Cubilin 147
   Endocytosis: RAP and Other Adaptor Molecules 148
   Megalin-RAP Binding 148
   Megalin-Shh-RAP Interactions 148
   Megalin in Development: More Links to Shh 149
   Proteoglycans and Megalin Interactions 149
   Lung Development and the Role of Megalin 149
   The Good, the Bad and the Ugly 151

Index 155

EDITORS

Carolyn E. Fisher
Immunobiology Group
MRC/UoE Centre for Inflammation Research
The Queen's Medical Research Institute
Edinburgh, U.K.
Email: carolyn.fisher@ed.ac.uk
Chapters 1, 14

Sarah E.M. Howie
Immunobiology Group
MRC/UoE Centre for Inflammation Research
The Queen's Medical Research Institute
Edinburgh, U.K.
Email: s.e.m.howie@ed.ac.uk
Chapter 1

CONTRIBUTORS

Marie-Andrée Akimenko
Department of Medicine and Cellular
 and Molecular Medicine
Ottawa Health Research Institute
University of Ottawa
Ottawa, Ontario, Canada
Email: makimenko@ohri.ca
Chapter 9

Fabien Avaron
Department of Medicine and Cellular
 and Molecular Medicine
Ottawa Health Research Institute
University of Ottawa
Ottawa, Ontario, Canada
Chapter 9

Wade Bushman
Department of Surgery
University of Wisconsin
Madison, Wisconsin, U.S.A.
Chapter 11

Andrés E. Carrasco
Laboratorio de Embriología Molecular
Instituto de Biología Celular
 y Neurociencias
Facultad de Medicina
Universidad de Buenos Aires - CONICET
Ciudad Autónoma de Buenos Aires,
 Argentina
Email: rqcarras@mail.retina.ar
Chapter 2

Chin Chiang
Department of Cell and Developmental
 Biology
Vanderbilt University Medical Center
Nashville, Tennessee, U.S.A.
Email: chin.chiang@vanderbilt.edu
Chapter 12

Martyn T. Cobourne
Department of Craniofacial
　Development and Orthodontics
GKT Dental Institute
King's College London
Guy's Hospital
London, U.K.
Chapter 7

Dwight Cordero
Department of Obstetrics
　and Gynecology
Brigham and Women's Hospital
Harvard Medical School
Boston, Massachusetts, U.S.A.
Chapter 5

Tessa Crompton
Department of Biological Sciences
South Kensington Campus
Imperial College London
London, U.K.
Chapter 10

Antonella Galli
Developmental Genetics
Centre for Biomedicine
University of Basel
Basel, Switzerland
Chapter 8

Ariadne L. Hager-Theodorides
Department of Biological Sciences
South Kensington Campus
Imperial College London
London, U.K.
Chapter 10

Jill A. Helms
Department of Plastic
　and Reconstructive Surgery
Stanford University
Stanford, California, U.S.A.
Email: jhelms@stanford.edu
Chapter 5

Marilyn L.G. Lamm
Department of Pediatrics
Children's Memorial Research Center
Northwestern University Feinberg
　School of Medicine
Chicago, Illinois, U.S.A.
Email: mlamm@northwestern.edu
Chapter 11

Ying Litingtung
Department of Cell and Developmental
　Biology
Vanderbilt University Medical Center
Nashville Tennessee, U.S.A.
Chapter 12

Silvia L. López
Laboratorio de Embriología Molecular
Instituto de Biología Celular
　y Neurociencias
Facultad de Medicina
Universidad de Buenos Aires - CONICET
Ciudad Autónoma de Buenos Aires,
　Argentina
Chapter 2

Ben Martynoga
Genes and Development IDG
Section of Biomedical Sciences
University of Edinburgh
Edinburgh, U.K.
Chapter 3

Isabelle Miletich
Department of Craniofacial
　Development
GKT Dental Institute
King's College London
Guy's Hospital
London, U.K.
Chapter 7

Susan Outram
Department of Biological Sciences
South Kensington Campus
Imperial College London
London, U.K.
Email: s.outram@ic.ac.uk
Chapter 10

Martin Post
Program in Lung Biology
The Hospital for Sick Children
  Research Institute
Institute of Medical Sciences
University of Toronto
Toronto, Ontario, Canada
Chapter 13

David J. Price
Genes and Development IDG
Section of Biomedical Sciences
University of Edinburgh
Edinburgh, U.K.
Chapter 3

Yingchuan Qi
Department of Anatomical Sciences
  and Neurobiology
School of Medicine
University of Louisville
Louisville, Kentucky, U.S.A.
Chapter 4

Mengsheng Qiu
Department of Anatomical Sciences
  and Neurobiology
School of Medicine
University of Louisville
Louisville, Kentucky, U.S.A.
Email: m0qiu001@gwise.louisville.edu
Chapter 4

Martin Rutter
Program in Lung Biology
The Hospital for Sick Children
  Research Institute
Institute of Medical Sciences
University of Toronto
Toronto, Ontario, Canada
Chapter 13

Paul T. Sharpe
Department of Craniofacial
  Development
GKT Dental Institute
King's College London
Guy's Hospital
London, U.K.
Email: paul.sharpe@kcl.ac.uk
Chapter 7

Amanda Smith
Department of Medicine and Cellular
  and Molecular Medicine
Ottawa Health Research Institute
University of Ottawa
Ottawa, Ontario, Canada
Chapter 9

Deborah L. Stenkamp
Department of Biological Sciences
University of Idaho
Moscow, Idaho, U.S.A.
Email: dstenkam@uidaho.edu
Chapter 6

Min Tan
Department of Anatomical Sciences
  and Neurobiology
School of Medicine
University of Louisville
Louisville, Kentucky, U.S.A.
Chapter 4

Minal Tapadia
Department of Plastic
  and Reconstructive Surgery
Stanford University
Stanford, California, U.S.A.
Chapter 5

Paulette A. Zaki
Genes and Development IDG
Section of Biomedical Sciences
University of Edinburgh
Edinburgh, U.K.
Email: pzaki@ed.ac.uk
Chapter 3

Huimin Zhang
Department of Cell and Developmental
  Biology
Vanderbilt University Medical Center
Nashville Tennessee, U.S.A.
Chapter 12

Aimée Zuniga
Developmental Genetics
Centre for Biomedicine
University of Basel
Basel, Switzerland
Email: Aimee.Zuniga@unibas.ch
Chapter 8

PREFACE

The hedgehog signalling pathway is highly conserved and seen in organisms ranging from *Drosophila* to humans. This pathway is critical in determining cell fate decisions in a variety of different cell types. There are several vertebrate analogues of the *Drosophila* hedgehog protein of which the most widely studied is Sonic hedgehog (Shh). Shh signalling classically involves the Gli family of zinc-finger transcription factors. The Shh signalling pathway is well characterised in the development of a number of vertebrate organ systems. It could indeed be argued that the Shh and Gli signalling may well be involved at some stage in the development of all the major organ systems in vertebrates. This volume represents a concerted drive to bring together 'state of the art' reviews by leading experts in the field of Shh and Gli signalling in development from all over the world. The chapters span vertebrate organisms from zebrafish to humans and cover development of the multiple organ systems in which the Shh signalling pathway is crucial for normal development. There are chapters on the development of the central nervous system, skeletal structures, visceral organs, prostate, lung, immune system and the structures of the human face. The authors themselves span three major continents and multiple nationalities which admirably illustrates the worldwide nature of the science. The international nature of the project has been very rewarding and the quality, depth and range of the reviews included speaks for itself. It is hoped that the reader will appreciate the wide variety of scientific approaches that have contributed to our current knowledge base of the importance of Shh and Gli signalling in vertebrate development and will at the same time realise that, as with all good science, there are still more questions than answers.

Sarah E.M. Howie, B.Sc. Hons., Ph.D.
Edinburgh
June 2006

CHAPTER 1

Introduction

Carolyn E. Fisher* and Sarah E.M. Howie

The Concept of Developmental Biology

Although no real insights into the mechanisms of development were obtained until after 1880, when experimental approaches to embryology were established, descriptive studies of embryo development have been around for millennia. Aristotle (384-322 BC) wrote a very detailed description of mammalian embryogenesis, similar to the picture we accept today, inferring that the process was driven by an *entelechy*, known as a "vital force" in later centuries. Descriptive studies continued after 1550 but there was no further serious discussion of the *mechanisms* of embryo development until the 18th and 19th centuries.

The anatomist Wilhelm Roux (1850–1924) pioneered experimental embryology, focusing on amphibian embryos, and was the first to suggest that chromosomes carry hereditary material. In 1882 he extended Darwin's theory of the struggle for existence to ontogenesis. He wrote that stronger cells leave more offspring than weaker cells, inferring that competition for space and nutrients governed development. We now know that cell reproduction is far from chaotic, and that competition for intercellular spaces is, in general, abnormal. Nevertheless, "neural Darwinism", the idea that neurites compete during growth and that only the first of the group to reach the target cell survives, is becoming established in developmental neurobiology.

Another pioneer of experimental embryology, Hans Driesch (1867–1941), discovered that cells of early sea urchin embryos "remembered" their individual locations in the cell mass—separated cells returned to their original positions—although there were no detectable physical or chemical differences among them. Lacking the understanding of the biochemistry of cell-cell interactions that we have today, Driesch concluded that a "vital force" drove embryogenesis – the idea proposed by Aristotle more than two millennia earlier. Modern-day biologists no longer believe in a "vital force"; biology is mechanistic in character.

Thanks to technological advances in the late 20th century, developmental genetics has grown in stature. The importance of these advances for understanding embryogenesis is recognised. Significantly, biologists now realise that the molecular components of many developmental pathways are present and active in adult organisms. They are not mere residues of morphogenesis; developmental pathways are important in maintaining as well as generating the adult form. In a sense, morphogenesis is never complete. As will be discussed in later chapters, developmental pathways are important in tissue repair and organ regeneration. In addition, it is now clear that these same pathways play a major role in some cancers, where mature cell types appear to "dedifferentiate", proliferating without adequate control and invading normal functioning organs. Cancer is another topic that will be covered later in the book.

*Corresponding Author: Carolyn E. Fisher—Immunobiology Group, MRC/UoE Centre for Inflammation Research, The Queen's Medical Research Institute, Little France Crescent, Edinburgh EH16 4TJ, Scotland, U.K. Email: carolyn.fisher@ed.ac.uk

Shh and Gli Signalling and Development, edited by Carolyn E. Fisher and Sarah E.M. Howie. ©2006 Landes Bioscience and Springer Science+Business Media.

Introduction to Morphogens: Shh

The term *morphogens* was coined by the mathematician Allan Turing in 1952 to denote graded signals released by 'organisers' such as the notochord and Zone of Polarising Activity (ZPA) in the developing limb bud. To qualify as a morphogen, a signal must fulfil two criteria: to form a concentration gradient, and to elicit distinct responses at different concentrations. Cells encounter different concentrations of a morphogen according to their distance from the organiser that secretes it. Different transcription factors are therefore induced, committing the cells to different fates.[1] At least four models of morphogen transport have been proposed.[2]

Chemoattractants and chemorepellents also form graded signals, guiding cell migration and various cellular processes, but they are "guidance cues" not morphogens. Cells respond to chemoattractant and chemorepellent *gradients* rather than absolute concentrations. Also, these signals act by regulating cytoskeletal and membrane dynamics, not by signalling to nuclei.[3]

The first morphogens identified were the transcription factors encoded by the *Drosophila* genes *bicoid* and *hunchback*, which operate in the embryo before cellularization, forming concentration gradients along the anterior-posterior axis.[1] Morphogenesis genes are highly conserved across species. They include members of the Wnt family (wingless in *Drosophila*) and *decapentaplegic* (Dpp) in *Drosophila* appendage development;[4,5] bone morphogenic proteins (BMPs); fibroblast growth factors (FGFs); members of the TGFβ family, such as Squint in early zebrafish embryogenesis;[6] and Hh genes. Sonic Hedgehog (Shh), one of three mammalian homologues of Hh, has been shown to act as both a morphogen and a guidance cue.[7]

In *Drosophila*, Hh functions as a short-range morphogen during wing development whereas Dpp acts over a long range. Imaginal discs (wings) comprise anterior (A) and posterior (P) compartments. Cells in the latter express *engrailed* (en), which induces Hh synthesis. Hh is secreted into the A compartment, inducing transcription of several genes including Patched (Ptc), Dpp and en.[8] In anterior cells bordering the A-P boundary (the disc lumen), Dpp organises the wing's A-P axis and is required for disc development and patterning.[5] After A-P subdivision the imaginal disc is divided into Dorsal-Ventral (DV) compartments, the border between which develops into the wing margin. DV patterning involves the Notch and wingless signal transduction pathways. Wg acts as a morphogen inducing target gene expression and patterning activities of the dorsal/ventral boundary.[9]

Morphogens also play a role during vertebrate development. For example, squint promotes the formation of mesoderm and endoderm in zebrafish embryos;[6] and Shh acts directly at long range to pattern the ventral neural tube in chicks. Shh is also involved in limb bud formation but whether it acts as a morphogen in this context is unclear.

The Hh Pathway in *Drosophila*

The Hh pathway was first recognised as important during segmentation in *Drosophila*.[10] An elegant study by Ingham and colleagues led to a now widely-accepted model of Hh signalling in *Drosophila*;[11] a simplified version is shown in Figure 1.

Hh signalling is absolutely dependent on smo. Smo is inhibited by the protein Ptc, which acts indirectly and substoichiometrically. The mechanism might involve the transport of an endogenous modulator of smo, but this has not been identified, nor has Ptc transport activity been characterised.[12] However, it is generally held that Hh removes the inhibition of smo by binding to Ptc. Hh stimulation of cells stabilises smo, which accumulates at least 10-fold and becomes more highly phosphorylated.[13]

Evidence suggests that intracellular localisation of smo-containing organelles depends partly on costal-2 protein (cos 2). Cos-2 tethers a group of segment polarity proteins to cytoskeletal microtubules, and full-length Ci is bound to these. Smo and cos-2 may interact directly.[14] Recruitment of cos-2 to smo causes Ci to dissociate from the cytoskeleton, preventing its cleavage to the transcriptional repressor form Ci[75] (CiR). When smo is activated, however, the Ci/protein complex dissociates and full-length Ci is translocated to the nucleus, where it activates target genes containing Ci-binding sites. A detailed analysis of smo has been published.[15]

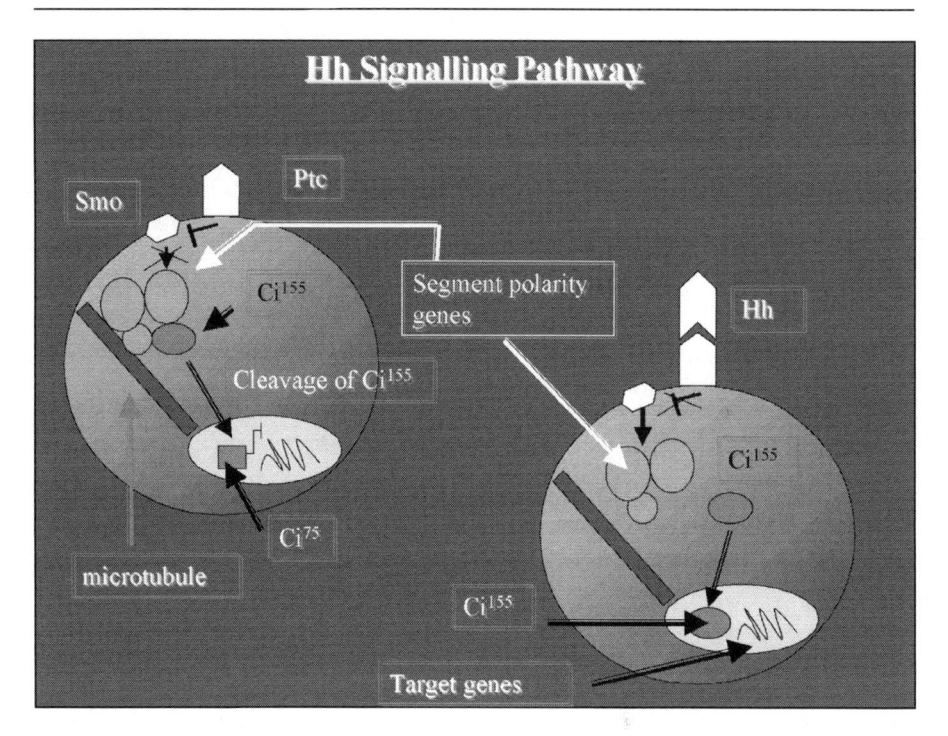

Figure 1. In the absence of ligand binding, Ptc-1 inhibits the activity of smo, allowing Ci to be cleaved to form a transcriptional repressor. When Hh binds to Ptc-1 this inhibition of smo is repressed. This allows full-length Ci to be translocated to the nucleus, where it acts as a transcription factor for various genes.

The smo-Cos-2 complex also contains Fu (Fused), and Fu kinase activity is needed for Hh signalling. Fu phosphorylates Cos-2 at the two positions induced by Hh stimulation.[16] A primary function of activated smo appears to be the inhibition of suppressor of fused (Su(fu)), activating Fu; this may happen indirectly via Cos-2.[15] The stability of Fu kinase is an absolute requirement for positive regulation by Cos-2. Therefore, the Hh-induced stabilisation of smo results in recruitment of both Fu and Cos-2.[13] Fu is dispensable if Su(fu) is lost. Su(fu) negatively regulates Ci by localising it in the cytoplasm, either through cytoplasmic anchoring or nuclear export; it might also inhibit Ci function in the nucleus.[17]

CiR (the N-terminal proteolytic fragment of Ci that suppresses transcription) retains the zinc finger-mediated DNA binding specificity but lacks nuclear export signals, a cytoplasmic anchoring sequence and a transcriptional activation domain.[17,18] *Drosophila* protein kinase A (dPKA) is required, along with Cos-2 and Fu, to process Ci^{155} to Ci^{75} in vivo. Intact Ci (Ci^{155}) is found in cells carrying mutations in these genes. It can activate the transcription of Hh target genes if normal Fu is present. Loss of Fu also causes accumulation of Ci, but in this situation Ci cannot activate Hh target genes.[19]

Although this Hh pathway has become widely accepted and has been mapped out in detail, some observations challenge it. In *Drosophila*, whilst smo protein is distributed throughout the imaginal disc, it accumulates in wing compartments and clones of cells lacking Ptc, but is reduced in cells overexpressing Ptc, even in the absence of Hh signalling. Also, cell-surface levels of smo increase in response to Hh stimulation whereas Ptc levels decrease. This suggests that most smo does not colocalise with Ptc, making it unlikely that Ptc-smo binding, if it occurs in vivo, is important in Hh signalling.[20] Some workers have gone so far as to suggest that the first step in the Hh pathway (modification of smo activity by Hh and Ptc) should be reconsidered. Currently it is hypothesised to involve changes in smo concentration, localisation, phosphorylation, conformation or binding to small molecules related to cyclopamine, i.e., changes in isolated smo molecules. Now it seems possible that Ptc and Hh might act primarily, or partly, through smo partners such as cos-2 instead of smo itself.[21]

In *Drosophila*, Hh regulates cell proliferation and differentiation in essential patterning events such as embryonic segmentation, appendage formation, and development of the eye and regions of the brain; either directly, or indirectly via recruitment of Dpp and wingless. Before they can execute such roles, Hh molecules are matured by autocatalytic cleavage. The products are Hh-Np (the N-terminal polypeptide), the functional signal, and a C-terminal polypeptide that appears to have no function other than catalysing the autoproteolysis. The signalling peptide (Hh-Np) is modified at its N- and C-termini by palmitoyl and cholesteryl adducts, respectively.[22] Although many proteins are lipid-modified, Hh and its vertebrate homologues are unique in being modified by cholesterol addition.[23]

The action of Hh on distant cells in developing tissues involves: (a) the transmembrane transporter-like protein Dispatched (Disp), which is required for releasing Hh from cells; (b) the heparan sulphate proteoglycans (HSPs) Dally-like (Dlp) and Dally, which are required for extracellular Hh transport; and (c) HSP biosynthesis enzymes such as Sulfateless and *tout velu*.[24-26] *Tout velu* is required for moving cholesterol-modified Hh.[27] The ability of Hh to attach to membranes via the C-terminal cholesterol may be critical for increasing the distance over which the morphogen acts.[23] *Dispatched*, a distant relative of Ptc, is predicted to encode a 12-pass transmembrane protein with a sterol-sensing domain. Its role in trafficking cholesterol-modified Hh might be executed through a secretory pathway, so that the active form arrives at the cell surface, or through the displacement of cholesterol-modified Hh from the lipid bilayer.[23] If *dispatched* is absent during the development of imaginal discs, normal levels of Hh are produced but it is not released from posterior cells and accumulates instead. Moreover, *Drosophila dispatched* mutants lacking both maternal and zygotic activity have a segment polarity phenotype identical to Hh mutants, demonstrating that this molecule is critical for proper Hh pathway signalling.[25]

The Shh Pathway in Vertebrates

The Hedgehog pathway in vertebrates parallels that in *Drosophila* but there are two or more homologues of some components, consistent with divergence of function. Mammals have two Ptc receptors (Ptc-1 and Ptc-2), though only the former is definitely involved in Hh signalling. It is confined to target cells and is upregulated in response to Hh. Ptc-2 is coexpressed with Hh but its transcription is independent of pathway activation.[28] Mammals also have three Hh proteins, *Sonic* (Shh), *Indian* (Ihh), *and Desert* (Dhh) *Hedgehog*, which differ in their tissue-specific expression patterns and in their roles during development. The mammalian homologues of *Drosophila* Ci are the three Gli molecules (Gli 1-3), which regulate the transcription of Hh-responsive genes both positively (Gli 2) and negatively (Gli 3).

The homologues of Hh, Ptch, smo and Ci are well conserved but those of Cos2 and Fu are less so. They have not been functionally linked to pathway regulation, suggesting that certain *Drosophila* routing mechanisms may be less important in mammals. SuFu, however, is conserved, and does have pathway regulatory functions. This is demonstrated by loss of function in zebrafish;[29] also, Cheng and Bishop (2002) showed that SuFu can enhance the binding of Gli proteins to DNA.[30]

As in *Drosophila*, Hh proteins undergo autocatalytic cleavage to an active 19kDa ligand with cholesterol covalently linked to the C-terminus. Caveolin-1 may be a Ptc-binding partner in *Drosophila*;[31] caveolins are the major constituents of caveolae, nonclathrin-coated membrane invaginations important in endocytosis and intracellular trafficking. This might imply that the cholesterol moiety is involved in directing intracellular transport, and cell culture experiments have shown that cholesterol-modified Hh remains bound to the cell surface, suggesting limited movement in vivo.[23] Nevertheless, cholesterol-modified Shh in vertebrates is thought to spread Shh activity rather than anchor it in one place; Lewis et al (2001) demonstrated that Shh-Nu (sonic that could not be cholesterol-modified) in mice had a restricted range of signalling in comparison to wild type Shh.[32]

This conflict of evidence might have been resolved by the discovery in vertebrates of inhibitors of Hh signalling, such as Hip1 (hedgehog interacting protein 1) and GAS-1 (growth arrest specific-1). These proteins have no *Drosophila* homologues. The former encodes a membrane-bound glycoprotein that binds Shh, and the latter is a Wnt-inducible mouse gene expressed in areas that respond to but do not express Shh.[33,34]

Hh proteins are involved in neural tube formation in vertebrates. In mammals, Shh activity at the midline patterns the ventral neural tube and somites, and is involved in the development of left-right asymmetry. It has polarising activity in the limb, acting at both short (posterior limb identities) and long (anterior limb identities) distances. It is involved in maintaining stem cells in postembryonic tissues and acts as a pathogenic mitogen in some endodermally-derived human cancers, which account for 25% of all cancer deaths.[35,36] Shh also regulates morphogenesis of many other organs (see below).

Gli Transcription Factors

Gli molecules are evolutionarily conserved, with homologues identified in invertebrates and in all vertebrate species analysed so far.[37] Humans and mice have three Gli genes that are candidates for mediating downstream activities of Shh but their precise roles are not fully determined.

Generally, expression of Gli1 is highly restricted compared to Gli2 and Gli3, and it is transcriptionally regulated by Hh signalling, whereas the others are less reliant on Hh for transcription. Gli1 only activates Shh transcription, whereas Gli2 and 3 are bi-functional and Hh signalling regulates their activities post-transcriptionally. Data from the many studies in mice with defective Gli genes show that Gli1 expression is tightly controlled by the activities of Gli2 and 3.[38] Gli genes are never expressed in Shh-expressing organiser cells during embryogenesis. Normally Gli1 is expressed in cells adjacent to the organiser, consistent with its role as a transcriptional activator of the Shh signal. Gli3 is usually situated opposite the organiser, possibly limiting its range.

First indications that transcription factors play a role in establishing cell fates in response to a morphogen came from studies on the spinal cord. Here, Gli 1-3 are expressed in partially overlapping patterns and establish the initial stripes of homeodomain transcription factor expression in the ventral neural tube in response to Shh produced by the notochord and floorplate, promoting the specification of several ventral cell types.[39] In the frog neural plate, widespread expression of Gli2/3R (repressors) abolishes neuronal differentiation.[40] In mice, inactivation of Gli2 results in absence of the floor plate, probably partly due to inefficient activation of the transcription factor HNF3β, which regulates floor plate identity.[41] Also, high expression of Gli3R in chick neural tube abolishes ventral cell differentiation.[42]

The importance of Gli factors during embryogenesis has been assessed in single and double knockout mice. Gli1-/- mice have no obvious defects, indicating that Gli1 is dispensable for embryogenesis.[43] Since Gli2-/- mice have phenotypes similar to but milder than Shh-/- mutants, it appears that Gli2 is the major transducer of Shh signalling.[38] These mice have severe skeletal abnormalities including no vertebral bodies or intervertebral discs, and shortened limbs.[44] Gli3-/- mutants have defects, such as polydactyly, distinct from those of Gli2-/- and Shh-/-. Xt mutant mice have alterations within the Gli3 locus, and Xt/Xt embryos display enhanced polydactyly in the fore and hind limbs. Heterozygotes show preaxial polydactyly of the hindlimbs.

Although deletion of the Gli1 zinc finger domain leads to no obvious abnormalities in the embryo, Gli1-/-Gli2+/- mice have reduced viability and exhibit lung and neural tube defects that are not found in either Gli1-/- or Gli2+/- mice.[43] This indicates that Gli1 has a physiological role in Shh signalling. Perhaps Gli2 and/or Gli3 can compensate for the lack of Gli1 function during embryogenesis.

Roles for Shh in Vertebrates

The importance of Shh signalling during development, in adult organisms, and in pathological processes, should not be underestimated. Although Shh signalling has been analysed in detail in relatively few organs/systems such as the CNS, limbs, lungs, eyes and the reproductive system, the pathway appears to have important roles in nearly every organ. Many of these are covered in detail in subsequent chapters.

CNS

Shh acts as a morphogen during development of the early vertebrate ventral neural tube. Later, in the dorsal brain, it acts as a mitogen on progenitors of the cerebellum, tectum, neocortex and hippocampus.[45] General consensus attributes dorsoventral specification of the neural tube to Shh secreted by the notochord inducing differentiation of the floor plate; the latter starts to express Shh in response to the notochordal signal.[46] An alternative proposal is that because the floor plate, notochord and dorsal endoderm share a common origin in Henson's node, all are sources of Shh.[47] Details notwithstanding, it is clear that Shh influences the development of, and many cell fates within, the CNS and associated structures.

A study on chick embryos by Ahlgren and Bronner-Fraser demonstrated the importance of Shh in craniofacial development, dealt with in a later chapter: branchial arch structures are lost and there are subsequent brain anomalies.[48] Somite development in Shh null mice has been investigated by Borycki et al, who demonstrated that Shh is critical in activating myogenic determination genes and that it is required for survival of sclerotome cells as well as ventral and dorsal neural tube cells.[49] Weschler-Reya and Scott implied a role for Shh during development of granule cells. They demonstrated that Shh, which is made by Purkinje cells, regulates the division of granule cell precursors.[50] A mitogenic action of Shh was also found by Rowitch et al, who suggested temporal restrictions on Shh-mediated cell proliferation.[51]

The three Gli genes are expressed in partially overlapping domains in the neural tube; Gli2 and 3 are proposed to mediate initial Hh signalling and to regulate Gli1. All have activator function but only Gli2 and 3 have potent repressor functions, and each appears to be regulated differently. Details of the role(s) of the Gli proteins during CNS development are dealt with in various subsequent chapters.

Limbs

Shh and Gli gene functions during limb bud formation have been studied extensively. Briefly, the ZPA (zone of polarising activity) signalling centre in the posterior limb bud is necessary for A-P patterning, and defects resulting from ZPA transplants can be mimicked by misexpression of Shh.[52]

Gli genes are expressed only in the mesenchyme during limb formation. However, only Gli3 appears to have a role in limb development, its major function being establishment of A-P asymmetry. It also represses Shh expression in the anterior margin of the limb bud; loss of Gli3 function results in ectopic Shh expression, induction of Gli1 in adjacent cells, and preaxial polydactyly. Despite the lack of limb defects in Gli1 mutant mice, Gli1 is always upregulated in the anterior region of limb buds adjacent to Shh-expressing cells in polydactylous animals, implying a mediating role in Shh signalling.[53]

All Gli genes are expressed in developing bones; Gli2 and 3 are essential for normal development. In Gli2-/- mice, bone ossification is delayed and long bones are shortened;[44] in Gli3-/-, the length and shape of most bones are altered and sometimes the radius and tibia are missing.[54]

Shh signalling is also involved in chondrogenesis and smooth muscle differentiation, with Shh and Ihh participating in the differentiation of chondrogenic precursor cells into chondrocytes.[55] The Hh family also plays a role in joint formation.[56]

Introduction 7

Reproductive Tract

Hh signalling is critical in the development and differentiation of the gonads and accessory sex glands.[57] In females, Ihh, rather than Shh, is the important molecule. In murine mammary gland development there appears to be a complete absence of Shh; Ihh is localised exclusively to the epithelium. During puberty it is found in undifferentiated epithelial 'body cells' at the tips of terminal end buds of elongating ducts.[58] The role of Hh in somatic and germline stem cell proliferation in adult Drosophila ovary is well-characterised,[59] but it is unclear whether Hh-signalling is involved in vertebrate ovaries.

In the adult male, Desert hedgehog (Dhh) signalling is essential for spermatogenesis and for development of Leydig cells, peritubular cells and seminiferous tubules; Shh appears to have no role. Male Dhh-/- mice lack mature sperm but no expression is observed in the female ovary during early or late stages of development.[60]

Shh is necessary for normal prostate development but not initial organogenesis. Specifically, it provides the signal for prostate ductal budding, a testosterone-dependent process, and is involved with ductal patterning.[61] All three Gli genes are expressed during ductal budding; their levels decline postnatally, becoming low in the adult.[62] Prostate development is covered in detail later in the book.

Lung and Visceral Organs

Lung bud morphogenesis begins in mice at E9.5 as an endodermal outbudding of the developing gut tube, the A-P patterning of which is governed by Shh. Normal lung development depends on Shh signalling and Gli transcription factors; Shh -/- murine embryos fail to form lungs, Gli3 is essential for proper pulmonary development, and Gli1 is known to act downstream of Shh signalling in lung.[63] Shh is essential during early stages of pulmonary branching morphogenesis but it does not appear to be important in the subsequent differentiation of specialised lung cells such as Clara cells. Shh signalling is also required for proper separation of the trachea and esophagus. It is also pivotal in digestive tract morphogenesis and differentiation; epithelial Shh regulates the formation of stomach glands, connective tissue and smooth muscle, and stratification of mesenchyme.[64] Lung development and the role of Shh in visceral organs are subjects of later chapters.

Eye

Much work has been done on eye development in *Drosophila*, *Xenopus*, chick, zebrafish and mouse, and in all cases Hh signalling regulates morphogenesis to some extent. The retinal determination gene in *Drosophila*, *eyes absent* (Eya), represents a crucial link between Hh signalling and photoreceptor differentiation: Hh acts as a binary switch, initiating retinal morphogenesis by inducing Eya expression.[65] In *Xenopus*, misexpression of Tbx2 and Tbx3 results in defective eye morphogenesis. Tbx2/3 expression is thought to be regulated by Gli-dependent Hh signal-transduction.[66] In zebrafish eye development (covered later in the book), the eye phenotype of the sonic-you (syu) mutant is consistent with multiple roles for Hh during retinal development.[67] Generally, Hh signalling regulates eye morphogenesis and photoreceptor differentiation and plays a role in defining the proximal-distal and dorsal-ventral axes in the eye.

Other Roles

Other roles of Shh in vertebrate morphogenesis include those in tooth development, covered in a later chapter. Attenuation of Shh signalling by means of a function-blocking Ab markedly delays tooth germ development and demonstrates that Shh is required for ameloblast and odontoblast maturation.[68] Shh is also vital for tongue formation; if signalling is disrupted early in rat embryogenesis (E12) then no tongue forms.[69] It is also important in renewing and maintaining tastebuds.[70] Liu et al[69] propose that high concentrations of Shh result in formation and maintenance of papillae, while low concentrations activate between-papillae genes that maintain a papilla-free epithelium. Shh signalling is essential for forming the olfactory pathway; disruption compromises distinct aspects of olfactory pathway patterning and differentiation.[71] It has a well-documented role in the formation of hair follicles and feather morphogenesis: Shh seems to be required for epithelial cell proliferation in the early development of hair follicles and for the morphogenetic movement of mesenchymal cells at later stages.[72] Gli2 is the key mediator of Shh responses in skin; Gli2(-/-) mouse mutants exhibit arrested hair follicle development.[73] Shh is involved in pituitary gland development; its role here seems to be largely conserved between fish and mice, despite the different modes of pituitary formation in the two vertebrate classes.[74] In the blood circulatory system, Shh has roles in heart morphogenesis, the induction of angiogenesis and blood cell development.[75-77] It plays roles in stem cell proliferation, thymocyte differentiation and, as discussed later, the development of lymphocytes. As developmental research continues, it seems inevitable that yet more roles for Shh and the Gli transcription factors will be uncovered. Conceivably, this pathway has functions in all aspects of vertebrate and invertebrate embryogenesis.

Clinical Aspects

Not only is Shh an indispensable developmental morphogen and mitogen with important roles in tissue repair in adult organisms, it is also linked to several human disease states. The Hh pathway may have an early and critical role in carcinogenesis; Shh-Gli signalling modulates normal dorsal brain growth by controlling precursor proliferation, which is deregulated in brain tumours.[78] Shh also seems to be involved in human pancreatic carcinoma. In vitro and in vivo experiments show that Shh is needed for the proliferation of some pancreatic cell lines, and it is suggested that maintenance of Hh-signalling is important for aberrant proliferation and tumourogenesis.[79] It is also a major determinant of skin tumourogenesis, most notably basal cell carcinoma (BCC).[80]

VACTERL (vertebral defects, atresia, tracheooesophageal fistula with esophageal atresia, radial and renal dysplasia, limb abnormalities) might also be linked to aberrant Shh-signalling:[38] defects in Gli2-/-, Gli3-/- and Gli2-/-; Gli3-/- mutant mice are associated with VACTERL and appear to represent the first animal model mimicking the human VACTERL syndrome.[81]

Foregut malformations such as oesophageal atresia, tracheo-oesophageal fistula, lung anomalies and congenital stenosis of the oesophagus and trachea account for 1 in every 2000-5000 live births. Experimental work in mice suggests that Gli 2 and 3 have specific and overlapping functions during foregut development, and that Gli gene mutations are involved in human foregut abnormalities.[82] Shh also has a role in lung hypoplasia CDH (congenital diaphragmatic hernia), probably affecting bronchiole development and causing thinning of the interstitium.[83]

Final Thoughts

Shh homologues are present in 'lower' animals such as sea urchins and leeches. Their roles are unknown, but involvement in patterning and symmetry seems likely. *C. elegans* contains a gene homologous to Gli and Ci that has an important role in sex determination, but there is no homologue of hedgehog in this species.

The accepted Hedgehog pathway has been elucidated from studies on *Drosophila* and shown to be involved in patterning during early embryogenesis across a wide range of species. It is important for the genesis of vertebrate lung, pancreas, prostate, eyes, limbs, CNS and other organs. It is clear from work done in species ranging from *Xenopus* to chick, mouse to human, that the specific pathways involved during organogenesis have been conserved, e.g., paired appendage formation in fish and tetrapods.

Recent work has shown that Shh and related proteins are heavily involved not only in early embryogenesis, and in the development of specific organs, but also in related functions such as wound healing and regeneration. Better understanding of the developmental role of Shh has also revealed that it may be vitally important in congenital human disorders such as congenital diaphragmatic hernia (CDH), as well as in tumour development.

References

1. Vincent J-P, Briscoe J. Morphogens. Curr Biol 2001; 11(21):R851-R854.
2. Tabata T, Takei Y. Morphogens, their identification and regulation. Development 2004; 131(4):703-712.
3. Tessier-Lavigne M. Axon guidance by diffusible repellants and attractants. Curr Opin Genet Dev 1994; 4(4):596-601.
4. Zecca M, Basler K, Struhl G. Direct and long-range action of a wingless morphogen gradient. Cell 1996; 87(5):833-844.
5. Lecuit T, Brook WJ, Ng M et al. Two distinct mechanisms for long-range patterning by Decapentaplegic in the Drosophila wing. Nature 1996; 381(6581):387-393.
6. Chen Y, Schier AF. The zebrafish nodal signal squint functions as a morphogen. Nature 2001; 411:607-610.
7. Charron F, Stein E, Jeong J et al. The morphogen sonic hedgehog is an axonal chemoattractant that collaborates with netrin-1 in midline axon guidance. Cell 2003; 113(1):11-23.
8. Tabata T, Kornberg TB. Hedgehog is a signaling protein with a key role in patterning Drosophila imaginal discs. Cell 1994; 76(1):89-102.
9. Neumann CJ, Cohen SM. A hierarchy of cross-regulation involving Notch, wingless, vestigial and cut organizes the dorsal/ventral axis of the Drosophila wing. Development 1996; 122(11):3477-3485.
10. Nusslein-Volhard C, Wieschaus E. Mutations affecting segment number and polarity in Drosophila. Nature 1980; 287(5785):795-801.
11. Ingham PW, Taylor AM, Nakano Y. Role of the Drosophila patched gene in positional signalling. Nature 1991; 353(6340):184-187.
12. Taipale J, Cooper MK, Maiti T et al. Patched acts catalytically to suppress the activity of Smoothened. Nature 2002; 418(6900):892-897.
13. Lum L, Zhang C, Oh S et al. Hedgehog signal transduction via Smoothened association with a cytoplasmic complex scaffolded by the atypical kinesin, Costal-2. Mol Cell 2003; 12(5):1261-1274.
14. Zhu AJ, Zheng L, Suyama K et al. Altered localization of Drosophila Smoothened protein activates Hedgehog signal transduction. Genes Dev 2003; 17(10):1240-1252.
15. Hooper JE. Smoothened translates Hedgehog levels into distinct responses. Development 2003; 130(17):3951-3963.
16. Nybakken KE, Turck CW, Robbins DJ et al. Hedgehog-stimulated phosphorylation of the kinesin-related protein Costal2 is mediated by the serine/threonine kinase fused. J Biol Chem 2002; 277(27):24638-24647.
17. Chen CH, von Kessler DP, Park W et al. Nuclear trafficking of Cubitus interruptus in the transcriptional regulation of Hedgehog target gene expression. Cell 1999; 98(3):305-316.
18. Aza-Blanc P, Ramirez-Weber FA, Laget MP et al. Proteolysis that is inhibited by hedgehog targets Cubitus interruptus protein to the nucleus and converts it to a repressor. Cell 1997; 89(7):1043-1053.
19. Alves G, Limbourg-Bouchon B, Tricoire H et al. Modulation of Hedgehog target gene expression by the Fused serine-threonine kinase in wing imaginal discs. Mech Dev 1998; 78(1-2):17-31.
20. Denef N, Neubuser D, Perez L et al. Hedgehog induces opposite changes in turnover and subcellular localization of patched and smoothened. Cell 2000; 102(4):521-531.
21. Kalderon D. Hedgehog signalling: Costal-2 bridges the transduction gap. Curr Biol 2004; 14:R67-R69.
22. Mann RK, Beachy PA. Novel lipid modifications of secreted protein signals. Annu Rev Biochem 2004; 73:891-923.
23. McMahon AP. More surprises in the hedgehog signaling pathway. Cell 2000; 100:185-188.
24. Han C, Belenkaya TY, Wang B et al. Drosophila glypicans control the cell-to-cell movement of Hedgehog by a dynamin-independent process. Development 2004; 131(3):601-611.
25. Burke R, Nellen D, Bellotto M et al. Dispatched, a novel sterol-sensing domain protein dedicated to the release of cholesterol-modified hedgehog from signaling cells. Cell 1999; 99(7):803-815.
26. Bellaiche Y, The I, Perrimon N. Tout-velu is a Drosophila homologue of the putative tumour suppressor EXT-1 and is needed for Hh diffusion. Nature 1998; 394:85-88.
27. Gallet A, Rodriguez R, Ruel L et al. Cholesterol modification of hedgehog is required for trafficking and movement, revealing an asymmetric cellular response to hedgehog. Dev Cell 2003; 4(2):191-204.
28. St-Jacques B, Dassule HR, Karavanova I et al. Sonic hedgehog signaling is essential for hair development. Curr Biol 1998; 8:1058-1068.
29. Wolff C, Roy S, Ingham PW. Multiple muscle cell identities induced by distinct levels and timing of hedgehog activity in the zebrafish embryo. Curr Biol 2003; 13(14):1169-1181.

30. Cheng SY, Bishop JM. Suppressor of Fused represses Gli-mediated transcription by recruiting the SAP18-mSin3 corepressor complex. Proc Natl Acad Sci USA 2002; 99(8):5442-5447.
31. Karpen HE, Bukowski JT, Hughes T et al. The sonic hedgehog receptor patched associates with caveolin-1 in cholesterol-rich microdomains of the plasma membrane. J Biol Chem 2001; 276(22):19503-19511.
32. Lewis PM, Dunn MP, McMahon JA et al. Cholesterol modification of sonic hedgehog is required for long-range signaling activity and effective modulation of signaling by Ptc1. Cell 2001; 105(5):599-612.
33. Chuang PT, McMahon AP. Vertebrate Hedgehog signalling modulated by induction of a Hedgehog-binding protein. Nature 1999; 397(6720):617-621.
34. Lee CS, Buttitta L, Fan CM. Evidence that the WNT-inducible growth arrest-specific gene 1 encodes an antagonist of sonic hedgehog signaling in the somite. Proc Natl Acad Sci USA 2001; 98(20):11347-11352.
35. Berman DM, Karhadkar SS, Maitra A et al. Widespread requirement for Hedgehog ligand stimulation in growth of digestive tract tumours. Nature 2003; 425(6960):846-851.
36. Watkins DN, Berman DM, Burkholder SG et al. Hedgehog signalling within airway epithelial progenitors and in small-cell lung cancer. Nature 2003; 422(6929):313-317.
37. Orenic TV, Slusarski DC, Kroll KL et al. Cloning and characterisation of the segment polarity gene cubitis interruptus dominant of Drosophila. Genes Dev 1990; 4:1053-1067.
38. Kim JH, Kim PCW, Hui C-C. The VACTERL association: Lessons from the sonic hedgehog pathway. Clin Genet 2001; 59:306-315.
39. Briscoe J, Ericson J. The specification of neuronal identity by graded Sonic Hedgehog signalling. Semin Cell Dev Biol 1999; 10(3):353-362.
40. Brewster R, Lee J, Ruiz i Altaba A. Gli/Zic factors pattern the neural plate by defining domains of cell differentiation. Nature 1998; 393(6685):579-583.
41. Sasaki H, Hui C, Nakafuku M et al. A binding site for Gli proteins is essential for HNF-3beta floor plate enhancer activity in transgenics and can respond to Shh in vitro. Development 1997; 124(7):1313-1322.
42. Meyer NP, Roelink H. The amino-terminal region of Gli3 antagonizes the Shh response and acts in dorsoventral fate specification in the developing spinal cord. Dev Biol 2003; 257(2):343-355.
43. Park HL, Bai C, Platt KA et al. Mouse Gli1 mutants are viable but have defects in SHH signaling in combination with a Gli2 mutation. Development 2000; 127(8):1593-1605.
44. Mo R, Freer AM, Zinyk DL et al. Specific and redundant functions of Gli2 and Gli3 zinc finger genes in skeletal patterning and development. Development 1997; 124(1):113-123.
45. Ruiz I Altaba A, Palma V, Dahmane N. Hedgehog-Gli signalling and the growth of the brain. Nat Rev Neurosci 2002; 3(1):24-33.
46. Echelard Y, Epstein DJ, St-Jacques B et al. Sonic hedgehog, a member of a family of putative signaling molecules, is implicated in the regulation of CNS polarity. Cell 1993; 75(7):1417-1430.
47. Teillet MA, Lapointe F, Le Douarin NM. The relationships between notochord and floor plate in vertebrate development revisited. Proc Natl Acad Sci USA 1998; 95(20):11733-11738.
48. Ahlgren SC, Bronner-Fraser M. Inhibition of sonic hedgehog signaling in vivo results in craniofacial neural crest cell death. Curr Biol 1999; 9(22):1304-1314.
49. Borycki AG, Brunk B, Tajbakhsh S et al. Sonic hedgehog controls epaxial muscle determination through Myf5 activation. Development 1999; 126(18):4053-4063.
50. Wechsler-Reya RJ, Scott MP. Control of neuronal precursor proliferation in the cerebellum by Sonic Hedgehog. Neuron 1999; 22(1):103-114.
51. Rowitch DH, S-Jacques B, Lee SM et al. Sonic hedgehog regulates proliferation and inhibits differentiation of CNS precursor cells. J Neurosci 1999; 19(20):8954-8965.
52. Riddle RD, Johnson RL, Laufer E et al. Sonic hedgehog mediates the polarizing activity of the ZPA. Cell 1993; 75(7):1401-1416.
53. Buscher D, Ruther U. Expression profile of Gli family members and Shh in normal and mutant mouse limb development. Dev Dyn 1998; 211(1):88-96.
54. Johnson DR. Extra-toes: Anew mutant gene causing multiple abnormalities in the mouse. J Embryol Exp Morphol 1967; 17(3):543-581.
55. Enomoto-Iwamoto M, Nakamura T, Aikawa T et al. Hedgehog proteins stimulate chondrogenic cell differentiation and cartilage formation. J Bone Miner Res 2000; 15(9):1659-1668.
56. Jorgensen C, Noel D, Gross G. Could inflammatory arthritis be triggered by progenitor cells in the joints? Ann Rheum Dis 2002; 61:6-9.
57. Walterhouse DO, Lamm MLG, Villavicencio E et al. Emerging roles for Hedgehog-Patched-Gli signal transduction in reproduction. Biol Reprod 2003; 69:8-14.

58. Lewis MT, Ross S, Strickland PA et al. Defects in mouse mammary gland development caused by conditional haploinsufficiency of Patched-1. Development 1999; 126(22):5181-5193.
59. Zhang Y, Kalderon D. Hedgehog acts as a somatic stem cell factor in the Drosophila ovary. Nature 2001; 410 (6828):599-604.
60. Bitgood MJ, Shen L, McMahon AP. Sertoli cell signaling by Desert hedgehog regulates the male germline. Curr Biol 1996; 6(3):298-304.
61. Berman DM, Desai N, Wang X et al. Roles for Hedgehog signaling in androgen production and prostate ductal morphogenesis. Dev Biol 2004; 267(2):387-398.
62. Lamm ML, Catbagan WS, Laciak RJ et al. Sonic hedgehog activates mesenchymal Gli1 expression during prostate ductal bud formation. Dev Biol 2002; 249(2):349-366.
63. Grindley JC, Bellusci S, Perkins D et al. Evidence for the involvement of the Gli gene family in embryonic mouse lung development. Dev Biol 1997; 188(2):337-348.
64. Fukuda K, Yasugi S. Versatile roles for sonic hedgehog in gut development. J Gastroenterol 2002; 37(4):239-246.
65. Pappu KS, Chen R, Middlebrooks BW et al. Mechanism of hedgehog signaling during Drosophila eye development. Development 2003; 130(13):3053-3062.
66. Takabatake Y, Takabatake T, Sasagawa S et al. Conserved expression control and shared activity between cognate T-box genes Tbx2 and Tbx3 in connection with Sonic hedgehog signaling during Xenopus eye development. Dev Growth Differ 2002; 44(4):257-271.
67. Stenkamp DL, Frey RA, Mallory DE et al. Embryonic retinal gene expression in sonic-you mutant zebrafish. Dev Dynamics 2002; 225(3):344-350.
68. Koyama E, Wu C, Shimo T et al. Chick limbs with mouse teeth: An effective in vivo culture system for tooth germ development and analysis. Dev Dynamics 2003; 226(1):149-154.
69. Liu HX, Maccallum DK, Edwards C et al. Sonic hedgehog exerts distinct, stage-specific effects on tongue and taste papilla development. Dev Biol 2004; 276(2):280-300.
70. Miura H, Kusakabe Y, Sugiyama C et al. Shh and Ptc are associated with taste bud maintenance in the adult mouse. Mech Dev 2001; 106(1-2):143-145.
71. LaMantia AS, Bhasin N, Rhodes K et al. Mesenchymal/epithelial induction mediates olfactory pathway formation. Neuron 2000; 28(2):411-425.
72. Nanba D, Nakanishi Y, Hieda Y. Role of Sonic hedgehog signaling in epithelial and mesenchymal development of hair follicles in an organ culture of embryonic mouse skin. Dev Growth Differ 2003; 45(3):231-239.
73. Mill P, Mo R, Fu H et al. Sonic hedgehog-dependent activation of Gli2 is essential for embryonic hair follicle development. Genes Dev 2003; 17(2):282-294.
74. Herzog W, Zeng X, Lele Z et al. Adenohypophysis formation in the zebrafish and its dependence on sonic hedgehog. Dev Biol 2003; 254(1):36-49.
75. Pola R, Ling LE, Silver M et al. The morphogen Sonic hedgehog is an indirect angiogenic agent upregulating two families of angiogenic growth factors. Nat Med 2001; 7(6):706-711.
76. Logan M, Pagan-Westphal SM, Smith DM et al. The transcription factor Pitx2 mediates situs-specific morphogenesis in response to left-right asymmetric signals. Cell 1998; 94(3):307-317.
77. Shah DK, Hager-Theodorides AL, Outram SV et al. Reduced thymocyte development in sonic hedgehog knockout embryos. J Immunol 2004; 172(4):2296-2306.
78. Dahmane N, Sanchez P, Gitton Y et al. The Sonic Hedgehog-Gli pathway regulates dorsal brain growth and tumorigenesis. Development 2001; 128(24):5201-5212.
79. Thayer SP, di Magliano MP, Heiser PW et al. Hedgehog is an early and late mediator of pancreatic cancer tumorigenesis. Nature 2003; 425(6960):851-856.
80. Ellis T, Smyth I, Riley E et al. Overexpression of Sonic Hedgehog suppresses embryonic hair follicle morphogenesis. Dev Biol 2003; 263(2):203-215.
81. Kim PC, Mo R, Hui CCC. Murine models of VACTERL syndrome: Role of sonic hedgehog signaling pathway. J Pediatr Surg 2001; 36(2):381-384.
82. Motoyama J, Liu J, Mo R et al. Essential function of Gli2 and Gli3 in the formation of lung, trachea and oesophagus. Nat Genet 1998; 20(1):54-57.
83. Unger S, Copland I, Tibboel D et al. Down-regulation of sonic hedgehog expression in pulmonary hypoplasia is associated with congenital diaphragmatic hernia. Am J Pathol 2003; 162(2):547-555.

CHAPTER 2

Sonic Hedgehog Signalling in Dorsal Midline and Neural Development

Silvia L. López and Andrés E. Carrasco*

Abstract

Sonic hedgehog is a secreted morphogen involved in patterning of a variety of structures and organs in vertebrates. In this chapter we focus on its role in the development of the floor plate and in the events that pattern and configure the shape and size of the central nervous system.

The Hedgehog Pathway

Hedgehog (Hh) proteins comprise a family of secreted morphogens that exert short and long range actions essential for patterning a variety of structures during animal embryogenesis.[1-5] In the course of their maturation process, Hh proteins undergo an autocatalytic cleavage that releases the active N-terminal polypeptide, which gains hydrophobicity by cholesterol and palmitate additions important for modulating the range of action. Short-range signalling involves tethering by cholesterol and up-regulation by Hh of its own receptor, Patched (Ptc), which is supposed to limit the range of action by ligand sequestration. Signalling at a distance depends on Dispatched, a transmembrane sterol-sensing protein necessary for release of Hh from the sending cell, and requires heparan sulfate proteoglycans and enzymes for heparan sulfate biosynthesis. After sensing the morphogen concentration, perhaps by perceiving the ratio of liganded to unliganded Ptc, the field of receiving cells modulate the activity of different forms of the latent cytoplasmic zinc-finger transcription factors Ci (Drosophila) and Gli (vertebrates), which ultimately turn-on different sets of target genes according to the distance to the morphogen source. In fact, Ci can display activating and repressing forms: the full-length transcriptional activator and the repressing N-terminal fragment generated by proteolysis (CiR). In vertebrates, the three homologues of Ci have activating properties, and only the proteolytic N-fragments of Gli2 and Gli3 appear to function as potent transcriptional repressors.

The mechanism by which Hh signalling is transduced is complex and subtly modulated, and actually involves a release of repression. In the absence of Hh signalling, the kinesin-like protein Costal-2 (Cos2) is stably associated with Ci. In this complex, Cos2 may mediate the scaffolding of a series of kinases that sequentially phosporylate Ci. Phosphorylated Ci then undergoes proteolysis, rendering the repressor form CiR. At the same time, Su(fu) (Suppressor of fused) inhibits the transcriptional activity of full-length Ci.

*Corresponding Author: Andrés E. Carrasco—Laboratorio de Embriología Molecular, Instituto de Biología Celular y Neurociencias, Facultad de Medicina, Universidad de Buenos Aires - CONICET. Paraguay 2155, Piso 3, Ciudad Autónoma de Buenos Aires, Argentina. Email: rqcarras@mail.retina.ar

Shh and Gli Signalling and Development, edited by Carolyn E. Fisher and Sarah E.M. Howie. ©2006 Landes Bioscience and Springer Science+Business Media.

When Hh binds Ptc, the transmembrane protein Smoothened (Smo) is released from a state of repression. If fully active, Smo recruits Cos2 and Fused (Fu). In this state, Fu is stabilized and inhibits Su(fu), leading to positive transcriptional activity of full-length Ci. Meanwhile, recruitment of Cos2 by active Smo results in blockage of Ci phosphorylation and ceasing of CiR production. Therefore, both actions result in the full range response of the transcriptional targets of Hh signalling. But if Smo is partially active, Fu is not stabilized, the activity of full-length Ci remains suppressed and only CiR formation is stopped, resulting in a partial response of a subset of target genes (for revision of Hh transport, release, reception and transduction, see refs. 6-10). Three distinct members of the Hh family have been characterized in vertebrates: *sonic (shh), indian (ihh)* and *desert (dhh)*. We will focus on the role of *shh* in floor plate and neural development.

Shh Signalling and Floor Plate Formation

The floor plate (FP) is a modified glial structure located in the ventral midline of the vertebrate neural tube. It constitutes an important source of signals involved in dorsoventral (D/V) neural patterning, proliferation and survival of neural precursors, and attraction and repulsion of axons en route to their destination.[11-16] It is generally accepted that as an anatomical structure, it extends from the ventral midbrain to the tail region. However, *shh* and *netrin-1*, typical FP markers, are also detectable in the diencephalon of all vertebrates, suggesting that the ventral midline of the anterior brain share FP properties.

Although vertebrates display some variation in the molecular dynamics, a consensus picture shows the FP as composed of three longitudinal regions, one medial (MFP) flanked by two lateral ones (LFP). In mouse and zebrafish, *shh* is only expressed by the MFP, while the winged-helix transcription factor *foxa2* (formerly known as *hnf3β*) is present in both populations. In chicken, MFP and LFP cells initially express both markers, but *foxa2* later becomes restricted to MFP, while some *shh* expression remains in the LFP.[17-18] However, an issue that cannot be circumvented is that apart from the variable expression of *shh* or *foxa2*, the LFP expresses neural markers, like the transcription factors *sox1* and *nkx2.2*. Sox1 is a general neuroepithelial marker necessary and sufficient to maintain panneural properties of neural progenitor cells.[19] Nkx2.2 is expressed by the progenitors of V3 interneurons and oligodendrocytes and it is necessary for their differentiation in the ventral neural tube.[20-22] Besides, the LFP is constituted by seudostratified neuroepithelium, different from the polarized cell structure characteristic of the MFP.[23] Therefore, the LFP could be seen as part of the ventral neural tube domain where neural progenitors are deciding their fate. Indeed, the feature that has been used to define the FP is the expression of *shh* or *foxa2*, which might be insufficient to determine a real functional unit. In conclusion, the division between LFP and MFP could be seen as a simplistic interpretation from the expression patterns of some markers.

The origin of the FP has been the subject of great controversy. The canonical model proposes that it is induced on the neural ectoderm by vertical signals from the underlying notochord. This was founded on the observations that the avian FP cannot develop after removal of the notochord but appears ectopically after grafting notochordal tissue onto the lateral or dorsal regions of the neural tube.[24] Evidence from different vertebrate species highlighted Shh as the signal responsible for inducing FP in a typical short-range action that requires direct contact with the notochord and exposition to high concentrations of the morphogen. Lower levels of Shh secreted from the FP would then induce diverse cell types in the neighbouring ventral spinal cord, including motor neurons and interneurons, in a dose-dependent way.[3,4,25-29]

In mouse, Shh signalling appears to be essential for FP development, since targeted disruption of *shh* blocks FP differentiation without impairing the early development of the notochord.[30] However, genetic manipulations in zebrafish undermined the protagonist role of Hh on FP development, because mutations of members of this pathway only impaired the development of the LFP. In this species, Nodal signalling was proposed to induce MFP and then, secreted Hh from the MFP would induce LFP. Nevertheless, in the absence of Nodal signals, some cells can acquire a mixed MFP-LFP character at later stages. It remains to be elucidated whether Hh could play a role in this late differentiation process. Strikingly, *smu (smo)* mutants, which have a general blockade of Hh signalling, ultimately show gaps in the MFP, suggesting that Hh, although not necessary for MFP specification in zebrafish, is later required for maintaining the phenotype of these cells or for their survival.[18,31,32]

In the last few years, a second model of FP development has challenged the idea of its neural origin. New experiments on birds led to the proposal that Hensen's node (equivalent to the amphibian's Spemann's organiser) generates both midline structures. The reason why the FP does not develop after the removal of the notochord is that the FP precursors are removed together.[23] Thus, the FP would be a mesodermal derivative rather than a neuroectodermal one. Indeed, the groundwork for this idea can be traced back to the pioneering experiments of Spemann and Mangold,[33] who clearly demonstrated that the implanted amphibian dorsal lip differentiates into notochord and FP in the trunk. Fate maps of the embryonic shield, the teleost equivalent of the Spemann's organiser, also established that this region contributes to both structures.[34,35]

The Hensen's node can be subdivided into three morphological and functional domains.[36] The caudal-most tip (named zone c by the authors) ends at the axial-paraxial hinge and contains *foxa2+ shh-* cells closely packed and randomly arranged. The medial part (zone b) lies on the median pit and contains *foxa2+ shh+* cells. An outline of two cell-layers becomes apparent in this zone: an epithelial-like layer, presumably containing the FP precursors, which already shows a columnar arrangement, and the deep layer, delineating the future notochord, with cells distributed at random; yet there is no clear separation between both structures. The rostral portion of the node (zone a) contains *foxa2+* cells and more *shh+* cells than zone b. Here, distinction of the notochord and FP is more clearly defined, but they are still in close association, although already separated by a basement membrane.

If zone b is removed, caudalward movement of zone c still occurs. However, the embryos then bear an interruption of midline cells (notochord and FP) at the trunk level. The stretch of neural tube formed consequently is smaller and is devoid of FP and motorneurons. Noteworthy, at more caudal levels, the midline cells and the neural tube resume normally. But if zone c is removed, caudalward movement of the node ceases. The neural tube formed posterior to the excision lacks notochord and FP, and it is completely dorsalised. On the basis of these observations, the authors suggest that zone c contains self-renewing cells with the potential to develop either as notochord or FP. However, grafting experiments demonstrated that although zone b contributes to all midline cells caudal to the level of the graft, zone c normally provides very few cells to the caudal FP.[26,36] Thus, although zone c, as a source of stem cells, can compensate for notochord and FP precursors when zone b is ablated, it is more likely that the bipotential, self-renewing precursors of FP and notochord are mainly found in zone b during normal development. This does not rule out that commitment to either fate also begins in this zone. Several groups have shown that Shh is a potent mitogen (see below). It would be interesting to test if Shh, whose expression is evident in zone b, is promoting mitosis of midline precursor cells. All together, the node could be considered as a functional cell niche.

The hypothesis that the notochord and the FP arise from a population of pluripotent precursors located in the vertebrate's organiser has been strongly supported by genetic evidence. In zebrafish and *Xenopus*, Delta-Notch signalling executes a binary cell-fate decision, promoting FP specification at the expense of the notochord.[37,38] In zebrafish *ntl* mutant embryos the notochord does not develop and the MFP is widened.[39]

In *Xenopus*, Notch signalling enhances *shh* expression in FP precursors (Fig. 1A,B), and secreted Shh represses notochord specification (Fig. 1C,D).[38] and expands the FP (Rosato-Siri et al, unpublished results). Therefore, Shh would amplify the effects of the binary decision initially promoted by active Notch, resulting in an even larger population of specified FP precursors in detriment of the notochord. This mechanism could in part underlie a previously unrecognised role of Shh as FP inducer.[38] This is consistent with the normal profile of *shh*

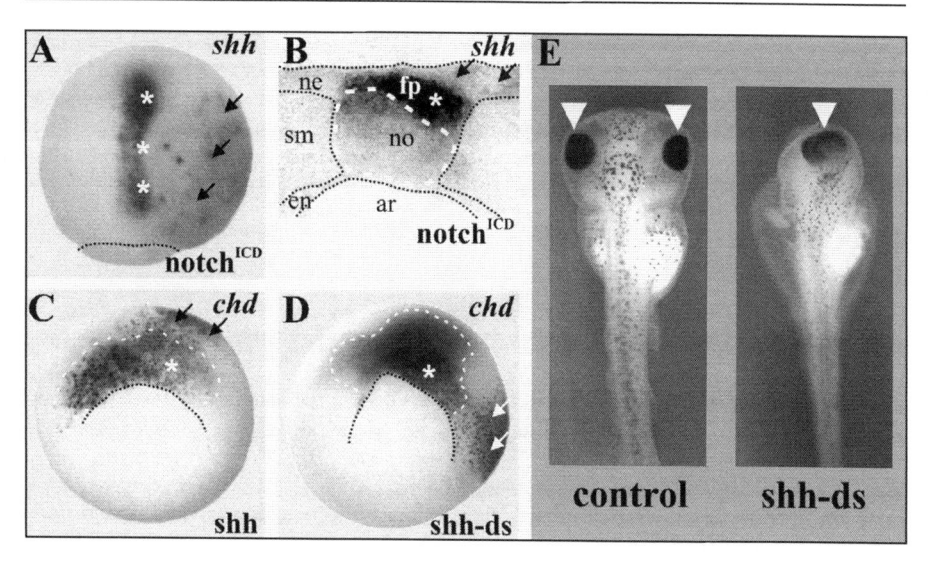

Figure 1. Notch and Shh signalling in *Xenopus* dorsal midline development. A,B) Notch increases *shh* expression in the FP, which is expanded at the expense of the notochord. A) Dorsal view of an early neurula injected with mRNA encoding the intracellular domain of Notch (notch^{ICD}), a constitutively active form independent of ligand binding. Anterior is oriented to the top. The level of *shh* transcripts (dark staining in the midline in black and white prints, purple in the original) is increased on the injected side (right, asterisks). The dotted line demarcates the nearly closed blastopore. B) Transverse section of the same embryo shown in (A). The *shh* expression domain in the FP (dark staining in black and white prints, purple in the original) is expanded (asterisk), while the notochord is reduced on the injected side (right). In A) and B), cells that inherited the injected mRNA were revealed by magenta-phos immunostaining of the myc-tag epitope fused to Notch^{ICD}, which is evidenced as the pale grey shadow indicated by arrows. The broken white line in B) demarcates the contour of the notochord (no). ne: neuroectoderm; sm: somitic mesoderm; en: endoderm; ar: archenteron. C,D) Shh signalling restricts the number of notochord precursors. C) Vegetal view of an early gastrula injected with *shh* mRNA. Notochordal precursors are decreased in the organiser on the injected side (right, asterisk), as evidenced by the down-regulation of the notochordal marker *chordin* (*chd*). The *chd* expression domain (purple staining in the original) is demarcated by the broken white line. D) vegetal view of a mid gastrula where *shh* function was knocked-down by injection of *shh* double-stranded RNA (shh-ds). *Chd* expression (purple staining in the original, demarcated by the broken white line) shows that the number of notochordal precursors is increased in the organiser on the injected side (right, asterisk). In (C) and (D), dorsal is oriented to the top, arrows point to the nuclear Xgal staining (turquoise in the original) revealing the co-injected nuc-lacZ lineage tracer, and the dotted line demarcates the dorsal blastopore lip. E) Dorsal views of a control tadpole (left) and a sibling embryo showing cyclopia as the result of knocking-down *shh* function with shh-ds RNA (right). Eyes are pointed by arrowheads. Embryos shown in (E) were extracted from López et al.[38]

expression during amphibian or avian development. Although *shh* transcripts are detectable both in notochord and FP precursors, expression becomes significantly higher in the latter during gastrulation.[38,40] Thus, the specification of the different midline fates may be linked to a differential regulation of the *shh* gene. If cells are committed to FP fates by active Notch, *shh* would be more actively transcribed. Interestingly, the dissection of the regulatory regions of the mouse *shh* gene has uncovered two enhancers that direct expression to FP, one upstream of the coding region (SFPE1) and the other within intron2 (SFPE2).[41] Strikingly, deletion of the proximal region of SFPE1 unmasks a potent notochord enhancer (SNE), whereas expression in the FP decreases substantially. It will be interesting to determine whether this region contains binding sites for repressors that restrict notochordal expression and if the complete SFPE1 underlies a mechanism for switching notochordal to FP expression. Besides, *foxa2* function is probably required by SNE but not by SFPE1 activity. On the other hand, SFPE2 contains a sequence of significant homology between mouse, chicken and zebrafish, which harbours binding sites for Foxa2 and homeodomain transcription factors.[42] When transgenes containing trimers of this sequence were assayed, expression of the reporter was then found in the notochord in addition to the FP. Intriguingly, while the Foxa2 binding sites are necessary for notochord and FP expression, the homeodomain binding site is required only for FP expression. The evidence suggests that expression of *shh* in the notochord and the FP is controlled by shared and independent mechanisms.

Despite growing support for the hypothesis that the vertebrate's organiser contributes significantly to FP formation, some disagreement still persists. In the mouse embryo, before *shh* expression begins in the central nervous system (CNS), transcripts are found in the ventral (mesodermal) layer of the node, and rostrally in the notochordal plate. Because expression was noticed neither in the dorsal layer of the node or in the ventral midline of the more rostral neural plate, it was argued that the mouse FP does not derive from the node but is induced by the canonical signal from the notochord.[42] However, the idea that the notochord and part of the FP share embryonic origin is not incompatible with different patterns of gene expression once both populations have been committed to their respective fates in the node. In fact, cell-lineage tracing has shown that descendants of the dorsal layer of the mouse node populate the FP[43,44] and loss-of-function of delta-1 in mouse results in an excess of FP cells, while the notochord is reduced.[45] Although the opposite activities of Delta signalling in dorsal midline development in mouse and anamniotes embryos are intriguing, in both cases they underscore the existence of a population of cells in the organiser with equal potential to develop either as FP or notochord.

Recent evidence from avian embryos have resolved important discrepancies between the two models of FP formation. Much of this understanding comes from considering the different cell populations that compose this structure, which allowed new questions about the role of *shh* to be addressed. Careful analysis of molecular markers in quail-chick chimeras, where the chick Hensen's node was replaced by its quail counterpart, demonstrated that the MFP derives from the node, while the LFP derives from the neural plate.[17] Utilising lineage tracing, two areas of FP precursors in the chick epiblast have been distinguished; one anterior to the Hensen's node (prenodal epiblast, originally designated as "area a" by the authors), which gives rise to the earliest-forming FP in the cephalic region, and the other one in the Hensen's node, whose descendants later populate the posterior FP and the notochord.[46] Thus, the anterior FP (AFP) would be of neural origin. Although the authors did not address the issue of whether the node-derived population becomes incorporated into the LFP or MFP, they are presumably observing the medial component, as demonstrated by Charrier et al.[17]

Notochord and MFP grafts are able to induce a complete ectopic FP in the avian neural tube but with temporal and spatial restrictions. While a supernumerary LFP appears at any stage of the window tested and throughout the full length of the caudal neural tube exposed to induction, MFP is induced only in the posterior-most region of host embryos younger than 15 ss. In addition, the MFP graft needs a very close contact with the host neural epithelium to induce MFP. In contrast, Shh can induce only LFP in the neural ectoderm. Thus, Shh alone is insufficient to transform neural cells into MFP in the avian embryo.[17] Other factors provided by the notochord, presumably BMP antagonists, may be acting in cooperation with Shh to fulfil this process.[47]

The AFP is rapidly induced on the neural ectoderm by a vertical contact with the nascent prechordal mesoderm while passing beneath the prenodal region of the epiblast. Prechordal mesoderm expresses *shh* and *nodal-1*. Shh alone is sufficient to induce FP markers in prenodal epiblast explants, but only at high concentrations; when the explants are exposed to low concentrations of Shh in the presence of Nodal 1, a robust induction of FP markers is observed. Thus, Nodal and Shh signalling may cooperate during the early and rapid induction of the AFP by the prechordal mesoderm.[46]

Interestingly, an area analogous to the prenodal epiblast may exist in *Xenopus*. It is conformed by an arc of *hairy-2a+* cells in the dorsal noninvoluting marginal zone and marks the earliest signs of FP induction.[48] It will be interesting to elucidate whether the FP phenotype of these cells is induced by Shh and Nodal signals secreted by the prechordal plate, as was proposed for the avian embryo.

At this point, a main conclusion can be raised: most of the embryonic models studied so far suggest that the specification of the FP and notochord start earlier than previously recognised, challenging the canonical model where notochord induces FP. The main disagreement resides in the absolute requirement of *shh* for FP development in mouse, as opposed to its sole role as LFP inducer in zebrafish. An integrative model for FP development can be synthesised as follows:

1. The anterior FP, of neural origin, would be induced early on the prenodal epiblast by Shh with the cooperation of Nodal, both emanating from the prechordal mesoderm that passes beneath.
2. The MFP, located posteriorly, and primarily of mesodermal origin, would be induced within the organiser, before the segregation of notochord and MFP precursors, in a binary switch triggered by Delta-Notch that favours FP fates at the expense of the notochord. This enhances *shh* expression in MFP precursors, and secreted Shh contributes to repress the notochordal fate and amplifies the switch. In turn, specified MFP precursors populate the midline of the neural plate. It remains to be elucidated whether Delta-Notch signalling modulates *shh* expression in amniote embryos, but *shh* is essential for the induction of FP in mouse. Other questions must be answered; for example, which are the molecules that pattern the scattered expression of *delta-1* in the organiser, which initially would define the distribution of MFP and notochord precursors.[48] It will be interesting to investigate whether Nodal signalling is involved in this process or acts independent of the Notch switch, given the absolute requirement of Nodal for MFP development in zebrafish. In addition, some MFP cells may be induced on the neural ectoderm by node derived MFP in close contact with the neural plate, but Shh alone is insufficient for this process and would require BMP antagonists derived from the notochord.
3. Consequent to Notch activation, MFP secretes high levels of Shh, leading to the short-range induction of the LFP on the neighbouring neural plate.

Shh in Neural Development

Ventral Neural Patterning

Shh has been classically considered as an inducer of different types of ventral neurons in the spinal cord, the phenotype of which varies according to the morphogen concentration, depending on the distance from the source in the FP. These kinds of neurons are arrayed from ventral to dorsal as follows: V3 interneurons (the closest to the FP), motorneurons (MN), V2, V1 and V0 interneurons, the latter at the level of the intermediate neural tube.[49] During the specification of ventral neural fates in the spinal cord, the expression of different combinations of homeodomain and basic-helix-loop-helix transcription factors determine the identity of neural progenitors. Shh promotes the expression of some of these molecules (Class II) and represses the expression of others (Class I). Cross-repression between both classes defines ventral spinal cord domains that will generate specific types of neurons.[50-51] Knock-out mice for *shh* neither develop FP nor most ventral neuronal types, including V2 interneurons.[9] *Smo* is essential for all Hh signalling, and its loss-of-function generates a more severe phenotype, where failure in the specification of ventral cell types is more dorsally extended. The differences with *shh* mutants most likely reflect a normal contribution of Ihh signalling from the underlying gut endoderm.[52]

The three known mediators of Hh signalling are expressed in partially overlapping domains in the neural epithelium. Although their patterns are arranged in gradients with more or less widespread distributions, Gli1 is predominantly ventral, Gli3 dorsal, and Gli2 intermediate and dorsal. In mouse, Gli2 is essential for FP formation and is required, together with Gli1, for V3 development.[9] In embryos lacking all Gli function the FP and V3 interneurons do not arise. Strikingly, these mutants develop MN and V0 to V2 interneurons, but their distribution is totally disorganised. Thus, Hh signalling is essential in mouse for the specification of FP and the most ventral interneurons. Although not necessary for the induction of MN and the remaining interneurons, it regulates their spatial pattern. Transcription factors other than Gli may be responsible for inducing differentiation of some MN and V0 to V2 interneurons.[53] Retinoids are good candidates for regulating the expression of such molecules,[51,54] and Shh signalling may be necessary for the selective survival and expansion of precursor pools.[9]

Shh as a Mitotic and Anti-Apoptotic Agent

The development of the CNS depends on the precise coordination of growth and patterning mechanisms. Although the latter are becoming well understood, less is known about the factors that govern the shape and size of the CNS. Recent studies indicate that Shh is involved in the control of growth and cell survival during early and late stages of development, providing cues for size and shape. Indeed, in 1950 it was already shown that the embryonic chick brain collapses if the notochord and anterior hindbrain are separated from the neuroepithelium. At first glance, these results were attributed to an "experimental overgrowth". However, although more cells were in mitosis because their cell cycle was longer, there was net cell loss. It was concluded that the notochord normally secretes a trophic factor important for the expansion of the brain vesicles. Recently, it was shown that when the notochord is transiently displaced from the midbrain FP, the brain vesicles also collapse and fold abnormally. Although patterning and differentiation is not impaired, proliferation decreases and apoptosis increases in the midbrain. This is explained by the reduction of Shh levels in the notochord and the FP, since an implant of Shh-secreting cells in the ventral midbrain reverts the effect, and the normal midbrain expansion is retarded by cyclopamine, an inhibitor of the Shh pathway.[55-56] In addition, the injection of antibodies against Shh into the chick cranial mesenchyme inhibits proliferation in the neural tube and induces massive apoptosis in cranial neural tube and neural crest.[57] This evidence supports the idea that the ventral midline of the neural tube, by secreting Shh, is involved in the process of three-dimensional shaping during the early growth of the brain by controlling proliferation and cell survival.

The anti-apoptotic role of Shh is also crucial for the development of more caudal regions of the CNS. Programmed cell death in the posterior neural tube of *shh* knock-out mice is restricted to discrete ventral and dorsal regions and occurs between E9.5 and E10.5.[58] When chick embryos are deprived of midline structures by ablation of zone c of Hensen's node, the neural tube posterior to the excised zone develops but it is completely dorsalised and displays massive cell death after 20 h of operation. However, apoptosis is prevented when a graft of midline cells (notochord or FP) or a Shh source is provided.[16,23] These results strengthen the idea that Shh secreted by midline cells, is absolutely required to keep the neural tube alive. However, it remains to be elucidated whether the apoptosis produced by the absence of midline structures can be entirely homologated to the effects produced by removing Shh signalling. Programmed cell death has the role of sculpting the shape and size of organs. The built-in suicide program, first demonstrated in *Caenorhabditis elegans*,[59] can be seen as a default condition that must be modulated to attain the correct form and shape of the neural tube, and Shh signalling has a crucial role in this balance. In fact it was recently determined that Ptc has a proapoptotic role, which is prevented by binding of Shh.[60]

Several findings from *Xenopus* embryos indicate that Shh signalling is involved in a balance between neuronal differentiation and the control of cell number, and this balance receives the input of retinoid signalling. Overexpression of *shh* in frog embryos inhibits primary neurogenesis in the spinal cord and thickens the neural plate but later, an increase of secondary motorneurons is evident. The expansion of the neural plate was attributed to an increase in proliferation. On the other hand, when *shh* was knocked-down, primary neurogenesis was enhanced, and the absence of midline signalling impaired the normal division of the brain into two hemispheres, which resulted in diverse grades of cyclopia (Fig. 1E). Retinoids inhibit *shh* expression and enhance primary neurogenesis. To explain the opposite effects on primary and secondary neurogenesis and the counterbalancing activity of retinoids, it was proposed that Shh withdraws neural precursors from premature differentiation by retinoid signalling, holding their proliferative state and reserving them for subsequent waves of differentiation.[38,61,62] These results are consistent with findings from mouse embryos. Ectopic expression of *shh* in the dorsal neural tube of transgenic mice induces proliferation of neural precursors and inhibits their differentiation.[14] In *shh* knock-out mice the telencephalon is 90% smaller than normal and consists of a single fused vesicle, strongly dorsalised; ventral and dorsal diencephalic structures are also reduced. This dramatic phenotype is due to the disruption of brain proliferation and to increased apoptosis.[30]

Shh is expressed in a layer-specific manner in the perinatal mouse neocortex and tectum, while *gli* transcripts are found in proliferative zones. *Shh* is required as a mitogen after stage E12 in the superficial layer of the tectum and neocortex (layer V) and also in ventricular and subventricular zones, where *gli* genes are expressed. This resembles the situation in the cerebellum, where Shh secreted by the Purkinje neurons induces proliferation of the granular layer. It is also similar to the mechanism in the hippocampus, where secreted Shh from cells in the hilus of the dentate gyrus induces proliferation of granular and septal cells.[63-65] Therefore, Shh signalling associated with proliferative regions could be part of a general mechanism of control of the cell number by regulating cell cycle and cell death.

Besides its role during CNS development, a growing line of evidence points to a crucial role of Shh signalling in the maintenance of postnatal and adult telencephalic stem cell niches. For example, the adult rat hippocampus expresses high levels of *ptc,* and when exposed to an adeno-associated viral vector delivering *shh* cDNA, a potent mitogenic effect is observed. Neural progenitors isolated from this region and cultured with Shh proliferate, retaining their multipotency.[66] Conditional null alleles of *shh* and *smo* display increased apoptosis of neural progenitors in the postnatal subventricular zone and reduced proliferation in the gyrus dentate.[67]

Therefore, Shh has been consolidated as an anti-apoptotic and mitogenic factor that controls growth and shape during the development of the CNS and it is also present in adult neurogenic niches, where the complex architecture requires premature differentiation to be inhibited on behalf of remodelling and plasticity.[68,69]

Closing the Idea

The midline structure composed by the FP and the notochord is formed by a crucial binary switch executed by Delta-1/Notch/hairy-2, which controls the distribution of cell fates in the organiser, and Shh contributes to refine the shape and size of both structures. In turn, their size provide the basis for the number of cells that secrete Shh, whose diffusion influences (1) the shape and size of the neural plate, by controlling cell number; (2) the correct patterning of the neural tube. Therefore, building of the midline is a crucial part of the program by which the organiser commands the distribution of signals and cell fates to insure the proper organising activity defined by Hilde Mangold and Hans Spemann in 1924.[33]

Acknowledgements

We are grateful to Paula Franco, Alejandra Paganelli and María Victoria Rosato-Siri for their permanent collaboration and helpful discussions. S. L. L. and A. E. C. are from CONICET.

References

1. Lee JJ, von Kessler DP, Parks S et al. Secretion and localized transcription suggest a role in positional signaling for products of the segmentation gene hedgehog. Cell 1992; 71:33-50.
2. Riddle RD, Johnson RL, Laufer E et al. Sonic hedgehog mediates the polarizing activity of the ZPA. Cell 1993; 75:1401-1416.
3. Echelard Y, Epstein DJ, St-Jacques B et al. Sonic hedgehog, a member of a family of putative signaling molecules, is implicated in the regulation of CNS polarity. Cell 1993; 75:1417-1430.
4. Krauss S, Concordet JP, Ingham PW. A functionally conserved homolog of the Drosophila segment polarity gene hh is expressed in tissues with polarizing activity in zebrafish embryos. Cell 1993; 75:1431-1444.
5. Ekker SC, McGrew LL, Lai CJ et al. Distinct expression and shared activities of members of the hedgehog gene family of Xenopus laevis. Development 1995; 121:2337-2347.
6. McMahon AP. More surprises in the Hedgehog signaling pathway. Cell 2000; 21:185-188.
7. Nybakken K, Perrimon N. Hedgehog signal transduction: Recent findings. Curr Opin Genet Dev 2002; 12:503-511.
8. Nusse R. Wnts and Hedgehogs: Lipid-modified proteins and similarities in signaling mechanisms at the cell surface. Development 2003; 130:5297-5305.
9. Ruiz i Altaba A, Nguyen V, Palma V. The emergent design of the neural tube: Prepattern, SHH morphogen and GLI code. Curr Opin Genet Dev 2003; 13:513-521.
10. Briscoe J. Hedgehog signaling: Measuring ligand concentrations with receptor ratios. Curr Biol 2004; 14:R889-R891.
11. Tanabe Y, Jessell TM. Diversity and pattern in the developing spinal cord. Science 1996; 274:1115-1123.
12. Colamarino SA, Tessier-Lavigne M. The role of the floor plate in axon guidance. Annu Rev Neurosci 1995; 18:497-529.
13. Stoeckli ET, Landmesser LT. Axon guidance at choice points. Curr Opin Neurobiol 1998; 8:73-79.
14. Rowitch DH, S-Jacques B, Lee SM et al. Sonic hedgehog regulates proliferation and inhibits differentiation of CNS precursor cells. J Neurosci 1999; 19:8954-8965.
15. Wechsler-Reya RJ, Scott MP. Control of neuronal precursor proliferation in the cerebellum by Sonic Hedgehog. Neuron 1999; 22:103-114.
16. Charrier JB, Lapointe F, Le Douarin NM et al. Anti-apoptotic role of Sonic hedgehog protein at the early stages of nervous system organogenesis. Development 2001; 128:4011-4020.
17. Charrier JB, Lapointe F, Le Douarin et al. Dual origin of the floor plate in the avian embryo. Development 2002; 129:4785-4796.
18. Strähle U, Lam CS, Ertzer R et al. Vertebrate floor-plate specification: Variations on common themes. Trends Genet 2004; 20:155-162.
19. Graham V, Khudyakov J, Ellis P et al. SOX2 functions to maintain neural progenitor identity. Neuron 2003; 39:749-765.
20. Briscoe J, Sussel L, Serup P et al. Homeobox gene Nkx2.2 and specification of neuronal identity by graded Sonic hedgehog signalling. Nature 1999; 398:622-627.
21. Soula C, Danesin C, Kan P et al. Distinct sites of origin of oligodendrocytes and somatic motoneurons in the chick spinal cord: Oligodendrocytes arise from Nkx2.2-expressing progenitors by a Shh-dependent mechanism. Development 2001; 128:1369-1379.
22. Qi Y, Cai J, Wu Y et al. Control of oligodendrocyte differentiation by the Nkx2.2 homeodomain transcription factor. Development 2001; 128:2723-2733.
23. Le Douarin NM, Halpern ME. Discussion point. Origin and specification of the neural tube floor plate: Insights from the chick and zebrafish. Curr Opin Neurobiol 2000; 10:23-30.
24. Placzek M, Dodd J, Jessell TM. Discussion point. The case for floor plate induction by the notochord. Curr Opin Neurobiol 2000; 10:15-22.
25. Roelink H, Augsburger A, Heemskerk J et al. Floor plate and motor neuron induction by vhh-1, a vertebrate homolog of hedgehog expressed by the notochord. Cell 1994; 76:761-775.
26. Roelink H, Porter JA, Chiang C et al. Floor plate and motor neuron induction by different concentrations of the amino-terminal cleavage product of sonic hedgehog autoproteolysis. Cell 1995; 81:445-455.
27. Marti E, Bumcrot DA, Takada R et al. Requirement of 19K form of Sonic hedgehog for induction of distinct ventral cell types in CNS explants. Nature 1995; 375:322-325.
28. Ericson J, Morton S, Kawakami A et al. Two critical periods of Sonic Hedgehog signaling required for the specification of motor neuron identity. Cell 1996; 87:661-673.
29. Ericson J, Rashbass P, Schedl A et al. Pax6 controls progenitor cell identity and neuronal fate in response to graded Shh signaling. Cell 1997; 90:169-180.

30. Chiang C, Litingtung Y, Lee E et al. Cyclopia and defective axial patterning in mice lacking Sonic hedgehog gene function. Nature 1996; 383:407-413.
31. Chen W, Burgess S, Hopkins N. Analysis of the zebrafish smoothened mutant reveals conserved and divergent functions of hedgehog activity. Development 2001; 128:2385-2396.
32. Varga ZM, Amores A, Lewis KE et al. Zebrafish smoothened functions in ventral neural tube specification and axon tract formation. Development 2001; 128:3497-3509.
33. Spemann H, Mangold H. Über Induktion von Embryonalanlagen durch Implantation artfremder Organisatoren. Arch f mikr Anat u Entw Mech 1924; 100:599-638.
34. Shih J, Fraser SE. Distribution of tissue progenitors within the shield region of the zebrafish gastrula. Development 1995; 121:2755-2765.
35. Melby AE, Warga RM, Kimmel CB. Specification of cell fates at the dorsal margin of the zebrafish gastrula. Development 1996; 122:2225-2237.
36. Charrier JB, Teillet MA, Lapointe F et al. Defining subregions of Hensen's node essential for caudalward movement, midline development and cell survival. Development 1999; 126:4771-4783.
37. Appel B, Fritz A, Westerfield M et al. Delta-mediated specification of midline cell fates in zebrafish embryos. Curr Biol 1999; 9:247-256.
38. López SL, Paganelli AR, Rosato-Siri MV et al. Notch activates sonic hedgehog and both are involved in the specification of dorsal midline cell-fates in Xenopus. Development 2003; 130:2225-2238.
39. Halpern ME, Hatta K, Amacher SL et al. Genetic interactions in zebrafish midline development. Dev Biol 1997; 187:154-170.
40. Teillet MA, Lapointe F, Le Douarin NM. The relationships between notochord and floor plate in vertebrate development revisited. Proc Natl Acad Sci USA 1998; 95:11733-11738.
41. Epstein DJ, McMahon AP, Joyner AL. Regionalization of Sonic hedgehog transcription along the anteroposterior axis of the mouse central nervous system is regulated by Hnf3-dependent and -independent mechanisms. Development 1999; 126:281-292.
42. Jeong Y, Epstein DJ. Distinct regulators of Shh transcription in the floor plate and notochord indicate separate origins for these tissues in the mouse node. Development 2003; 130:3891-3902.
43. Sulik K, Dehart DB, Iangaki T et al. Morphogenesis of the murine node and notochordal plate. Dev Dyn 1994; 201:260-278.
44. Wilson V, Beddington RS. Cell fate and morphogenetic movement in the late mouse primitive streak. Mech Dev 1996; 55:79-89.
45. Przemeck GK, Heinzmann U, Beckers J et al. Node and midline defects are associated with left-right development in Delta1 mutant embryos. Development 2003; 130:3-13.
46. Patten I, Kulesa P, Shen MM et al. Distinct modes of floor plate induction in the chick embryo. Development 2003; 130:4809-4821.
47. Patten I, Placzek M. Opponent activities of Shh and BMP signaling during floor plate induction in vivo. Curr Biol 2002; 12:47-52.
48. Lopez SL, Rosato Siri MV, Franco PG et al. The Notch-target gene hairy-2a impedes the involution of notochord cells promoting floor plate fates in Xenopus embryos. Development 2005; 132(5):1035-46.
49. Jessell TM. Neuronal specification in the spinal cord: Inductive signals and transcriptional codes. Nat Rev Genet 2000; 1:20-29.
50. Briscoe J, Pierani A, Jessell TM et al. A homeodomain protein code specifies progenitor cell identity and neuronal fate in the ventral neural tube. Cell 2000; 101:435-445.
51. Appel B, Eisen JS. Retinoids run rampant: Multiple roles during spinal cord and motor neuron development. Neuron 2003; 40:461-464.
52. Wijgerde M, McMahon JA, Rule M et al. A direct requirement for Hedgehog signaling for normal specification of all ventral progenitor domains in the presumptive mammalian spinal cord. Genes Dev 2002; 16:2849-2864.
53. Bai CB, Stephen D, Joyner AL. All mouse ventral spinal cord patterning by hedgehog is Gli dependent and involves an activator function of Gli3. Dev Cell 2004; 6:103-115.
54. Pierani A, Brenner-Morton S, Chiang C et al. A sonic hedgehog-independent, retinoid-activated pathway of neurogenesis in the ventral spinal cord. Cell 1999; 97:903-915.
55. Kaellen B. Proliferation in the embryonic brain with special reference to the overgrowth phenomenon an its possible relationship to neoplasia. Prog Brain Res 1965; 14:263-278.
56. Britto J, Tannahill D, Keynes R. A critical role for sonic hedgehog signaling in the early expansion of the developing brain. Nat Neurosci 2002; 5:103-110.
57. Ahlgren SC, Bronner-Fraser M. Inhibition of sonic hedgehog signaling in vivo results in craniofacial neural crest cell death. Curr Biol 1999; 9:1304-1314.

58. Borycki AG, Brunk B, Tajbakhsh S et al. Sonic hedgehog controls epaxial muscle determination through Myf5 activation. Development 1999; 126:4053-4063.
59. Ellis HM, Horvitz HR. Genetic control of programmed cell death in the nematode C. elegans. Cell 1986; 44:817-829.
60. Thibert C, Teillet MA, Lapointe F et al. Inhibition of neuroepithelial patched-induced apoptosis by sonic hedgehog. Science 2003; 301:843-846.
61. Franco PG, Paganelli AR, López SL et al. Functional association of retinoic acid and hedgehog signaling in Xenopus primary neurogenesis. Development 1999; 126:4257-4265.
62. Paganelli AR, Ocaña OH, Prat MI et al. The Alzheimer-related gene presenilin-1 facilitates sonic hedgehog expression in Xenopus primary neurogenesis. Mech Dev 2001; 107:119-131.
63. Dahmane N, Ruiz i Altaba A. Sonic hedgehog regulates the growth and patterning of the cerebellum. Development 1999; 126:3089-3100.
64. Dahmane N, Sánchez P, Gitton Y et al. The Sonic Hedgehog-Gli pathway regulates dorsal brain growth and tumorigenesis. Development 2001; 128:5201-5212.
65. Palma V, Ruiz i Altaba A. Hedgehog-Gli signaling regulates the behaviour of cells with stem cell properties in the developing neocortex. Development 2003; 131:337-345.
66. Lai K, Kaspar BK, Gage FH et al. Sonic hedgehog regulates adult neural progenitor proliferation in vitro and in vivo. Nat Neurosci 2003; 6:21-27.
67. Machold R, Hayashi S, Rutlin M et al. Sonic hedgehog is required for progenitor cell maintenance in telencephalic stem cell niches. Neuron 2003; 39:937-950.
68. Alvarez-Buylla A, Lim DA. For the long run: Maintaining germinal niches in the adult brain. Neuron 2004; 41:683-686.
69. Kempermann G, Jessberger S, Steiner B et al. Milestones of neuronal development in the adult hippocampus. Trends Neurosci 2004; 27:447-452.

CHAPTER 3

Role of Hedgehog and Gli Signalling in Telencephalic Development

Paulette A. Zaki,* Ben Martynoga and David J. Price

Abstract

Studies performed over the last decade have significantly increased our understanding of the role of Hedgehog (Hh) signalling in brain development. Here, we review the various in vitro and in vivo studies demonstrating the importance of Hh signalling for dorsoventral patterning of the telencephalon. The use of conditional knockouts has been particularly helpful in defining the spatial and temporal requirements of Hh signalling during telencephalic development. We also discuss the primary effectors of Hh signalling, the Gli family of transcription factors, and focus on Gli3, which is particularly important for telencephalic development, as reflected in the severe telencephalic phenotype of *Gli3* mutant mice. The presence of some dorsoventral patterning in animals lacking both *Shh* and *Gli3* implies that, although these molecules are major players in patterning the telencephalon, other patterning factors exist.

Introduction

The secreted morphogen, Sonic hedgehog (Shh), is vital for ventral patterning along the entire rostrocaudal extent of the neural tube.[1-3] Although most work has concentrated on the role of Shh in patterning of the caudal part of the neural tube, the spinal cord, studies are beginning to elucidate the role that Shh plays in the development of the most rostral part of the neural tube, the telencephalon.

The Hedgehog Signalling Pathway

The *Shh* gene, along with genes for *Indian hedgehog (Ihh)* and *Desert hedgehog (Dhh)*, are homologues of the *Drosophila* gene *hedgehog* and code for ~45-kD precursor proteins.[4] When Hh binds to the transmembrane receptor, Patched (Ptc), an inhibitory effect on Smoothened (Smo) is relieved and the pathway is activated (for a thorough review of these interactions, see ref. 5). In *Drosophila*, Hh signalling is transduced by one protein, the zinc-finger transcription factor cubitus interruptus (Ci) (reviewed in refs. 5, 6). In the absence of Hh, Ci is cleaved to form an N-terminal fragment which acts as a transcriptional repressor. When Hh is present, the cleavage of Ci is inhibited and the full-length form of Ci is able to act as a transcriptional activator.

The Hh signalling pathway is more complex in vertebrates. One important difference between *Drosophila* and vertebrates is that there are three proteins in vertebrates which are

---

*Corresponding Author: Paulette A. Zaki—Genes and Development IDG, Section of Biomedical Sciences, University of Edinburgh Hugh Robson Building George Square, Edinburgh, EH8 9XD, U.K. Email: pzaki@ed.ac.uk

Shh and Gli Signalling and Development, edited by Carolyn E. Fisher and Sarah E.M. Howie. ©2006 Landes Bioscience and Springer Science+Business Media.

homologous to Ci: Gli1, Gli2 and Gli3.[7,8] It has been proposed that repressor and activator functions of Ci have been distributed among the three Gli proteins. For instance, expression of *Gli1* and *Gli2* results in activation of Hh target genes when expressed in *Drosophila* (similar to the actions of full-length Ci) whereas expression of *Gli3* results in the repression of target genes (similar to the actions of cleaved Ci).[9,10] Indeed, combined expression of *Gli1* and *Gli3* is able to substitute for *Ci* during *Drosophila* development.[9]

Based on these results in *Drosophila*, it is tempting to postulate that Gli1 and Gli2 act as transcriptional activators and Gli3 acts as a transcriptional repressor of Hh target genes in vertebrates. However, the situation is far more complex than this. For instance, although expression of *Gli1* results in transcriptional activation of various genes (*cyclin D2*;[11] *Ptch1*;[11,12] *Gli1*;[13] *Bcl-2*;[14] *Bmp4*;[15] *Bmp7*;[15] *HNF3β*[16,17]), it is also able to cause down-regulation of gene transcription (*plakoglobin*[11]). Furthermore, although expression of *Gli3* can result in transcriptional repression,[13,16,18,19] Gli3 has also been shown to mediate Shh-induced activation of the *Gli1* promoter[13] and expression of *Gli3* can result in an increase in transcription of *Ptch1*,[12,20] *Bmp4*[15] and *Bmp7*.[15] Also, Gli3 has been shown to have activator function in vivo.[21-23] Of course, it must be taken into account that many of the studies looking at the transcriptional properties of the Glis have been performed in artificial over-expression systems in vitro and that the transcriptional activities of the Gli proteins may be very different in vivo. Furthermore, the Glis may function differently depending on location and time of action.

Understanding the transcriptional repertoire of the Gli proteins is further complicated by the fact that not all Gli proteins are processed in a similar fashion to Ci. For instance, Gli3, but not Gli1, is cleaved in the absence of Hh.[10,13,24-26] Furthermore, the shorter form of Gli3 has been shown to be a more potent repressor of transcription than full-length Gli3.[12,24] Because it is unclear in most studies whether Gli3 is cleaved, or the relative amounts of full-length and short forms present in the system, it is difficult to determine whether the transcriptional effects of Gli3 expression are mediated by the full-length or short form of the protein.

Defining the relationship between Shh and the Glis is made even more difficult by the observation that (unlike Hh and Ci in *Drosophila*) Shh can affect the transcription of *Gli1* and *Gli3*. Shh has been shown to increase *Gli1*[16,27] and decrease *Gli3* transcription in various systems.[19,28-30] Furthermore, it has been suggested that Gli3 represses *Shh* transcription based on observations of ectopic *Shh* expression in the limb and spinal cord of *Gli3* mutant animals.[19,31,32] However, whether Shh and Gli3 are cross-repressive in all tissues is unclear.

Overview of Telencephalic Development

The neural plate is formed from the ectodermal layer of the gastrulating embryo and gives rise to the entire central nervous system (CNS). Neural folds arise in the neural plate (Fig. 1A), appose and fuse to form the neural tube (Fig. 1B). The brain develops from the most anterior region of the neural tube and is divided into three primary vesicles: the hindbrain vesicle (rhombencephalon), the midbrain vesicle (mesencephalon) and the forebrain vesicle (prosencephalon). The forebrain becomes divided into the diencephalon caudally and telencephalon rostrally. Rapid proliferation of telencephalic cells results in the disproportionate swelling of the telencephalon which forms a pair of fluid-filled vesicles (telencephalic vesicles). The telencephalon eventually differentiates to become the olfactory bulbs anteriorly, the cerebral cortex dorsally and the basal ganglia ventrally.

During the second half of embryogenesis (~E11 onward in mouse), distinct telencephalic progenitor zones are morphologically apparent (Fig. 2). For example, two physically distinguishable eminences are found in the ventral region of the telencephalon: the lateral ganglionic eminence (LGE), the precursor to the adult striatum, and the more ventrally positioned medial ganglionic eminence (MGE), which gives rise to the globus pallidus. The striatum and globus pallidus comprise the basal ganglia, which are important for motor function. Cells from the MGE and LGE, as well as from the recently described caudal ganglionic eminence, produce GABAergic interneurons which migrate to populate a wide range of mature telencephalic structures.[33] Around the time the LGE and MGE become physically recognisable, the dorsal

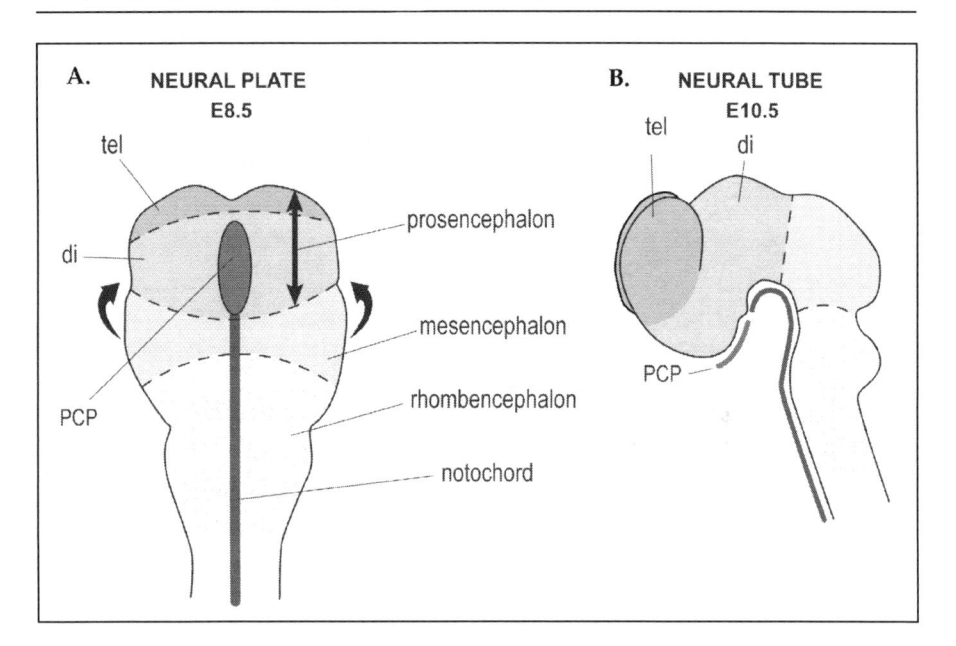

Figure 1. Neural plate and neural tube stages in mouse. A) The anterior neural plate at around E8.5. The neural plate folds in the direction of the arrows to form the neural tube. B) The brain viewed from the side after neural tube closure (at around E10.5). The brain vesicles are the prosencephalon (comprised of the telencephalon (tel) and diencephalon (di)), mesencephalon and rhombencephalon. The prechordal plate (pcp) underlies the rostral part of the neural tube (at the level of the diencephalon) whereas the notochord underlies the caudal neural tube.

midline of the telencephalon invaginates, leading to the separation of the telencephalic vesicles. This dorsal midline structure gives rise to the hippocampus, a structure crucial for learning and memory, as well as choroid plexus, which generates cerebrospinal fluid. The neocortex, which underpins complex cognitive functions, arises from the dorsolateral area of the telencephalon.

In addition to their distinguishable morphology, embryonic telencephalic progenitor domains have unique gene expression profiles (Fig. 2). For example, the MGE uniquely expresses the transcription factor, *Nkx2.1*, whereas transcription factors such as *Emx1* and *Pax6* are expressed in the cerebral cortex. Characterising these gene expression patterns has facilitated analyses of telencephalic regional specification in various mutant embryos, as described below.

Role of Shh in Telencephalic Dorsoventral Patterning

One of the first studies to implicate Shh in telencephalic regional specification showed that Shh induces the expression of the MGE marker Nkx2.1 in telencephalic neural plate explants.[34] Genetic evidence for the involvement of Shh in telencephalic development came from the discovery that humans heterozygous for mutations in the *SHH* gene suffer from holoprosencephaly (HPE).[35,36] Rather than becoming cleaved into distinct left and right hemispheres, the holoprosencephalic telencephalon develops as a single unpaired vesicle and, in extreme cases, ventral structures including the striatum and globus pallidus are completely absent. As a consequence of the lack of ventral diencephalic structures, the optic primordia fail to separate, resulting in a single cyclopic eye.

Around the same time as the human *SHH* gene was implicated in HPE, researchers generated transgenic mice mutant for *Shh*.[37] *Shh*$^{-/-}$ animals die at birth, have cyclopic eyes, lack olfactory bulbs and exhibit defects in the development of ventral structures along the entire neuraxis. The forebrain is particularly affected and strikingly reminiscent of human HPE. The

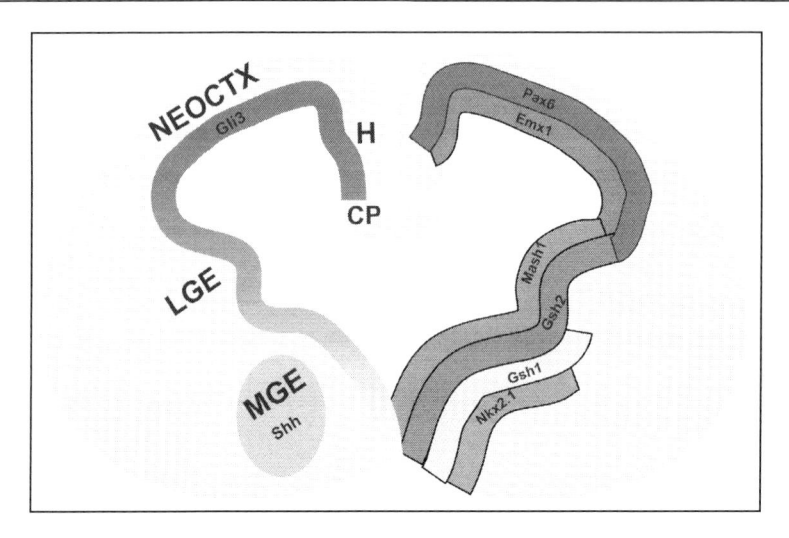

Figure 2. Coronal section of a midgestional (E12.5) mouse telencephalon illustrating major telencephalic subdivisions, a selection of gene expression patterns which are regionally restricted and the expression pattern of *Shh* and *Gli3*. Genes such as *Pax6* and *Emx1* are expressed in the dorsal telencephalon, which will give rise to the neocortex (neoctx), hippocampus (H) and choroid plexus (CP). The ventral telencephalon contains the precursors for the adult striatum (lateral ganglionic eminence, LGE) and globus pallidus (medial ganglionic eminence, MGE). Genes such as *Mash2* and *Gsh2* are expressed in both the LGE and MGE whereas genes such as *Gsh1* and *Nkx2.1* are primarily restricted to the MGE. *Shh* expression is confined to the MGE, while *Gli3* is expressed throughout the entire telencephalon, with high levels in the dorsal telencephalon and LGE and lower levels in the MGE.

telencephalon is severely hypoplastic, uninvaginated and the ganglionic eminences are not morphologically identifiable. Consistent with the lack of ventral telencephalic structures, expression of genes characteristic of the most ventral region of the telencephalon, such as *Nkx2.1*, *Lhx6* and *Gsh1*, is completely absent.[38-40] In concert with the reduction of ventral gene expression, genes such as *Emx1* and *Pax6*, normally restricted to the dorsal telencephalon, are expressed throughout the majority of the remaining telencephalic tissue.[37,41,42]

It has recently been observed, however, that ventral gene expression is not totally absent in the *Shh*[-/-] telencephalon. In less severely affected *Shh*[-/-] embryos, a small ventral telencephalic domain continues to express genes such as *Gsh2*, *Mash1* and *Dlx2*.[39,40] The gene expression profile of this ventral domain is reminiscent of wild-type LGE. Consequently, whilst providing good evidence for the importance of Shh in setting up a correctly patterned ventral telencephalon, the *Shh*[-/-] phenotype demonstrates that Shh is not wholly necessary for the specification of all ventral cell types in the telencephalon.

It remains an open question as to which factors induce the residual ventral gene expression in *Shh* mutants. It is possible that other Hh homologues can pattern the telencephalon or can partially compensate for the absence of Shh. In support of this possibility, mice mutant for both *Shh* and *Ihh* appear to lack all ventral character throughout the CNS.[43] This phenotype is essentially indistinguishable from *Smo*[-/-] mutants, which are unable to transduce any Hh signal.[43] It is also possible that Hh-independent signalling pathways can induce ventral gene expression. Indeed, the ability of both *Shh*[-/-] and *Smo*[-/-] telencephalic cells to express ventral telencephalic markers when *Gli3* is removed[39] (see below) strongly supports the idea that Hh signalling is not the only inducer of ventral telencephalic fate. There is evidence that retinoids, acting in a pathway parallel to that of Shh, induce ventral interneurons in the spinal cord.[44,45] It is likely that retinoids also play a role in patterning the telencephalon.[46-48]

Although *Shh* and *Smo* mutants clearly demonstrate the importance of these factors during development, they do not define the spatial and temporal requirements of Hh signalling during telencephalic development. For example, because the appearance of MGE-specific gene expression occurs before *Shh* is expressed in this region,[49] it is most likely that sources of Shh outside the telencephalon itself influence its patterning. But exactly where and when is Hh signalling required for telencephalic patterning? A range of cellular and genetic approaches has provided insight into these issues. For instance, Gunhaga et al[50] demonstrate that blocking Shh signalling in epiblast explants from gastrula stage embryos results in failure of ventral telencephalic cell specification. Because *Shh* is expressed in the anterior primitive streak and Henson's node at gastrula stages,[4,50,51] it is believed that these sources of Shh are crucial for specification of the MGE.

Shh signalling from the prechordal plate (mesendodermal tissue underlying the prospective rostral diencephalon) (Fig. 1),[52] may also be required for specification of the ventral telencephalon. Rostral neural plate explants lacking prechordal plate do not express Nkx2.1, whilst transplantation of prechordal plate results in expression of Nkx2.1 and repression of lateral neural plate markers.[34,53-55] Furthermore, Shh can induce Nkx2.1 expression in neural explants lacking prechordal plate.[34,53,55] Although the prechordal plate does not lie directly under the telencephalon, it may still be an important source of Shh with regard to telencephalic patterning due to the proposed long-range actions of Shh.[56-58]

Later in telencephalic development, the MGE itself becomes a source of Shh[59,60] (Fig. 2) and in vitro studies have demonstrated that Shh can induce gene expression characteristic of the LGE and inhibit dorsal marker expression in telencephalic explants.[61] Interestingly, even at high concentrations, Shh is unable to induce expression of the MGE marker, Nkx2.1, at this developmental stage.[61] These experiments suggest that the role of Shh in telencephalic development is regulated temporally by changes in responsiveness to Shh. Thus, early signalling from extra-telencephalic sources appears to induce MGE fates and later signalling from within the telencephalon itself seems to induce LGE fates. It will be very interesting to determine, at a molecular level, what underlies these changes and whether they involve context-dependent alteration in *Gli* target genes.

Recent work using conditional gene ablation has also attempted to unravel the temporal and spatial requirements for Hh signalling. Two studies involving the conditional ablation of *Smo* (in order to abolish all Hh signalling) or *Shh* reveal strikingly different telencephalic phenotypes depending on the timing of gene excision. Machold et al[62] used Cre recombinase under the control of the *Nestin* promoter to remove 'floxed' alleles of either *Shh* (Shh^{cl-};$Nestin^{Cre}$) or *Smo* (Smo^{cl-};$Nestin^{Cre}$) in neural progenitors. In these mutants, target gene transcription is reduced by E10.5 and abolished by E12.5. Removal of Shh or Hh signalling by these means results in a surprisingly normal telencephalon, although the olfactory bulbs are reduced in size. In stark contrast to the $Shh^{-/-}$ telencephalon, both the MGE and LGE in Shh^{cl-};$Nestin^{Cre}$ and Smo^{cl-};$Nestin^{Cre}$ animals are morphologically present and exhibit largely appropriate gene expression. The MGE is variably reduced in size and contains considerably fewer oligodendrocyte precursors, prefiguring the paucity of oligodendrocytes observed later in development. More severe defects were observed postnatally, where there were significantly reduced numbers of progenitors in the neocortical subventricular zone and hippocampal proliferative zones, supporting the idea that Shh is required in adult mammals to maintain telencephalic stem cell niches.[63]

Fuccillo et al[64] used a floxed allele of *Smo* to ablate Hh signalling earlier in telencephalic development using Cre under the control of the *Foxg1* promoter (Smo^{cl-};$Foxg1^{Cre}$). *Foxg1* is expressed throughout the telencephalic neuroepithelium from its inception at neural plate stages (~E7.5 in mouse).[53] This early ablation of *Smo*, which is estimated to be complete by E9, results in a much more severe phenotype than animals where *Shh* or *Smo* is excised using Nestin-Cre. Smo^{cl-};$Foxg1^{Cre}$ embryos lack all trace of the ventral telencephalon, as assessed by morphology and gene expression, and all remaining telencephalic tissue expresses dorsal markers. As in the Shh^{cl-};$Nestin^{Cre}$ and Smo^{cl-};$Nestin^{Cre}$ animals, the olfactory bulbs of Smo^{cl-};$Foxg1^{Cre}$

embryos are reduced in size. Later in development, the majority of telencephalic GABAergic interneurons and oligodendrocytes, two ventrally-derived cell types, are absent.

From these two studies, we can surmise that early Hh signalling, between the activation of *Foxg1-Cre* expression (~E7.5) and *Nestin-Cre* expression (~E10.5), is crucial in setting up ventral telencephalic progenitor domains and the cells derived from them. Furthermore, the relatively mild phenotype of the $Shh^{c/-};Nestin^{Cre}$ and $Smo^{c/-};Nestin^{Cre}$ mice suggests that Hh signalling after approximately E12.5 is not required for the maintenance of ventral telencephalic territories which, as suggested above, are specified earlier in development.

The absence of ventral telencephalic fate specification observed in the $Smo^{c/-};Foxg1^{Cre}$ mutant would appear to contradict in vitro studies suggesting that early Shh signalling during gastrulation (before significant *Foxg1* expression) is necessary and sufficient for induction of ventral telencephalic cell fates.[50] The $Smo^{c/-};Foxg1^{Cre}$ mutant presumably has intact Hh signalling at gastrulation, which should be sufficient for the induction of at least some ventral fate. Some of the contradictions between these studies may simply reflect the inherent differences that exist between in vitro and in vivo studies. It is possible that, whilst very early Hh signalling may indeed specify ventral lineages, maintenance of ventral fates in the absence of persistent Hh signalling can only occur in the rarefied environment of the tissue culture dish. In vivo, continued Hh signalling may be required to maintain ventral fate and removing Hh signalling during this phase may expose ventrally specified cells to dorsalising factors, which are likely to be absent in vitro.

It is also interesting to note that the ventral patterning defects in $Smo^{c/-};Foxg1^{Cre}$ mutants are more severe than those found in the constitutive *Shh* knockouts. This might best be explained by activity of other Hh ligands in the embryo. Indeed, a low level of Smo-dependent Hh signalling has been reported to be present in the $Shh^{-/-}$ neural tube.[65] The ability of $Shh^{-/-}$ (but not $Smo^{-/-}$) telencephalic cells to respond to other Hh ligands (if present) might contribute to the different phenotypes of the $Shh^{-/-}$ and $Smo^{c/-};Foxg1^{Cre}$ mutants. Thus, it will be important to determine whether the ventral telencephalic phenotype of embryos where *Shh* is excised by Foxg1-Cre is similar to or less severe than that of the Foxg1-Cre excised *Smo* mutants. It is also possible that heterozygosity at the *Foxg1* locus (due to insertion of Cre) synergises with the absence of *Smo* to contribute to the severe ventral phenotype observed.

The role of Shh in the development of dorsomedial telencephalic structures remains more ambiguous than its role in patterning the ventral telencephalon. In contrast to the increased severity of ventral patterning defects of $Smo^{c/-};Foxg1^{Cre}$ mutants compared to *Shh* mutants, the dorsal telencephalic midline of $Smo^{c/-};Foxg1^{Cre}$ mutants appears to be largely unaffected, whereas it is morphologically absent in *Shh* mutant mice and holoprosencephalic humans. This discrepancy may suggest that very early Hh signalling (before Foxg1-Cre expression) is required to pattern the dorsal midline. However, Ohkubo et al[42] demonstrate that *Bmp2* and *7* and *Msx1* and *2*, genes expressed in the dorsal-most regions of the telencephalon, are still expressed, and may even be over-expressed, in the $Shh^{-/-}$ telencephalon. As such, the requirement for Shh in dorsal midline development may be one of morphological induction rather than cell fate specification. It is also possible that Shh has some Smo-independent activity in this region of the telencephalon.

Role of Shh in Cell Death and Proliferation

In addition to affecting telencephalic dorsoventral patterning, Hh signalling likely influences other processes during telencephalic development. The small size of the telencephalon in various *Shh* and *Smo* mutants[37,42,62,64,66] suggests that cell death and proliferation may be affected. As Bmps and their effectors, Msx transcription factors, have been shown to mediate cell death in many regions of the developing embryo,[67,68] including the brain,[42,69-72] the increased *Bmp* and *Msx* expression in the *Shh* mutant[42] may mediate some of the increase in cell death observed, although it is not known whether Hh signalling is directly required to repress *Bmp* expression. A more direct mechanism could involve the pro-apoptotic function of Ptc.

Thibert et al[73] have shown that Ptc induces apoptotic cell death in neuroepithelial cells that can be prevented by binding of Shh to Ptc. In the absence of Shh, this property of Ptc may contribute to the increased cell death observed in the *Shh* mutant.[42] Because this type of Ptc-induced cell death does not involve the Ptc/Smo transducing module,[73] the cell death observed in the *Smo^(c/-);Foxg1^(Cre)* telencephalon may also be effected by this pathway.[64]

The role of Shh in proliferation is supported by the observations of decreased proliferation in the *Shh^(-/-)* telencephalon[66] and the increased proliferation of neocortical precursors after Shh treatment in vitro.[66] It has also been observed that telencephalic vesicles are enlarged after ectopic expression of Shh in vivo.[39,74] One possible mechanism for the mitogenic effect of Shh is its ability to relieve the inhibition of proliferation caused by Ptc's interaction with cyclin B1.[75] However, as no obvious proliferation defects have been observed in the telencephalon of *Smo^(c/-); Foxg1^(Cre)*,[64] *Shh^(c/-);Nestin^(Cre)*[62] and *Smo^(c/-);Nestin^(Cre)* mice,[62] further work is warranted in order to determine when and where Hh signalling is required for telencephalic cell proliferation.

Role of Shh in Cell Type Specification

Hh signalling is important for the specification of two cell types derived from the ventral telencephalon: oligodendrocytes and GABAergic interneurons. As mentioned above, oligodendrocytes are depleted in Nestin-Cre and Foxg1-Cre excised Hh signalling mutants[62,64] and the *Shh^(-/-)* telencephalon lacks oligodendrocyte precursors altogether.[76] Although various in vitro and in vivo studies suggest that Shh is necessary and sufficient for telencephalic oligodendrocyte generation,[59,76-78] the ability of *Shh^(-/-)* telencephalic tissue to generate oligodendrocytes in vitro[59] suggests that Shh is not required in vitro for oligodendrocyte generation and/or that there exists a pathway parallel to that of Shh for oligodendrogenesis. Shh also plays a role in the generation of GABAergic interneurons. As mentioned earlier, most GABAergic interneurons are absent in the *Smo^(c/-);Foxg1^(Cre)* mutant.[64] Moreover, Shh induces dorsomedial telencephalic cells to produce more GABAergic interneurons than normal in vitro.[79]

Telencephalic Phenotypes of Gli Mutants

Given the strong telencephalic phenotype of mice mutant for *Shh* or *Smo* and that the Gli proteins are transducers of Hh signalling in vertebrates, it is reasonable to assume that mice mutant for *Glis* would also have strong telencephalic phenotypes. Interestingly, mice mutant for *Gli1* do not show any obvious abnormalities,[80,81] demonstrating that *Gli1* is dispensable for normal development and/or may be compensated for by the presence of other *Glis*. Mice mutant for *Gli2* were initially reported to have a grossly normal telencephalon,[81,82] although, on an outbred background, these mice display a variably penetrant incidence of exencephaly.[83] In nonexencephalic *Gli2^(-/-)* mice, the telencephalic vesicles are expanded but have a thinner proliferative zone.[83]

The *Gli3* mouse mutant has the most dramatic telencephalic phenotype of all three *Gli* mutants. *Gli3* is widely expressed very early in mouse development in both the mesoderm and ectoderm.[8] It is then expressed throughout the telencephalon, with high expression in the cortex and LGE and lower expression in the MGE (Fig. 2).[8,84] The most widely studied strain of mice with mutation in the *Gli3* gene is referred to as extra toes (*Xt*) due to heterozygotes demonstrating polydactyly.[85,86] The *Xt* deletion results in a *Gli3* transcript lacking the sequence coding for the DNA binding element,[87,88] presumably resulting in a functionally null *Gli3* allele. Mice homozygous for the *Xt* allele die perinatally, display extreme polydactyly and are often exencephalic.[86] In nonexencephalic *Gli3^(Xt/Xt)* mice, the telencephalon is highly abnormal. *Gli3^(Xt/Xt)* embryos have no olfactory bulbs and do not develop dorsomedial telencephalic structures such as the hippocampus and choroid plexus.[39,86,89-92] The tissue of the putative neocortex is severely disorganised and heterotopic clusters of cells are observed in this area.[90]

Gene expression patterns in the *Gli3^(Xt/Xt)* dorsal telencephalon are distinctly abnormal. Genes such as *Emx1* and *2* have been reported to be reduced or absent in the *Gli3^(Xt/Xt)* dorsal telencephalon, although the telencephalon retains dorsal character as reflected by the perdurance of some dorsal marker expression.[41,90,92-94] Furthermore, the genes *Dlx2* and *Mash1*, which are normally ventrally-restricted, are expressed in the dorsal region of the telencephalon, particularly rostrally.[92,94] The boundary between the dorsal telencephalon and the ventral telencephalon is also compromised.[39,92] Gene expression within the ventral telencephalon appears relatively normal with *Nkx2.1* being expressed in the area of the MGE and *Gli1* expressed at the boundary between the MGE and LGE.[90,92,94]

Based on studies observing ectopic expression of *Shh* in *Gli3$^{Xt/Xt}$* limbs (and with lower penetrance in the *Gli3$^{Xt/Xt}$* spinal cord[19]),[31,32] it was thought that ectopic expression of *Shh* might be observed in the dorsal telencephalon of *Gli3$^{Xt/Xt}$* mice, contributing to some of the telencephalic defects present in these embryos. Somewhat surprisingly, *Shh* expression appears to be normal in the *Gli3$^{Xt/Xt}$* ventral telencephalon.[41,90,92] Hh target genes, such as *Gli1* and *Ptc1*, also appear to be normally expressed,[90,92] providing further evidence that Hh signalling is not aberrantly activated in the dorsal region of the *Gli3$^{Xt/Xt}$* telencephalon.

Abnormal expression of genes for signalling molecules other than Shh is, however, observed in the *Gli3$^{Xt/Xt}$* telencephalon. For instance, expression of various *Bmps* are decreased or absent in the dorsal telencephalon[90,92] and the cortical hem, a *Bmp*- and *Wnt*-rich signalling center in the dorsal midline important for formation of the hippocampus and choroid plexus, does not form in the *Gli3$^{Xt/Xt}$* mutant.[91] Furthermore, *Fgf8* expression is expanded in the anterior neural ridge[41] and dorsomedial telencephalon.[94] Because these signalling molecules are crucial for the proper development of the telencephalon,[95-97] it is likely that the abnormal expression of these molecules contributes to the severe phenotype of the *Gli3$^{Xt/Xt}$* telencephalon.

Role of Gli3 in Cell Death

Whereas increased cell death is observed in Hh signalling mutants, decreased cell death is observed in the forebrain of *Gli3* mutants.[41] As Bmps mediate cell death in many regions of the developing embryo, one possible mechanism through which Gli3 might regulate cell death is via modulation of Bmp signalling. This is supported by findings that expression of several *Bmps* is lost or reduced in the *Gli3$^{Xt/Xt}$* telencephalon.[90,92] Reduced expression of *Bmp* genes is consistent with the ability of Gli3 to enhance promoter activity of *Bmp4* and *Bmp7*.[15] Gli3 may also decrease cell death by directly affecting genes such as the anti-apoptotic factor *Bcl2*.[98] Because the repressor form of Gli3 is able to inhibit transactivation of the *Bcl2* gene in vitro,[14] it is possible that loss of *Gli3* function results in an overall increase in Bcl2 activity, resulting in decreased levels of cell death.

Loss of Gli3 Partially Rescues *Shh$^{-/-}$* Telencephalic Phenotypes

Based on work in the limb[25,99] and spinal cord,[65,100] it has been proposed that Shh acts to antagonise the actions of Gli3. For example, Litingtung and Chiang[100] were the first to demonstrate that many of the ventral spinal cord defects found in *Shh$^{-/-}$* animals were partially rescued in *Shh$^{-/-}$;Gli3$^{Xt/+}$* animals and further rescued in *Shh$^{-/-}$;Gli3$^{Xt/Xt}$* animals. Based on these findings, it was suggested that *Gli3* normally represses ventral fates and that Shh is required to counteract Gli3 function in order to allow ventral fate specification in the spinal cord. Could Hh signalling play a similar role with respect to Gli3 in the telencephalon?

It has been shown that loss of *Gli3* can partially rescue the telencephalic phenotype of *Shh$^{-/-}$* mutants.[39] For example, formation of two telencephalic vesicles is restored when one copy of *Gli3* is removed from *Shh$^{-/-}$* embryos. Furthermore, correct regional expression of ventral markers *Mash1*, *Dlx2* and *Gsh2* appears to be restored in the *Shh$^{-/-}$;Gli3$^{Xt/+}$* mutant compared to the aberrant expression of these genes in the *Shh$^{-/-}$* mutant. There is even a small amount of *Nkx2.1* expression present in the *Shh$^{-/-}$;Gli3$^{Xt/+}$* mutant, which is never seen in the *Shh$^{-/-}$* telencephalon, suggesting that some MGE character is restored in the *Shh$^{-/-}$;Gli3$^{Xt/+}$* mutant. Unfortunately, the high incidence of exencephaly in double homozygous mutants precluded a thorough analysis of the dorsoventral patterning of these animals. However, it appears that the MGE is more fully specified in the *Shh$^{-/-}$;Gli3$^{Xt/Xt}$* mutant than in the *Shh$^{-/-}$;Gli3$^{Xt/+}$* and *Shh$^{-/-}$* mutants. Thus, ventral patterning is able to occur in the absence of both Shh and Gli3, demonstrating that other pathways are capable of dorsoventral patterning in the telencephalon.

Conclusion

It has been a decade since Hh signalling was first implicated in development of the telencephalon and the Hh pathway now has a unique and undisputed position as a key regulator of ventral fate specification. Nevertheless, many issues remain to be addressed regarding the mechanism of the Hh-Gli signalling pathway. For example, it will be important to define what the respective contributions of the full-length and cleaved forms of the Gli3 protein are during telencephalic development. One group has already begun to address this by generating a mouse ($Gli3^{\Delta 699/\Delta 699}$ mutant) that only expresses a truncated Gli3 protein similar to the cleaved form of Gli3.[101] This mutant form of Gli3 would thus have DNA binding capability, unlike the potential protein product resulting from the Xt allele. Interestingly, these $Gli3^{\Delta 699/\Delta 699}$ mutants exhibit a very different phenotype to that of $Gli3^{Xt/Xt}$ mice. They exhibit a variety of defects, such as imperforate anus and absence of adrenal glands,[101] which are not present in $Gli3^{Xt/Xt}$ mice. Furthermore, $Gli3^{\Delta 699/\Delta 699}$ mice do not display the spinal cord defects found in $Gli3^{Xt/Xt}$ mice.[102] Although analysis of the rostral portion of the nervous system in these mice has not been published, it appears unlikely that these mice have a similar telencephalic phenotype to $Gli3^{Xt/Xt}$ mice. If the telencephalon of the $Gli3^{\Delta 699/\Delta 699}$ mutant is correctly patterned, it would suggest that full-length Gli3 is either not necessary for telencephalic development, or other proteins, presumably Gli1 or Gli2, are able to compensate for its absence. The generation of a mouse expressing a cleavage-resistant form of Gli3 would be of great help in defining the relative importance of full-length and cleaved forms of Gli3 during development.

Questions regarding the relationship between Hh ligands and the Gli proteins also remain. For example, do Hhs have Gli-independent action in telencephalic development? Conversely, to what extent do Gli proteins have roles independent of their Hh transducing functions? With regard to the first issue, the identification of a Shh-response element in the COUP-TFII promoter that is distinct from the Gli-response element suggests that factors other than Gli can transduce the Shh signal.[103] This is particularly relevant to telencephalic development since COUP-TFII is thought to be involved in the migration of neurons from the ventral telencephalon.[104] Furthermore, the Hh receptor Ptc has been shown to modulate both cell death[73] and proliferation[75] independent of the Ptc/Smo/Gli transducing module and these actions are regulated by binding of Shh, adding further support to the notion that Shh can act without Gli proteins. Regarding whether Gli proteins have roles independent of their Hh transducing functions, there is evidence that C-terminally truncated Gli3 is able to interact with Smads,[105] transducers of Bmp signalling. This, in addition to the ability of Glis to activate the $Bmp4$ and $Bmp7$ promoters,[15] suggests that Glis are able to influence Bmp signalling at both a transcriptional and post-translational level. Thus, it is important to keep in mind that not all functions of Shh and Gli proteins are confined to the well-described linear Shh-Smo-Gli pathway and that future models will need to accommodate these actions.

Acknowledgements

The authors are supported by the National Institutes of Health (NEI individual NRSA postdoctoral fellowship, PAZ), the Wellcome Trust (BM, DJP), the Medical Research Council (DJP) and the Biotechnology and Biological Sciences Research Council (DJP).

References

1. Ericson J, Muhr J, Jessell TM et al. Sonic hedgehog: A common signal for ventral patterning along the rostrocaudal axis of the neural tube. Int J Dev Biol 1995; 39(5):809-816.
2. Lumsden A, Krumlauf R. Patterning the vertebrate neuraxis. Science 1996; 274(5290):1109-1115.
3. Patten I, Placzek M. The role of Sonic hedgehog in neural tube patterning. Cell Mol Life Sci 2000; 57(12):1695-1708.
4. Echelard Y, Epstein DJ, St-Jacques B et al. Sonic hedgehog, a member of a family of putative signaling molecules, is implicated in the regulation of CNS polarity. Cell 1993; 75(7):1417-1430.
5. Ingham PW, McMahon AP. Hedgehog signaling in animal development: Paradigms and principles. Genes Dev 2001; 15(23):3059-3087.

6. Lum L, Beachy PA. The Hedgehog response network: Sensors, switches, and routers. Science 2004; 304(5678):1755-1759.
7. Walterhouse D, Ahmed M, Slusarski D et al. Gli, a zinc finger transcription factor and oncogene, is expressed during normal mouse development. Dev Dyn 1993; 196(2):91-102.
8. Hui CC, Slusarski D, Platt KA et al. Expression of three mouse homologs of the Drosophila segment polarity gene cubitus interruptus, Gli, Gli-2, and Gli-3, in ectoderm- and mesoderm-derived tissues suggests multiple roles during postimplantation development. Dev Biol 1994; 162(2):402-413.
9. von Mering C, Basler K. Distinct and regulated activities of human Gli proteins in Drosophila. Curr Biol 1999; 9(22):1319-1322.
10. Aza-Blanc P, Lin HY, Ruiz i Altaba A et al. Expression of the vertebrate Gli proteins in Drosophila reveals a distribution of activator and repressor activities. Development 2000; 127(19):4293-4301.
11. Yoon JW, Kita Y, Frank DJ et al. Gene expression profiling leads to identification of GLI1-binding elements in target genes and a role for multiple downstream pathways in GLI1-induced cell transformation. J Biol Chem 2002; 277(7):5548-5555.
12. Agren M, Kogerman P, Kleman MI et al. Expression of the PTCH1 tumor suppressor gene is regulated by alternative promoters and a single functional Gli-binding site. Gene 2004; 330:101-114.
13. Dai P, Akimaru H, Tanaka Y et al. Sonic Hedgehog-induced activation of the Gli1 promoter is mediated by GLI3. J Biol Chem 1999; 274(12):8143-8152.
14. Bigelow RL, Chari NS, Unden AB et al. Transcriptional regulation of bcl-2 mediated by the sonic hedgehog signaling pathway through gli-1. J Biol Chem 2004; 279(2):1197-1205.
15. Kawai S, Sugiura T. Characterization of human bone morphogenetic protein (BMP)-4 and -7 gene promoters: Activation of BMP promoters by Gli, a sonic hedgehog mediator. Bone 2001; 29(1):54-61.
16. Sasaki H, Hui C, Nakafuku M et al. A binding site for Gli proteins is essential for HNF-3beta floor plate enhancer activity in transgenics and can respond to Shh in vitro. Development 1997; 124(7):1313-1322.
17. Hynes M, Stone DM, Dowd M et al. Control of cell pattern in the neural tube by the zinc finger transcription factor and oncogene Gli-1. Neuron 1997; 19(1):15-26.
18. Sasaki H, Nishizaki Y, Hui C et al. Regulation of Gli2 and Gli3 activities by an amino-terminal repression domain: Implication of Gli2 and Gli3 as primary mediators of Shh signaling. Development 1999; 126(17):3915-3924.
19. Ruiz i Altaba A. Combinatorial Gli gene function in floor plate and neuronal inductions by Sonic hedgehog. Development 1998; 125(12):2203-2212.
20. Shin SH, Kogerman P, Lindstrom E et al. GLI3 mutations in human disorders mimic Drosophila cubitus interruptus protein functions and localization. Proc Natl Acad Sci USA 1999; 96(6):2880-2884.
21. Motoyama J, Milenkovic L, Iwama M et al. Differential requirement for Gli2 and Gli3 in ventral neural cell fate specification. Dev Biol 2003; 259(1):150-161.
22. Bai CB, Stephen D, Joyner AL. All mouse ventral spinal cord patterning by hedgehog is Gli dependent and involves an activator function of Gli3. Dev Cell 2004; 6(1):103-115.
23. Tyurina OV, Guner B, Popova E et al. Zebrafish Gli3 functions as both an activator and a repressor in Hedgehog signaling. Dev Biol 2005; 277(2):537-556.
24. Wang B, Fallon JF, Beachy PA. Hedgehog-regulated processing of Gli3 produces an anterior/posterior repressor gradient in the developing vertebrate limb. Cell 2000; 100(4):423-434.
25. Litingtung Y, Dahn RD, Li Y et al. Shh and Gli3 are dispensable for limb skeleton formation but regulate digit number and identity. Nature 2002; 418(6901):979-983.
26. Bastida MF, Delgado MD, Wang B et al. Levels of Gli3 repressor correlate with Bmp4 expression and apoptosis during limb development. Dev Dyn 2004; 231(1):148-160.
27. Lee J, Platt KA, Censullo P et al. Gli1 is a target of Sonic hedgehog that induces ventral neural tube development. Development 1997; 124(13):2537-2552.
28. Marigo V, Johnson RL, Vortkamp A et al. Sonic hedgehog differentially regulates expression of GLI and GLI3 during limb development. Dev Biol 1996; 180(1):273-283.
29. Takahashi M, Tamura K, Buscher D et al. The role of Alx-4 in the establishment of anteroposterior polarity during vertebrate limb development. Development 1998; 125(22):4417-4425.
30. Schweitzer R, Vogan KJ, Tabin CJ. Similar expression and regulation of Gli2 and Gli3 in the chick limb bud. Mech Dev 2000; 98(1-2):171-174.
31. Buscher D, Bosse B, Heymer J et al. Evidence for genetic control of Sonic hedgehog by Gli3 in mouse limb development. Mech Dev 1997; 62(2):175-182.
32. Masuya H, Sagai T, Moriwaki K et al. Multigenic control of the localization of the zone of polarizing activity in limb morphogenesis in the mouse. Dev Biol 1997; 182(1):42-51.

33. Corbin JG, Nery S, Fishell G. Telencephalic cells take a tangent: Nonradial migration in the mammalian forebrain. Nat Neurosci 2001; (4 Suppl):1177-1182.
34. Ericson J, Muhr J, Placzek M et al. Sonic hedgehog induces the differentiation of ventral forebrain neurons: A common signal for ventral patterning within the neural tube. Cell 1995; 81(5):747-756.
35. Roessler E, Belloni E, Gaudenz K et al. Mutations in the human Sonic Hedgehog gene cause holoprosencephaly. Nat Genet 1996; 14(3):357-360.
36. Belloni E, Muenke M, Roessler E et al. Identification of Sonic hedgehog as a candidate gene responsible for holoprosencephaly. Nat Genet 1996; 14(3):353-356.
37. Chiang C, Litingtung Y, Lee E et al. Cyclopia and defective axial patterning in mice lacking Sonic hedgehog gene function. Nature 1996; 383(6599):407-413.
38. Pabst O, Herbrand H, Takuma N et al. NKX2 gene expression in neuroectoderm but not in mesendodermally derived structures depends on sonic hedgehog in mouse embryos. Dev Genes Evol 2000; 210(1):47-50.
39. Rallu M, Machold R, Gaiano N et al. Dorsoventral patterning is established in the telencephalon of mutants lacking both Gli3 and Hedgehog signaling. Development 2002; 129(21):4963-4974.
40. Corbin JG, Rutlin M, Gaiano N et al. Combinatorial function of the homeodomain proteins Nkx2.1 and Gsh2 in ventral telencephalic patterning. Development 2003; 130(20):4895-4906.
41. Aoto K, Nishimura T, Eto K et al. Mouse GLI3 regulates Fgf8 expression and apoptosis in the developing neural tube, face, and limb bud. Dev Biol 2002; 251(2):320-332.
42. Ohkubo Y, Chiang C, Rubenstein JL. Coordinate regulation and synergistic actions of BMP4, SHH and FGF8 in the rostral prosencephalon regulate morphogenesis of the telencephalic and optic vesicles. Neuroscience 2002; 111(1):1-17.
43. Zhang XM, Ramalho-Santos M, McMahon AP. Smoothened mutants reveal redundant roles for Shh and Ihh signaling including regulation of L/R asymmetry by the mouse node. Cell 2001; 105(6):781-792.
44. Pierani A, Brenner-Morton S, Chiang C et al. A sonic hedgehog-independent, retinoid-activated pathway of neurogenesis in the ventral spinal cord. Cell 1999; 97(7):903-915.
45. Novitch BG, Wichterle H, Jessell TM et al. A requirement for retinoic acid-mediated transcriptional activation in ventral neural patterning and motor neuron specification. Neuron 2003; 40(1):81-95.
46. LaMantia AS, Colbert MC, Linney E. Retinoic acid induction and regional differentiation prefigure olfactory pathway formation in the mammalian forebrain. Neuron 1993; 10(6):1035-1048.
47. Toresson H, Mata de Urquiza A, Fagerstrom C et al. Retinoids are produced by glia in the lateral ganglionic eminence and regulate striatal neuron differentiation. Development 1999; 126(6):1317-1326.
48. Schneider RA, Hu D, Rubenstein JL et al. Local retinoid signaling coordinates forebrain and facial morphogenesis by maintaining FGF8 and SHH. Development 2001; 128(14):2755-2767.
49. Rubenstein JL, Shimamura K, Martinez S et al. Regionalization of the prosencephalic neural plate. Annu Rev Neurosci 1998; 21:445-477.
50. Gunhaga L, Jessell TM, Edlund T. Sonic hedgehog signaling at gastrula stages specifies ventral telencephalic cells in the chick embryo. Development 2000; 127(15):3283-3293.
51. Epstein DJ, McMahon AP, Joyner AL. Regionalization of Sonic hedgehog transcription along the anteroposterior axis of the mouse central nervous system is regulated by Hnf3-dependent and -independent mechanisms. Development 1999; 126(2):281-292.
52. Kiecker C, Niehrs C. The role of prechordal mesendoderm in neural patterning. Curr Opin Neurobiol 2001; 11(1):27-33.
53. Shimamura K, Rubenstein JL. Inductive interactions direct early regionalization of the mouse forebrain. Development 1997; 124(14):2709-2718.
54. Pera EM, Kessel M. Patterning of the chick forebrain anlage by the prechordal plate. Development 1997; 124(20):4153-4162.
55. Dale JK, Vesque C, Lints TJ et al. Cooperation of BMP7 and SHH in the induction of forebrain ventral midline cells by prechordal mesoderm. Cell 1997; 90(2):257-269.
56. Briscoe J, Chen Y, Jessell TM et al. A hedgehog-insensitive form of patched provides evidence for direct long-range morphogen activity of sonic hedgehog in the neural tube. Mol Cell 2001; 7(6):1279-1291.
57. Lewis PM, Dunn MP, McMahon JA et al. Cholesterol modification of sonic hedgehog is required for long-range signaling activity and effective modulation of signaling by Ptc1. Cell 2001; 105(5):599-612.
58. Zeng X, Goetz JA, Suber LM et al. A freely diffusible form of Sonic hedgehog mediates long-range signalling. Nature 2001; 411(6838):716-720.
59. Nery S, Wichterle H, Fishell G. Sonic hedgehog contributes to oligodendrocyte specification in the mammalian forebrain. Development 2001; 128(4):527-540.

60. Platt KA, Michaud J, Joyner AL. Expression of the mouse Gli and Ptc genes is adjacent to embryonic sources of hedgehog signals suggesting a conservation of pathways between flies and mice. Mech Dev 1997; 62(2):121-135.
61. Kohtz JD, Baker DP, Corte G et al. Regionalization within the mammalian telencephalon is mediated by changes in responsiveness to Sonic Hedgehog. Development 1998; 125(24):5079-5089.
62. Machold R, Hayashi S, Rutlin M et al. Sonic hedgehog is required for progenitor cell maintenance in telencephalic stem cell niches. Neuron 2003; 39(6):937-950.
63. Palma V, Lim DA, Dahmane N et al. Sonic hedgehog controls stem cell behavior in the postnatal and adult brain. Development 2005; 132(2):335-344.
64. Fuccillo M, Rallu M, McMahon AP et al. Temporal requirement for hedgehog signaling in ventral telencephalic patterning. Development 2004; 131(20):5031-5040.
65. Wijgerde M, McMahon JA, Rule M et al. A direct requirement for Hedgehog signaling for normal specification of all ventral progenitor domains in the presumptive mammalian spinal cord. Genes Dev 2002; 16(22):2849-2864.
66. Dahmane N, Sanchez P, Gitton Y et al. The Sonic Hedgehog-Gli pathway regulates dorsal brain growth and tumorigenesis. Development 2001; 128(24):5201-5212.
67. Graham A, Koentges G, Lumsden A. Neural crest apoptosis and the establishment of craniofacial pattern: An honorable death. Mol Cell Neurosci 1996; 8(2-3):76-83.
68. Merino R, Ganan Y, Macias D et al. Bone morphogenetic proteins regulate interdigital cell death in the avian embryo. Ann NY Acad Sci 1999; 887:120-132.
69. Golden JA, Bracilovic A, McFadden KA et al. Ectopic bone morphogenetic proteins 5 and 4 in the chicken forebrain lead to cyclopia and holoprosencephaly. Proc Natl Acad Sci USA 1999; 96(5):2439-2444.
70. Furuta Y, Piston DW, Hogan BL. Bone morphogenetic proteins (BMPs) as regulators of dorsal forebrain development. Development 1997; 124(11):2203-2212.
71. Mabie PC, Mehler MF, Kessler JA. Multiple roles of bone morphogenetic protein signaling in the regulation of cortical cell number and phenotype. J Neurosci 1999; 19(16):7077-7088.
72. Israsena N, Kessler JA. Msx2 and p21(CIP1/WAF1) mediate the proapoptotic effects of bone morphogenetic protein-4 on ventricular zone progenitor cells. J Neurosci Res 2002; 69(6):803-809.
73. Thibert C, Teillet MA, Lapointe F et al. Inhibition of neuroepithelial patched-induced apoptosis by sonic hedgehog. Science 2003; 301(5634):843-846.
74. Gaiano N, Kohtz JD, Turnbull DH et al. A method for rapid gain-of-function studies in the mouse embryonic nervous system. Nat Neurosci 1999; 2(9):812-819.
75. Barnes EA, Kong M, Ollendorff V et al. Patched1 interacts with cyclin B1 to regulate cell cycle progression. EMBO J 2001; 20(9):2214-2223.
76. Lu QR, Yuk D, Alberta JA et al. Sonic hedgehog—regulated oligodendrocyte lineage genes encoding bHLH proteins in the mammalian central nervous system. Neuron 2000; 25(2):317-329.
77. Tekki-Kessaris N, Woodruff R, Hall AC et al. Hedgehog-dependent oligodendrocyte lineage specification in the telencephalon. Development 2001; 128(13):2545-2554.
78. Spassky N, Heydon K, Mangatal A et al. Sonic hedgehog-dependent emergence of oligodendrocytes in the telencephalon: Evidence for a source of oligodendrocytes in the olfactory bulb that is independent of PDGFRalpha signaling. Development 2001; 128(24):4993-5004.
79. Gulacsi A, Lillien L. Sonic hedgehog and bone morphogenetic protein regulate interneuron development from dorsal telencephalic progenitors in vitro. J Neurosci 2003; 23(30):9862-9872.
80. Matise MP, Epstein DJ, Park HL et al. Gli2 is required for induction of floor plate and adjacent cells, but not most ventral neurons in the mouse central nervous system. Development 1998; 125(15):2759-2770.
81. Park HL, Bai C, Platt KA et al. Mouse Gli1 mutants are viable but have defects in SHH signaling in combination with a Gli2 mutation. Development 2000; 127(8):1593-1605.
82. Ding Q, Motoyama J, Gasca S et al. Diminished Sonic hedgehog signaling and lack of floor plate differentiation in Gli2 mutant mice. Development 1998; 125(14):2533-2543.
83. Palma V, Ruiz i Altaba A. Hedgehog-GLI signaling regulates the behavior of cells with stem cell properties in the developing neocortex. Development 2004; 131(2):337-345.
84. Shinozaki K, Yoshida M, Nakamura M et al. Emx1 and Emx2 cooperate in initial phase of archipallium development. Mech Dev 2004; 121(5):475-489.
85. Schimmang T, Lemaistre M, Vortkamp A et al. Expression of the zinc finger gene Gli3 is affected in the morphogenetic mouse mutant extra-toes (Xt). Development 1992; 116(3):799-804.
86. Hui CC, Joyner AL. A mouse model of greig cephalopolysyndactyly syndrome: The extra-toesJ mutation contains an intragenic deletion of the Gli3 gene. Nat Genet 1993; 3(3):241-246.
87. Buscher D, Grotewold L, Ruther U. The XtJ allele generates a Gli3 fusion transcript. Mamm Genome 1998; 9(8):676-678.

88. Maynard TM, Jain MD, Balmer CW et al. High-resolution mapping of the Gli3 mutation extra-toes reveals a 51.5-kb deletion. Mamm Genome 2002; 13(1):58-61.
89. Franz T. Extra-toes (Xt) homozygous mutant mice demonstrate a role for the Gli-3 gene in the development of the forebrain. Acta Anat (Basel) 1994; 150(1):38-44.
90. Theil T, Alvarez-Bolado G, Walter A et al. Gli3 is required for Emx gene expression during dorsal telencephalon development. Development 1999; 126(16):3561-3571.
91. Grove EA, Tole S, Limon J et al. The hem of the embryonic cerebral cortex is defined by the expression of multiple Wnt genes and is compromised in Gli3-deficient mice. Development 1998; 125(12):2315-2325.
92. Tole S, Ragsdale CW, Grove EA. Dorsoventral patterning of the telencephalon is disrupted in the mouse mutant extra-toes(J). Dev Biol 2000; 217(2):254-265.
93. Theil T, Aydin S, Koch S et al. Wnt and Bmp signalling cooperatively regulate graded Emx2 expression in the dorsal telencephalon. Development 2002; 129(13):3045-3054.
94. Kuschel S, Ruther U, Theil T. A disrupted balance between Bmp/Wnt and Fgf signaling underlies the ventralization of the Gli3 mutant telencephalon. Dev Biol 2003; 260(2):484-495.
95. Rubenstein JL, Beachy PA. Patterning of the embryonic forebrain. Curr Opin Neurobiol 1998; 8(1):18-26.
96. Monuki ES, Walsh CA. Mechanisms of cerebral cortical patterning in mice and humans. Nat Neurosci 2001; (4 Suppl):1199-1206.
97. Zaki PA, Quinn JC, Price DJ. Mouse models of telencephalic development. Curr Opin Genet Dev 2003; 13(4):423-437.
98. Roth KA, D'Sa C. Apoptosis and brain development. Ment Retard Dev Disabil Res Rev 2001; 7(4):261-266.
99. te Welscher P, Zuniga A, Kuijper S et al. Progression of vertebrate limb development through SHH-mediated counteraction of GLI3. Science 2002; 298(5594):827-830.
100. Litingtung Y, Chiang C. Specification of ventral neuron types is mediated by an antagonistic interaction between Shh and Gli3. Nat Neurosci 2000; 3(10):979-985.
101. Bose J, Grotewold L, Ruther U. Pallister-Hall syndrome phenotype in mice mutant for Gli3. Hum Mol Genet 2002; 11(9):1129-1135.
102. Persson M, Stamataki D, te Welscher P et al. Dorsal-ventral patterning of the spinal cord requires Gli3 transcriptional repressor activity. Genes Dev 2002; 16(22):2865-2878.
103. Krishnan V, Pereira FA, Qiu Y et al. Mediation of Sonic hedgehog-induced expression of COUP-TFII by a protein phosphatase. Science 1997; 278(5345):1947-1950.
104. Tripodi M, Filosa A, Armentano M et al. The COUP-TF nuclear receptors regulate cell migration in the mammalian basal forebrain. Development 2004; 131(24):6119-6129.
105. Liu F, Massague J, Ruiz i Altaba A. Carboxy-terminally truncated Gli3 proteins associate with Smads. Nat Genet 1998; 20(4):325-326.

CHAPTER 4

Role of *Shh* and *Gli* Signalling in Oligodendroglial Development

Min Tan, Yingchuan Qi and Mengsheng Qiu*

Abstract

Recent molecular and genetic studies have demonstrated that early oligodendrocyte progenitor cells are induced from the ventral neural tube by the Sonic hedgehog (Shh) protein produced in the ventral midline structures. Whilst *Shh* signalling is required for ventral oligodendrogenesis in the entire central nervous system, *Gli2* activity only regulates oligodendrocyte development in the ventral spinal cord. *Gli3* plays a nonessential role in ventral oligodendrogenesis during normal development. However, in the absence of *Shh* signalling, *Gli3* functions as a repressor of ventral oligodendrogenesis. In addition, there is growing evidence that a separate population of oligodendrocyte progenitor cells is also produced from the dorsal region of the neural tube independent of *Shh* signalling.

Early Oligodendrocyte Precursors Originate from the *Olig1/2*+ Ventral Neuroepithelium and Share the Same Lineage with Motor Neurons

Oligodendrocytes are myelinating macroglial cells found in all regions of the central nervous system (CNS). Despite their widespread distribution, recent studies suggest that early oligodendrocyte progenitors (OPCs or OLPs) are derived from specific loci in the ventral neuroepithelium in the developing CNS. For instance, expression of several early oligodendrocyte marker genes, such as *PDGFRα* and *Sox10*, is initially observed in the ventral ventricular zone in the entire CNS.[1-3] Moreover, in neural explant culture and chick-quail transplantation studies, only the ventral spinal cord tissues gave rise to oligodendrocytes, whereas the dorsal tissue largely produced astrocytes.[4-6]

The origin and molecular specification of oligodendrocytes have been studied most extensively in the developing spinal cord. Recently it was established that early OPC cells in the spinal cord specifically originate from the motor neuron progenitor domain (pMN domain) of the ventral neuroepithelium.[7] During early neural development, the pMN domain expresses the *Olig2* bHLH transcription factor.[8-10] From the *Olig2*+ pMN domain sequentially arise the HB9+ motor neurons and *Olig2*+ OPC cells.[7,11] Loss of *Olig2* function disrupts the development of both motor neurons and oligodendrocytes.[12-15] Based on these observations, it has been proposed that in the ventral spinal cord, motor neurons and oligodendrocytes are derived from the same pool of neural progenitor cells, with motor neurons being generated first followed by oligodendrocytes.[16,17]

---

*Corresponding Author: Mengsheng Qiu—Department of Anatomical Sciences and Neurobiology, School of Medicine, University of Louisville, Louisville, Kentucky 40292, U.S.A. Email: m0qiu001@louisville.edu

Shh and Gli Signalling and Development, edited by Carolyn E. Fisher and Sarah E.M. Howie. ©2006 Landes Bioscience and Springer Science+Business Media.

During oligodendrogenesis stages, the *Olig1* gene is also expressed in the pMN domain. Recent studies have suggested that the closely related *Olig1* and *Olig2* genes have distinct functions in the development of the oligodendrocyte lineage. While *Olig2* activity is essential for the fate specification of OPCs, *Olig1* appears to have a crucial role in oligodendrocyte maturation and remyelination. Mutation of the *Olig1* gene leads to delayed oligodendrocyte differentiation[13] and impaired remeylination in the insult-induced demyelination model.[18]

Ventral Oligodendrogenesis Is Induced by Sonic Hedgehog Signalling

During early neural development, Shh protein functions as a morphogen to induce various types of ventral neurons.[19] It has been proposed that Shh protein produced from the ventral midline structures (notochord and floor plate) sets up a concentration gradient in the ventral neural tube, and different concentrations of the protein can induce different subtypes of ventral neurons.[20,21] Several lines of evidence suggest that the production of OPCs from the ventral spinal cord is also a *Shh*-dependent process.[16,17,22] First, activation of the *Shh* pathway by Shh recombinant protein is sufficient to induce oligodendrocyte development from dorsal spinal cord explants.[23-26] The induction of oligodendrocytes occurs with a similar concentration of Shh protein that is required for motor neuron induction, consistent with the notion that oligodendrocytes and motor neurons share the same lineage.[25,26] Second, blockade of *Shh* activity can inhibit oligodendrogenesis in spinal cord explant culture.[26,27] Consistently, in *Shh* mutants, oligodendrocyte generation in the ventral spinal cord is completely abolished.[28]

Similarly, early OPCs are also generated from the ventral region of the brain in a *Shh*-dependent mechanism.[29] In the developing forebrain, early OPCs originate from the ventral telencephalon, specifically the anterior entopeduncular area (AEP).[30-32] *Shh* is expressed in the ventricular and subventricular zone of the AEP as well as the adjacent median ganglionic eminence (MGE) and anterior preoptic area (POA). Loss of *Shh* expression in the basal forebrain in *Nkx2.1* mutants and in *Shh* mutants is associated with an inhibition of early oligodendrocyte development in the telecephalon.[31-33] There is also evidence that early OPCs are generated from the ventral hindbrain in a *Shh*-dependent mechanism.[3,34] Therefore, *Shh*-dependent ventral oligodendrogenesis appears to be a universal phenomenon in the CNS.

A *Shh*-Independent Pathway for Oligodendrogenesis in the Developing Spinal Cord

Although it is generally accepted that early OPCs are produced from the ventral neural tube by a *Shh*-dependent mechanism, there is emerging evidence that dorsal neural progenitor cells also contribute to oligodendrocyte formation during development. Earlier studies demonstrated that prolonged culture of dorsal spinal tissues were capable of producing oligodendrocytes,[6] indicating that dorsal neural progenitor cells have the potential to generate oligodendrocytes in vitro under certain circumstances. However, it has not been clear until recently whether this potential is realised during animal development. Our recent studies revealed that a small number of OPCs are indeed generated from the dorsal spinal cord at E14.5, about two days later than the commencement of ventral oligodendrogenesis (E12.5) (Fig. 1). The generation of dorsal OPCs is particularly evident in *Nkx6.1-/-Nkx6.2-/-* double mutants, in which ventral oligodendrogenesis is inhibited due to the lack of the pMN domain. The dorsally-derived *Olig2+* OPCs in both wild-type and *Nkx6* mutants coexpress several dorsal neural progenitor genes including *Pax7, Mash1* and *Gsh1*.[28,35] However, the time window for the late phase of oligodendrogenesis from dorsal neural progenitor cells is relatively short (from E14.5-E15.5) as compared to that for the early phase of ventral oligodendrogenesis (from E12.5 to E15.5), suggesting that dorsal contribution to the OPC population is likely to be limited. Due to the lack of traceable markers for this population of OPCs (expression of dorsal progenitor genes is quickly down-regulated), it is difficult to estimate what percentage of spinal cord OPCs have a dorsal origin.[28,35] For the same reason, the fate and function of dorsal OPCs in adult spinal tissue are unknown. It is possible that dorsal OPCs may differ functionally from their ventral

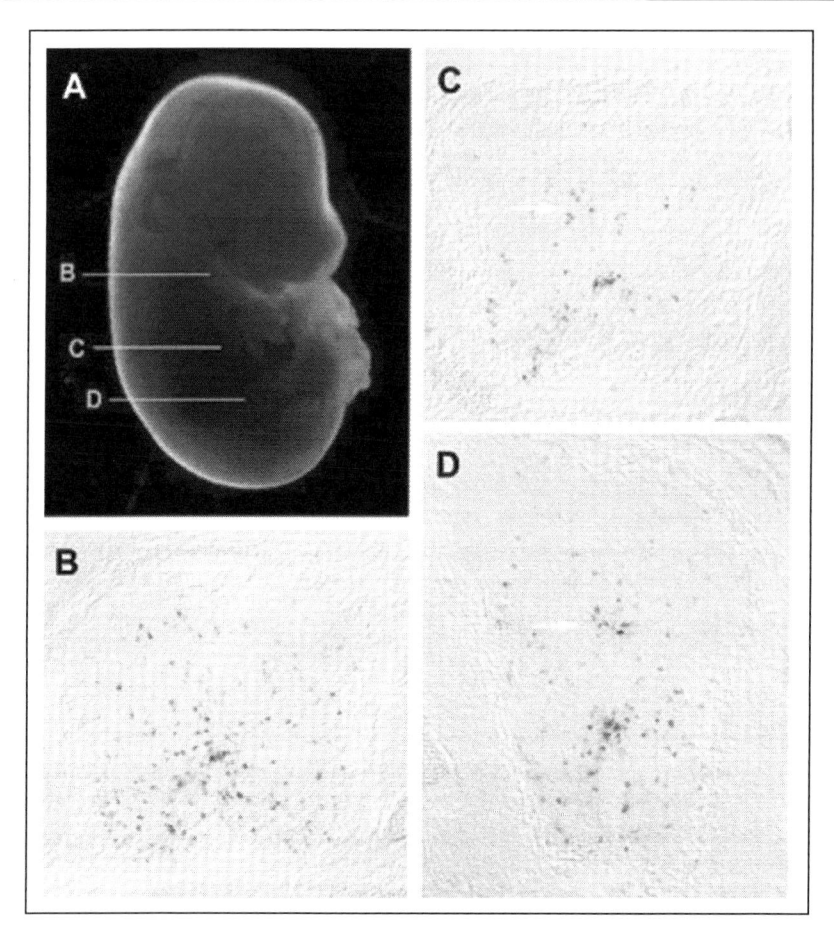

Figure 1. *Olig2*+ cells are generated transiently from restricted sites of E14.5 dorsal spinal cord. Serial cross sections of E14.5 wild-type mouse embryos were subjected to ISH with *Olig2* riboprobe. A) Photograph of an E14.5 mouse embryo to indicate the positions of transverse sections in B-D. B-D) *Olig2* expression in E14.5 spinal cord along the rostral-caudal axis as indicated in A. The dorsal OPC population is more evident in the caudal spinal cord, as indicated by white arrows.

counterparts. For instance, dorsal OPCs may remain undifferentiated and become adult progenitor cells, whereas ventral OPCs proceed to become myelinating cells. Even if dorsal OPCs do differentiate into myelinating cells, as suggested by the observation that dorsally-derived OPCs can form myelin sheets in culture,[28] it is conceivable that these two different pools of OPCs may be targeted to myelinate different populations of axons.[36] Definite answers to these important questions need to await future fate mapping studies employing the contemporary molecular and genetic approaches such as the CreLoxP system.

Similar to the fate specification of dorsal interneurons, the generation of oligodendrocytes from the dorsal neural progenitor cells also appears to be a *Shh*-independent process. In *Shh* -/- mutants, *Olig1/2*+ OPC cells emerge from the dorsal region of the spinal cord at E14.5 (Fig. 2), indicating that dorsal oligodendrogenesis proceeds as normal in the absence of *Shh* signalling. Although it is conceivable that the loss of *Shh* function could be compensated for by the expression of other hedgehog members (*Ihh* and *Dhh*) in the surrounding tissues, there is both pharmacological and genetic evidence that oligodendrocyte development occurs in the absence

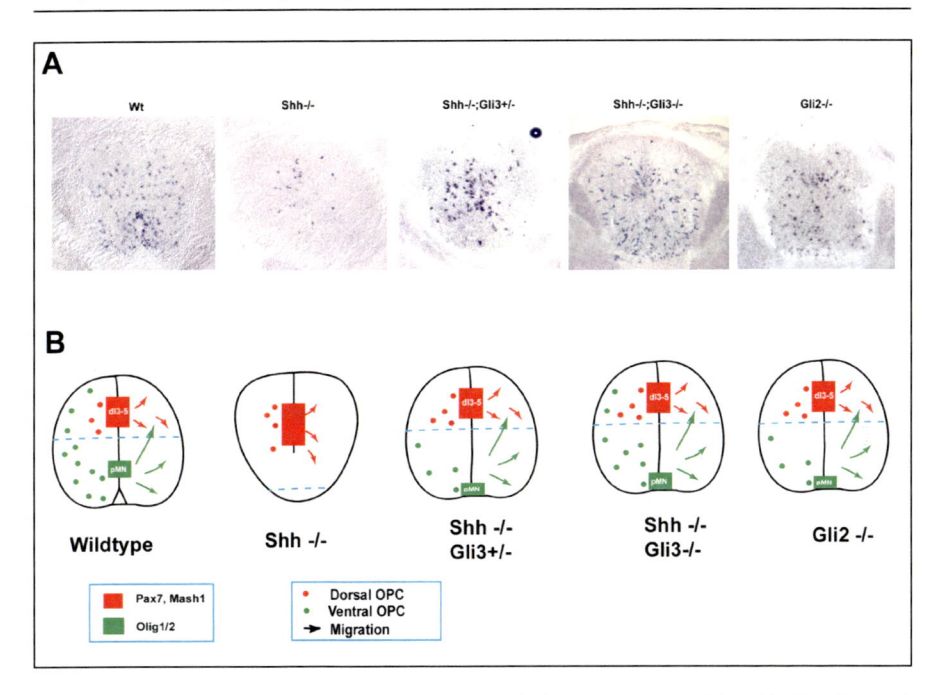

Figure 2. A) Expression of *Olig1* gene in E14.5 spinal cord of various mutants of the *Shh-Gli* pathway. B) A hypothetical model for the origin of spinal cord oligodendrocytes in these mutants. In the wild-type, a majority of OPCs are derived from the ventral pMN domain in a *Shh*-dependent mechanism. A small population of OPCs is also generated from the dorsal dI3-5 domains independent of *Shh* signalling. Only ventral, but not dorsal, oligodendrogenesis is affected by mutations in the *Shh-Gli* pathway. The floor plate is missing in all four mutants. The arrows represent the possible migratory directions of OPC cells.

of all *hedgehog* signalling in vitro. First, oligodendrocytes can be induced from dorsal neural progenitor cells by FGF in the presence of the pan-hedgehog inhibitor cyclopamine.[27,37] Second, oligodendrocytes can develop from embryonic stem (ES) cells deficient in the pan-hedgehog receptor *Smoothened*,[28] which is required for all *hedgehog* signalling.[38]

The signalling mechanism underlying the *Shh*-independent late phase of dorsal oligodendrogenesis in the spinal cord remains unknown at this time. Since FGF signalling can induce oligodendrocyte development in dissociated dorsal neural progenitor cells[39] independent of *Shh* signalling,[27,37] it is possible that FGF signalling could be partially responsible for the late production of OPCs in the dorsal spinal cord. In addition, the progressive reduction of BMP (Bone Morphogenetic Protein) signalling over time may also contribute to dorsal oligodendrogenesis.[35] It is known that BMP can antagonize *Shh*-induced oligodendrocyte specification, and experimental inhibition of BMP signalling is sufficient to induce oligodendrocyte production both in vivo and in vitro.[35,36,40] It is possible that dorsal oligodendrogenesis may result from a combination of increased FGF and decreased BMP, signalling.

Differential Roles of *Gli* Genes in Ventral Oligodendrogenesis

The intracellular mechanisms underlying *Shh* induction of motor neurons and oligodendrocytes in the ventral spinal cord are not well understood. Previous studies in *Drosophila* have identified a zinc-finger transcription factor, *Cubitus Interruptus* (*Ci*), as the key mediator of *hedgehog* signalling. Three homologues of *Ci* have been identified in vertebrates; these are the *Gli* genes (*Gli1*, *Gli2* and *Gli3*). Although all three *Gli* genes are expressed in the developing spinal cord,[41,42] they appear to have distinct roles in mediating the *Shh* induction of various cell types in the ventral neural tube.[43,44] Although both *Gli1* and *Gli2* can function as activators of *Shh* target genes in overexpression studies,[45-47] only *Gli2* is involved in high-level *Shh* signalling for the induction of floor plate and V3 ventral interneurons.[48,49] In contrast, *Gli1* activity is not required for normal D-V patterning of the neural tube, and there are no discernable neuronal defects in *Gli1* null mutants.[49,50]

The *Gli3* transducer appears to function both as a repressor and activator of *Shh* signalling.[42,51,52] The primary function of *Gli3* appears to be repression of fate specification of motor neurons and interneurons at more dorsal positions. In *Gli3* single mutant, there is a marked dorsal expansion of V0, V1 and dI6 interneurons in the intermediate region of the spinal cord.[53] In addition, *Gli3* mutation can rescue the development of motor neurons and V2 interneurons in *Shh* mutants in a dose-dependent manner, indicating that *Gli3* functions as a repressor of these two ventral neuronal cell types in the absence of *Shh* signalling.[51,53] However, the development of the floor plate and V3 neurons are not restored in *Shh-/-Gli3-/-* double mutants, similar to the phenotypes observed in *Gli2* mutants. Thus, it has been proposed that *Gli3* acts as a coactivator of *Gli2* in the *Shh*-mediated induction of these two ventral-most cell types.[42,51,54]

Gli2 Activity Regulates *Olig* Gene Expression in the Ventral Spinal Cord and the Initial Production of Oligodendrocyte Progenitors

Whilst no role for *Gli1* in oligodendrogenesis has been reported, the role of the *Gli2* transducer in ventral oligodendroglial development has recently been investigated in our laboratory.[55] In *Gli2* mutant embryos, the early expression of *Olig2* gene in the ventral spinal cord during neurogenesis is not affected, and the production of motor neurons appears to be normal.[48,49] However, *Olig2* expression in the ventral neural progenitor cells is not up-regulated and maintained during the oligodendrogenesis stage in these mutants. Consequently, the production of OPC cells from the ventral spinal cord is significantly delayed and reduced, but not completely inhibited.[55] Therefore, *Gli2* activity regulates the late phase of *Olig2* gene expression in the ventral neuroepithelium and its subsequent production of OPC cells. One plausible explanation for this mutant phenotype is that the ventricular expression of the *Olig* genes during oligodendrogenesis depends on a late supply of *Shh* protein from the floor plate, which is absent in *Gli2* mutants. It is known that oligodendrogenesis requires continued *Shh* signalling[26] and that *Shh* is expressed in the floor plate.[26,56] If this is the case, *Gli2* regulates ventral oligodendrogenesis indirectly through its effect on floor plate formation. In support of the nonautonomous role of *Gli2* in ventral oligodendrogenesis, *Shh* expression is not affected in the ventral forebrain in *Gli2* mutants and oligodendrogenesis proceeds normally in this region.[55] Although *Gli2* is not absolutely required for ventral oligodendrogenesis, it is still possible that *Gli2* is normally involved in this *Shh*-dependent process, but loss of its function is compensated for by *Gli3* or *Gli1*.

As expected, the generation of OPCs from the dorsal spinal cord does not seem to be compromised in *Gli2* mutants. At E14.5, a small number of *Olig*+ OPC are generated and located immediately adjacent to the dorsal neuroepithelium in the mutant spinal cords (Fig. 2). Despite delayed and reduced ventral OPC production, a similar steady-state number or density of OPCs is eventually achieved in the wild-type and *Gli2* mutant spinal cords at late gestation stages, possibly due to increased OPC proliferation in the mutants.

Interestingly, in spite of the similar number of OPC cells in the wild-type and *Gli2* mutant spinal cords at late gestation stages, oligodendrocyte differentiation is severely reduced and delayed in the mutants. However, this delay is also observed in other mutants (e.g., *Nkx6.1-/-* and *Shh-/-Gli3-/-* mutants) in which the initial production of ventral OPCs is also delayed but *Gli2* activity is reserved.[28,57] Therefore, the delay of OPC terminal differentiation is unlikely to be due to the loss of *Gli2* function itself. One possible mechanism for the parallel delay of OPC generation and differentiation in *Gli2* and other mutants is that an intrinsic timing mechanism may be responsible for regulating the onset of oligodendrocyte differentiation and maturation.

Previous studies showed that OPC cells in culture conditions withdrew from the cell cycle and differentiated after a certain number of cell divisions,[58] or a fixed amount of time.[59] However, alternative mechanisms are also conceivable, and therefore the mechanism underlying the parallel delay of OPC generation and maturation in the mutants remains to be determined.

Gli3 Functions as a Repressor of Ventral Oligodendrogenesis in the Absence of Shh Signalling

The role of the *Gli3* gene in ventral oligodendrogenesis has been investigated in our laboratory. In *Gli3* single mutants, there are no obvious phenotypes related to oligodendrocyte specification and differentiation in the spinal cord. The generation of OPCs from both ventral and dorsal spinal cord appears to be normal and on schedule (Fig. 2). The lack of an oligodendrocyte phenotype in *Gli3* single mutants is not surprising, given that *Gli3* mutation does not affect the specification of neural progenitor cells that give rise to oligodendrocytes in both ventral and dorsal spinal cord, i.e., the pMN domain and the dI3-5 domains, respectively.[51,53]

Similar to the scenario in ventral neurogenesis, *Gli3* mutation can also rescue ventral oligodendrogenesis in *Shh* mutants in a dose-dependent manner. In *Shh-/-* single mutants, OPCs are only produced from the dorsal, but not ventral, spinal cord, due to the lack of pMN domain (Fig. 2).[28] However, in *Shh-/-Gli+/-* embryos, a small number of OPCs start to appear in the ventral spinal cord at E14.5 (Fig. 2), indicating that ventral oligodendrogenesis is partially restored in these mutants. In *Shh-/-Gli3-/-* double mutants, the number of OPCs derived from the ventral spinal neuroepithelium is comparable to that observed in the wild-type embryos (Fig. 2). In all cases, the generation of OPCs from dorsal neuroepithelial cells does not appear to be affected. Together, these observations suggest that *Gli3* plays a nonessential role in both ventral and dorsal oligodendrogenesis during normal development. However, in the absence of *Shh* signalling, *Gli3* functions as a repressor of ventral oligodendrogenesis.

Acknowledgements

This work is supported by NIH (NS37717) and by National Multiple Sclerosis Society (RG 3275).

References

1. Noll E, Miller R. Oligodendrocyte precursors originate at the ventral ventricular zone dorsal to the ventral midline region in the embryonic rat spinal cord. Development 1993; 118:563-573.
2. Pringle N, Richardson W. A singularity of PDGF alpha-receptor expression in the dorsoventral axis of the neural tube may define the origin of the oligodendrocyte lineage. Development 1993; 117:525-533.
3. Timsit S, Martinez S, Allinquant B et al. Oligodendrocytes originate in a restricted zone of the embryonic ventral neural tube defined by DM-20 mRNA expression. J Neurosci 1995; 15:1012-1024.
4. Warf B, Fok-Seang J, Miller R. Evidence for the ventral origin of oligodendrocyte precursors in the rat spinal cord. J Neurosci 1991; 11:2477-2488.
5. Pringle N, Guthrie S, Lumsden A et al. Dorsal spinal cord neuroepithelium generates astrocytes but not oligodendrocytes. Neuron 1998; 20:883-93.
6. Sussman CR, Dyer KL, Marchionni M et al. Local control of oligodendrocyte development in isolated dorsal mouse spinal cord. J Neurosci Res 2000; 59:413-420.
7. Sun T, Pringle NP, Hardy AP et al. Pax-6 influences the time and site of origin of glial precursors in the ventral neural tube. Mol Cell Neurosci 1998; 12:228-239.
8. Lu Q, Yuk D, Alberta J et al. Sonic Hedgehog-regulated oligodendrocyte lineage genes encoding bHLH proteins in the mammalian central nervous system. Neuron 2000; 25:317-329.
9. Takebayashi H, Yoshida S, Sugimori M et al. Dynamic expression of basic helix-loop-helix Olig family members: Implication of Olig2 in neuron and oligodendrocyte differentiation and identification of a new member, Olig3. Mech Dev 2000; 99:143-8.
10. Zhou Q, Wang S, Anderson DJ. Identification of a novel family of oligodendrocyte lineage-specific basic helix-loop-helix transcription factors. Neuron 2000; 25:331-343.

11. Fu H, Qi Y, Tan M et al. Dual origin of spinal oligodendrocyte progenitors and evidence for the cooperative role of Olig2 and Nkx2.2 in the control of oligodendrocyte differentiation. Development 2002; 129:681-693.
12. Zhou Q, Choi G, Anderson DJ. The bHLH transcription factor Olig2 promotes oligodendrocyte differentiation in collaboration with Nkx2.2. Neuron 2001; 31:791-807.
13. Lu Q, Sun T, Zhu Z et al. Common developmental requirement for Olig function indicates a motor neuron/oligodendrocyte connection. Cell 2002; 109:75-86.
14. Takebayashi H, Nabeshima Y, Yoshida S et al. The basic helix-loop-helix factor olig2 is essential for the development of motoneuron and oligodendrocyte lineages. Curr Biol 2002; 12:1157-1163.
15. Zhou Q, Anderson DJ. The bHLH transcription factors OLIG2 and OLIG1 couple neuronal and glial subtype specification. Cell 2002; 109:61-73.
16. Richardson W, Pringle N, Yu W et al. Origins of spinal cord oligodendrocytes: Possible developmental and evolutionary relationships with motor neurons. Dev Neurosci 1997; 19:58-68.
17. Richardson W, Smith H, Sun T et al. Oligodendrocyte lineage and the motor neuron connection. Glia 2000; 29:136-142.
18. Arnett H, Fancy S, Alberta J et al. bHLH transcription factor Olig1 is required to repair demyelinated lesions in the CNS. Science 2004; 306:2111-2115.
19. Jessell TM. Neuronal specification in the spinal cord: Inductive signals and transcriptional codes. Nat Rev Genet 2000; 1:20-29.
20. Echelard Y, Epstein D, St-Jacques B et al. Sonic hedgehog, a member of a family of putative signaling molecules, is implicated in the regulation of CNS polarity. Cell 1993; 75:1417-1430.
21. Roelink H, Porter J, Chiang C et al. Floor plate and motor neuron induction by different concentrations of the amino-terminal cleavage product of sonic hedgehog autoproteolysis. Cell 1995; 81:445-455.
22. Orentas D, Miller R. Regulation of oligodendrocyte development. Mol Neurobiol 1999; 18:247-259.
23. Trousse F, Giess M, Soula C et al. Notochord and floor plate stimulate oligodendrocyte differentiation in cultures of the chick dorsal neural tube. J Neurosci Res 1995; 41:552-560.
24. Poncet C, Soula C, Trousse F et al. Induction of oligodendrocyte progenitors in the trunk neural tube by ventralizing signals: Effects of notochord and floor plate grafts, and of sonic hedgehog. Mech Dev 1996; 60:13-32.
25. Pringle N, Yu W, Guthrie S et al. Determination of neuroepithelial cell fate: Induction of the oligodendrocyte lineage by ventral midline cells and sonic hedgehog. Dev Biol 1996; 177:30-42.
26. Orentas D, Hayes J, Dyer K et al. Sonic hedgehog signaling is required during the appearance of spinal cord oligodendrocyte precursors. Development 1999; 126:2419-29.
27. Chandran S, Kato H, Gerreli D et al. FGF-dependent generation of oligodendrocytes by a hedgehog-independent pathway. Development 2003; 130:6599-6609.
28. Cai J, Qi Y, Hu X et al. Generation of oligodendrocyte precursor cells from mouse dorsal spinal cord independent of Nkx6-regulation and Shh signaling. Neuron 2005; 45:41-53.
29. Qi Y, Stapp D, Qiu M. Origin and molecular specification of oligodendrocytes in the telencephalon. TINS 2002; 25:223-225.
30. Olivier C, Cobos I, Villegas E et al. Monofocal origin of telencephalic oligodendrocytes in the anterior entopeduncular area of the chick embryo. Development 2001; 128:1757-1769.
31. Nery S, Wichterle H, Fishell G. Sonic hedgehog contributes to oligodendrocyte specification in the mammalian forebrain. Development 2001; 128:527-540.
32. Tekki-Kessaris N, Woodruff R, Hall A et al. Hedgehog-dependent oligodendrocyte lineage specification in the telencephalon. Development 2001; 128:2545-2554.
33. Alberta J, Park S-K, Mora J et al. Sonic hedgehog is required during an early phase of oligodendrocyte development in mammalian brain. Mol Cell Neurosci 2001; 18:434-441.
34. Davies JE, Miller RH. Local sonic hedgehog signaling regulates oligodendrocyte precursor appearance in multiple ventricular zone domains in the chick metencephalon. Dev Biol 2001; 233:513-525.
35. Vallstedt A, Klos J, Ericson J. Multiple dorsoventral origins of oligodendrocyte generation in the spinal cord and hindbrain. Neuron 2005; 45:55-67.
36. Miller RH. Dorsally derived oligodendrocytes come of age. Neuron 2005; 45:1-3.
37. Kessaris N, Jamen F, Rubin L et al. Cooperation between sonic hedgehog and fibroblast growth factor/MAPK signaling pathways in neocortical precursors. Development 2004; 131:1289-1298.
38. Wijgerde M, McMahon J, Rule M et al. A direct requirement for Hedgehog signaling for normal specification of all ventral progenitor domains in the presumptive mammalian spinal cord. Genes Dev 2002; 16:2849-2864.
39. Gabay L, Lowell S, Rubin L et al. Deregulation of dorsoventral patterning by FGF confers trilineage differentiation capacity on CNS stem cells in vitro. Neuron 2003; 40:485-499.

40. Mekki-Dauriac S, Agius E, Kan P et al. Bone morphogenetic proteins negatively control oligodendrocyte precursor specification in the chick spinal cord. Development 2002; 129:5117-5130.
41. Hui CC, Slusarski D, Platt K et al. Expression of three mouse homologs of the Drosophila segment polarity gene cubitus interruptus, Gli, Gli-2, and Gli-3, in ectoderm- and mesoderm-derived tissues suggests multiple roles during postimplantation development. Dev Biol 1994; 162:402-13.
42. Lei Q, Zelman AK, Kuang E et al. Transduction of graded Hedgehog signaling by a combination of Gli2 and Gli3 activator functions in the developing spinal cord. Development 2004; 131:3593-3604.
43. Bai C, Stephen D, Joyner AL. All mouse ventral spinal cord patterning by hedgehog is Gli dependent and involves an activator function of Gli3. Dev Cell 2004; 6:103-115.
44. Ruiz I Altaba A, Palma V, Dahmane N. Hedgehog-Gli signalling and the growth of the brain. Nat Rev Neurosci 2002; 3:24-33.
45. Lee J, Platt KA, Censullo P et al. Gli1 is a target of Sonic hedgehog that induces ventral neural tube development. Development 1997; 124:2537-2552.
46. Ruiz I Altaba A. Combinatorial Gli gene function in floor plate and neuronal inductions by Sonic hedgehog. Development 1998; 125:2203-2212.
47. Bai C, Joyner AL. Gli1 can rescue the in vivo function of Gli2. Development 2001; 128:5161-5172.
48. Ding Q, Motoyama J, Gasca S et al. Diminished Sonic hedgehog signaling and lack of floor plate differentiation in Gli2 mutant mice. Development 1998; 125:2533-2543.
49. Matise M, Epstein D, Park H et al. Gli2 is required for induction of floor plate and adjacent cells, but not most ventral neurons in the mouse central nervous system. Development 1998; 125:2759-2770.
50. Park H, Bai C, Platt K et al. Mouse Gli1 mutants are viable but have defects in SHH signaling in combination with a Gli2 mutation. Development 2000; 127:1593-605.
51. Litingtung Y, Chiang C. Specification of ventral neuron types is mediated by an antagonistic interaction between Shh and Gli3. Nat Neurosci 2000; 3:979-985.
52. Tyurina O, Guner B, Popova E et al. Zebrafish Gli3 functions as both an activator and a repressor in Hedgehog signaling. Dev Biol 2005; 277:537-556.
53. Persson M, Stamataki D, Welscher P et al. Dorsal-ventral patterning of the spinal cord requires Gli3 transcriptional repressor activity. Genes Dev 2002; 16:2865-2878.
54. Motoyama J, Milenskovic L, Iwama M et al. Differential requirement for Gli2 and Gli3 in ventral neural cell fate specification. Dev Biol 2003; 259:150-161.
55. Qi Y, Tan M, Hui C-C et al. Gli2 activity is required for normal Shh signaling and oligodendrocyte development. Mol Cell Neurosci 2003; 23:440-450.
56. Bitgood MJ, McMahon AP. Hedgehog and Bmp genes are coexpressed at many diverse sites of cell-cell interaction in the mouse embryo. Dev Biol 1995; 172:126-138.
57. Liu R, Cai J, Hu X et al. Region-specific and stage-dependent regulation of Olig gene expression and oligodendrogenesis by Nkx6.1 homeodomain transcription factor. Development 2003; 130:6221-6231.
58. Temple S, Raff M. Clonal analysis of oligodendrocyte development in culture: Evidence for a developmental clock that counts cell divisions. Cell 1986; 44:773-779.
59. Gao FB, Durand B, Raff M. Oligodendrocyte precursor cells count time but not cell divisions before differentiation. Curr Biol 1997; 7:152-155.

CHAPTER 5

The Role of Sonic Hedgehog Signalling in Craniofacial Development

Dwight Cordero, Minal Tapadia and Jill A. Helms*

Introduction

The unique characteristics of our face contribute to individuality, distinguishing us from other human beings as well as other species. This has led to the face being thought of as an isolated entity, in terms of both embryonic development and postnatal physical characteristics. The artistic intricacy of facial features is a reflection of multiple sophisticated spatial and temporal developmental events and interactions, not only within tissues that give rise to the face but also between these and other tissues such as the brain. The culmination of such interactions transforms planar tissue into readily recognizable complex three-dimensional structures with unique characteristics that we identify as our face. Complexity not simplicity, and interactions not seclusion, are the axioms in craniofacial development.

The Sonic Hedgehog (Shh) signalling pathway is involved in a number of tissue interactions during craniofacial development, and is integral in providing information to cells that give rise to facial features. This chapter provides an overview of craniofacial development and our present understanding of the roles Shh plays in the genesis of the face. We also discuss how mutations in this pathway, and environmental agents, may lead to craniofacial dysmorphologies.

Overview of the Anatomy of Craniofacial Development

In mammals and birds, facial structures develop from the facial primordia: a single frontonasal primordium and paired maxillary and mandibular processes (Figs. 1A-C). These primordia, also referred to as the facial mesenchyme, consist of an epithelium that encloses undifferentiated neural crest cells (Figs. 1D-G), and are active centers of mesenchymal cell proliferation, condensation, differentiation and apoptosis. They may share fundamental similarities in terms of structural organization but the molecular mechanisms controlling their patterned outgrowth appear to be distinct. This may be due to differing axial origins of the neural crest cells from the neural tube, or to regional differences in the overlying ectoderm.[1]

The forebrain and the epithelia of facial primordia originate from the same ectoderm (Fig. 1D). Neural cell fate is thought to result from the presence of bone morphogenic protein (Bmp) antagonists and fibroblast growth factors (Fgfs) (Fig. 1D) inducing neural character [reviewed in ref. 2] and therefore formation of the forebrain [reviewed in ref. 3,4]. During development, the neuroectoderm of the ventral forebrain is in intimate contact with the mesenchyme within the epithelium-covered frontonasal process (Figs. 1F-G), allowing communication between the three tissues (the neuroectoderm, mesenchyme and facial ectoderm). In the

---

*Corresponding Author: Jill A. Helms—Department of Plastic and Reconstructive Surgery, Stanford University, 257 Campus Drive, Stanford, California 94305, U.S.A. Email: jhelms@stanford.edu.

Shh and Gli Signalling and Development, edited by Carolyn E. Fisher and Sarah E.M. Howie. ©2006 Landes Bioscience and Springer Science+Business Media.

The Role of Sonic Hedgehog Signalling in Craniofacial Development

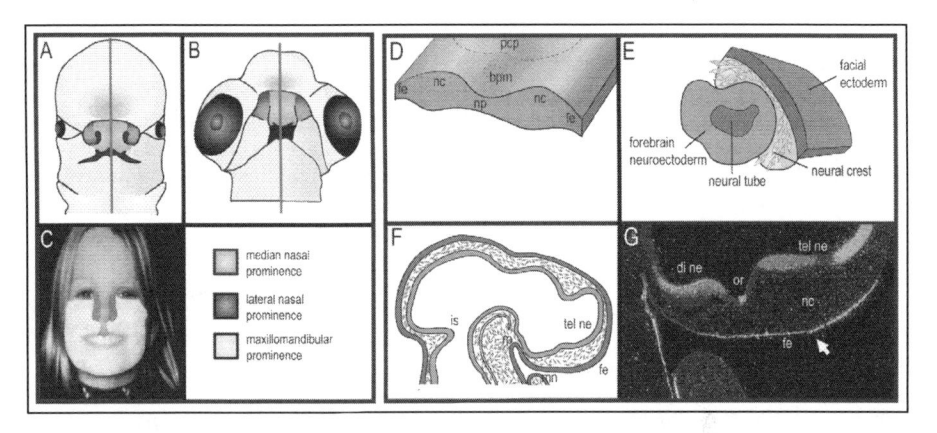

Figure 1. Development of the facial prominences. Craniofacial structures develop from facial primordia and consist of neural crest mesenchyme enclosed by facial ectoderm. Frontal view representations of (A) murine and (B) chick embryo facial primordia showing the physical location and relationships of the primordia. C) Photograph of a child depicting the facial structures that arise from the respective primordia. Color codes represent the primordia of structural derivation. The forehead and nose (beak in the chick) are derived from the frontonasal primordium (forehead and medial nasal prominence in orange, lateral nasal prominence in purple). The maxillary and mandibular processes (maxillomandibular prominences) (yellow) give rise to the midface, lateral aspects of the lips and secondary palate (maxillary prominences) and the lower jaw or beak in chicks (mandibular prominences). D) The epithelia comprising the primordia originate from a unified sheet of ectoderm, which is subdivided into neural and nonneural regions that are influenced by the concentration of Bone morphogenetic proteins (Bmps). E) The ectoderm folds upward and becomes the neural folds. As the neural folds fuse, creating the neural tube, distinct tissue layers of neuroectoderm (ne, green) and facial ectoderm (fe, blue) are seen. Neural crest cells delaminate from the border region between the neuroectoderm and surface ectoderm, and migrate into specific areas of the face to give rise to the facial prominences depicted in A-C. F-G) Following migration, the neural crest cells lie between the neuroectoderm and facial ectoderm, and receive developmental cues from both the neuro- and facial ectoderm such as Sonic Hedgehog (Shh) and Fibroblast Growth Factor 8 (Fgf8). G) In situ hybridization performed on sagittal sections of chick embryos, where red (pseudocolored using photoshop) represents *Shh* expression and green represents *Fgf8* expression. Note the neural crest cell (NC) relationship to both the ne and fe. The Facial Ectodermal Zone (FEZ) consists of a boundary between *Shh* expression (red) and *Fgf8* expression (green) in the fe (arrowhead), which is an organizing centre for proper outgrowth and patterning of structures derived from the frontonasal process. Abbreviations: di ne: diencephalic neuroectoderm; is: isthmus; mn: mandible; PA: pharyngeal arch; pe: pharyngeal endoderm; or: optic recess; PCP: prechordal plate; tel ne: telencephalic neuroectoderm. A-G) Reprinted courtesy of *Development*.[12]

maxillary and mandibular primordia, signalling occurs between the facial ectoderm, mesenchyme and endoderm (Fig. 1G), resulting in maturation of these tissues into facial structures (Fig. 1C).

Cranial neural crest cells (CNCCs) give rise to the facial mesenchyme, and originate from specific axial positions along the dorsal neural tube. They migrate into specific regions of the facial primordia where they have the pluripotential to form pericytes, which are components of blood vessels,[5] cartilage and bones of the face [reviewed in ref. 11].[6-10] There is debate over what determines the cell fate decision of CNCCs. One view is that they are preprogrammed with all the information needed to determine their cell fate but others argue that their fate is determined by responses to developmental cues from the local environment after they arrive in the facial primordia.[12] Evidence for preprogramming comes from transplantation experiments.[13-15] Transplanting presumptive second and third arch neural crest with presumptive first arch neural crest results in ectopic skeletal elements of the first arch growing in locations usually associated with the second and third arches.[13] However, transplantation experiments from other laboratories suggest that CNCCs are not preprogrammed but interpret information within the new local environment of the facial primordia and are capable of responding in a manner that depends on their developmental history [reviewed in ref. 12].[7,9,13,15-19] Opponents of this view suggest that signals originating in the overlying facial epithelium or adjacent tissues control cell proliferation, survival, patterning and differentiation of the mesenchyme.[20,21] However, the generation of information does not appear to be unidirectional i.e., from ectoderm to mesenchyme; signals emanating from the mesenchyme are likely to influence the character of facial ectoderm and neuroectoderm, which are in contact with the frontonasal process (FNP). Integration of and responses to such signalling determine patterning, cell proliferation and outgrowth, leading to the fusion of facial structures, and thereby creating the intricate morphologies characterizing the human face.

Sonic Hedgehog in Development of the Upper Face

The Dynamic Spatial and Temporal Expression of Shh in the Brain and Face

Physicians have recognized a clinical association between forebrain and facial development for more than forty years[22] but it is only recently that the molecular basis of this relationship has begun to be revealed.[23-25] Studies have shown that a number of molecules including Shh,[26] Fgf8,[26,27] BMPs[28-31] and retinoic acid (RA)[23] play important roles in patterning, growth and morphogenesis of the forebrain and face, and that Shh and RA are two of possibly many molecules that may mediate the transmission of developmental information between the forebrain and face during embryogenesis.[23]

Shh expressed in the developing central nervous system (CNS), ectoderm of the first pharyngeal arch, FNP and endoderm[32] mediates ectodermal-mesenchymal interactions, which are necessary for the appropriate patterning and growth of the facial primordia.[33] *Shh* is expressed in the rostral head in the midline of the neural plate.[34] The mesoendoderm (prechordal plate) is a source of Shh required for normal ventral forebrain development.[35] After neurulation in the avian embryo, this midline region gives rise to the ventral prosencephalon which subsequently divides into the telencephalon (future cerebral cortex) and diencephalon (future thalamus, hypothalamus, subthalamus and epithalamus).

Recently, chick models have revealed that *Shh* is dynamically expressed in the forebrain and face. *Shh* transcripts are restricted to the ventral diencephalon at HH[36] stage 15 (Figs. 2C-D), and at HH stage 17 it is induced in the ventral telencephalon, which is separated from the diencephalic domain by the *Shh*-negative optic recess (Figs. 2E-F).[24] At HH stage 20, *Shh* is induced in ventral ectoderm of the FNP (Figs. 2G-H)[24] but is not expressed in intervening CNCCs (Figs. 2G-H). Shh is required for normal skeletal development of the craniofacial region,[37] and once it has been established the facial domain of *Shh* persists[38] but is limited to the ectoderm of the FNP and maxillary processes.[39] The spatial and temporal expression of *Shh* in the CNS and face suggests that Shh may be important for the coordinated development of the forebrain and face.

The Clinical Implications of the Spatial-Temporal Relationship between the Brain and Face

Shh is required for normal forebrain[40-42] and facial development[26,32] in many species including humans.[43,44] *Shh* null mutations in mice result in holoprosencephaly (HPE) and severe facial manifestations such as cyclopia, a proboscis, and hypoplastic maxillary and mandibular derivatives (Figs. 3A-B).[35] Unfortunately interruption of Shh signalling early in gestation affects neural plate patterning and thereby precludes analyzing the direct contribution of Shh to facial morphogenesis at later developmental stages. The chick model system, which allows for manipulation of Shh signalling by physical, biochemical and other means, has in part bypassed this limitation and provided insights into the roles of Shh during patterning and outgrowth of the craniofacial complex.[23-25,32]

The Role of Sonic Hedgehog Signalling in Craniofacial Development   47

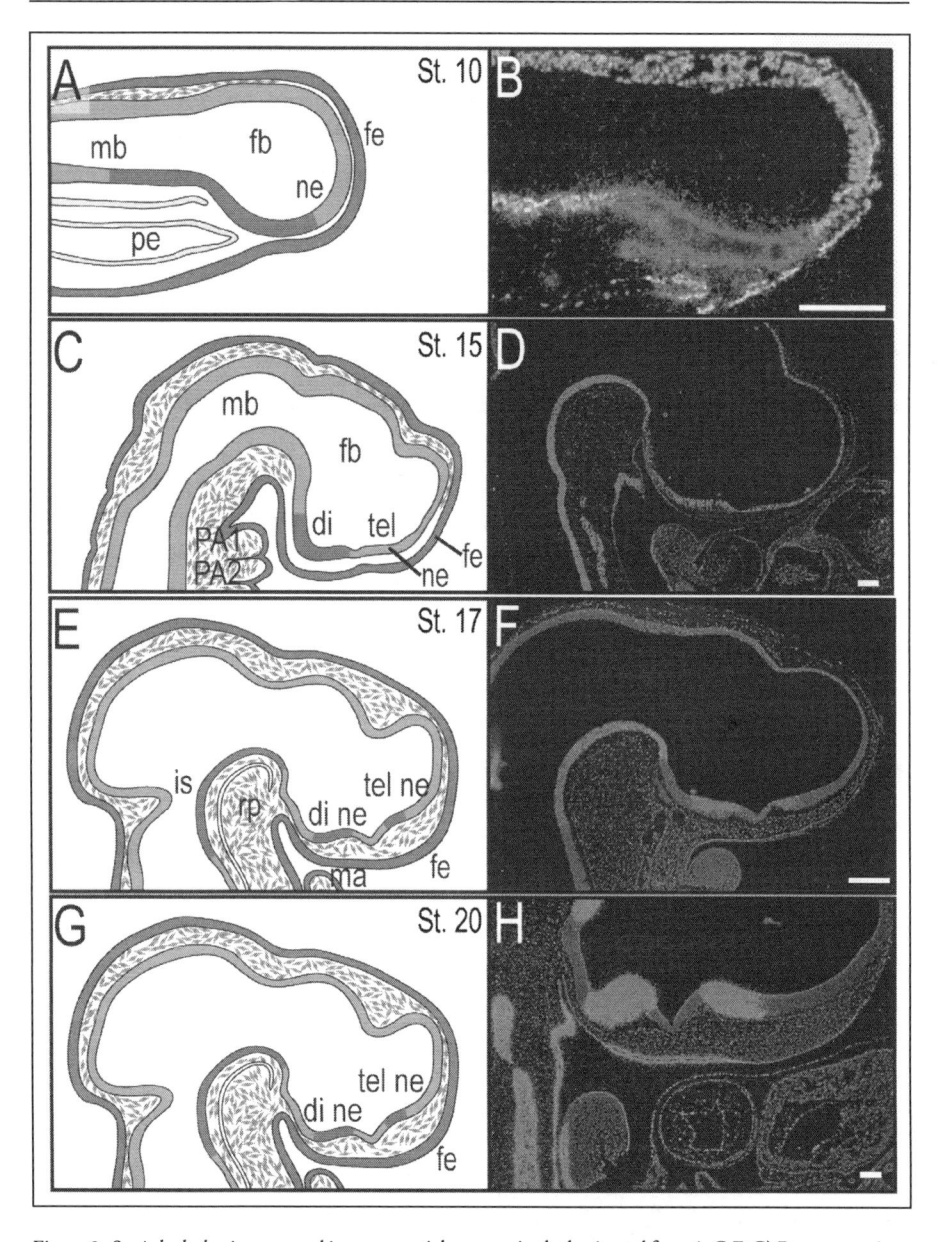

Figure 2. *Sonic hedgehog* is expressed in a sequential manner in the brain and face. A,C,E,G) Representations of midline sagittal sections of the chick craniofacial complex at various developmental stages, next to (B,D,F,H) actual midline sagittal sections at corresponding stages (red represents *Shh* expression). A,B) At stage 10, *Shh* is expressed in the forebrain (fb), in the ventral prosencephalon (vp), and pharyngeal endoderm (pe). C,D) At stage 15, the forebrain (fb) has divided into the telencephalic (tel) and the diencephalic (di) domains. At this stage *Shh* transcripts are localized to the neuroectoderm of the di. E,F) At stage 17, *Shh* is expressed in telencephalic neuroectoderm (tel ne). G,H) By stage 20, *Shh* is expressed in the diencephalic (di ne) and telencephalic neuroectoderm (tel ne) and in the facial ectoderm (fe). Abbreviations: is: isthmus; ma: maxillary process; mb: midbrain; PA: pharyngeal arch; rp: Rathke's pouch. A,C,E,G) Reprinted courtesy of *Development*,[12] *Drug Discov Today: Disease Mech*,[99] *J Anatomy*.[101] B,F) Reprinted courtesy of *J Clin Invest*.[24] D) Reprinted courtesy of *J Anatomy*.[101] H) Reprinted courtesy of *Development*.[12]

The first experiment directly to address the contribution of Shh at later stages of craniofacial development involved the excision of a *Shh*-positive region of facial ectoderm corresponding to the presumptive FNP in HH stage 25 chick embryos.[32] Loss of *Shh* from the facial epithelium (without disturbing the underlying mesenchyme) resulted in decreased outgrowth of the FNP along the anterior-posterior axis and an inability to fuse with the maxillary primordia (Figs. 3C-D). In humans, this failure of fusion causes cleft lip and palate.[39,45] The addition of Shh recombinant protein (Shh-N)-containing beads in the presumptive FNP region of HH stage 25 chick embryos caused an increase in the width of the FNP,[32] which is reminiscent of hypertelorism, observed in a number of human craniofacial disorders. Over-expressing *Shh* in the facial ectoderm of chicks using RCAS-Shh results in similar phenotypes,[46] as do the murine gain-of-function mutations in *Gli3*[-/-].[47,48] These data suggest that Shh is involved in the medial-lateral growth (axis) of the FNP. These 'extra-toe' mice also have polydactyly.

In chicks, *Shh* expression in the FNP appears to be modulated by the vitamin A derivative RA.[26] RALDH6, a member of the aldehyde dehydrogenase family involved in the synthesis of RA,[49] is localized to the ventral epithelium of the presumptive FNP in chick embryos,[23] and RALDH3 is found in the neuroepithelium of the telencephalon and olfactory placode.[50,51] Two nuclear receptors, RARβ and RXRγ, which bind to RA, are present in the FNP mesenchyme.[23] The spatial relationship between the ligands and their receptors suggests a possible link between Shh production in the developing brain and face, and/or that facial mesenchyme receives inputs from dual sources. To investigate the possible relationship between RA signalling and development of the forebrain and FNP, chick embryos were treated at HH stage 10 with a synthetic pan-specific retinoid antagonist that transiently inhibits the ability of retinoid receptors to bind RA in the rostral head.[23] Treated embryos exhibited hypoplastic forebrains, fused eyes, and no FNP derived structures such as the upper beak (Figs. 3E-F). These defects were caused by a down-regulation of *Shh* and *Fgf8* in the forebrain and FNP ectoderm, leading to increased apoptosis and decreased cell proliferation in both the forebrain and FNP primordia.[23] The malformations were rescued by reintroducing all-trans RA, Fgf2, or Shh protein[23] [reviewed in ref. 12] to embryos treated with the antagonists at HH stage 10, which were removed 8-10 hours later. The forebrain and FNP are linked developmentally; both structures depend upon the same local retinoid signalling during early morphogenesis, and Fgf8 and Shh signalling pathways are downstream targets of RA in the rostral head.

Investigating the possible role(s) of Shh in the communication between the forebrain and face utilized the steroidal alkaloid cyclopamine, a teratogenic agent extracted from the *Veratrum californicum* plant[52-54] that inhibits Shh signal transduction by binding to the heptahelical bundle of Smoothened (SMO) and altering its protein conformation.[55] This biochemical approach has the advantage of allowing the interruption of Shh signalling at multiple select embryonic stages, which is not possible with gene targeting.

Chick embryos were exposed to cyclopamine at select developmental time points governed by the dynamic *Shh* induction pattern described above,[24] and produced a variety of facial malformations reminiscent of the human HPE phenotypic spectrum. The severity of the craniofacial malformations correlated with the temporal and spatial inhibition of Shh signal transduction. Cyclopamine administration during gastrulation produced severe malformations involving the forebrain and face, including cyclopia with a proboscis as described previously,[54] and as found in *Shh* null mice.[35] When cyclopamine was administered prior to the initiation of *Shh* expression in the telencephalon (stage 15), embryos exhibited abnormal forebrain morphology consisting of incomplete division of the cerebral hemispheres.[24] The craniofacial abnormalities were less severe than those found in embryos treated at gastrulation, consisting of microcephaly, microptthalmia, a moderate degree of hypotelorism, and hypoplasia of the maxillary primordia.[24] Inhibiting Shh after induction of *Shh* in the telencephalon but prior to induction in the facial ectoderm (HH stage 17) yielded a grossly morphologically normal forebrain (two cerebral hemispheres) with facial dysmorphologies consisting of mild hypotelorism and distal upper beak truncation (consistant with cleft lip and palate in humans).[24] Inhibition

The Role of Sonic Hedgehog Signalling in Craniofacial Development 49

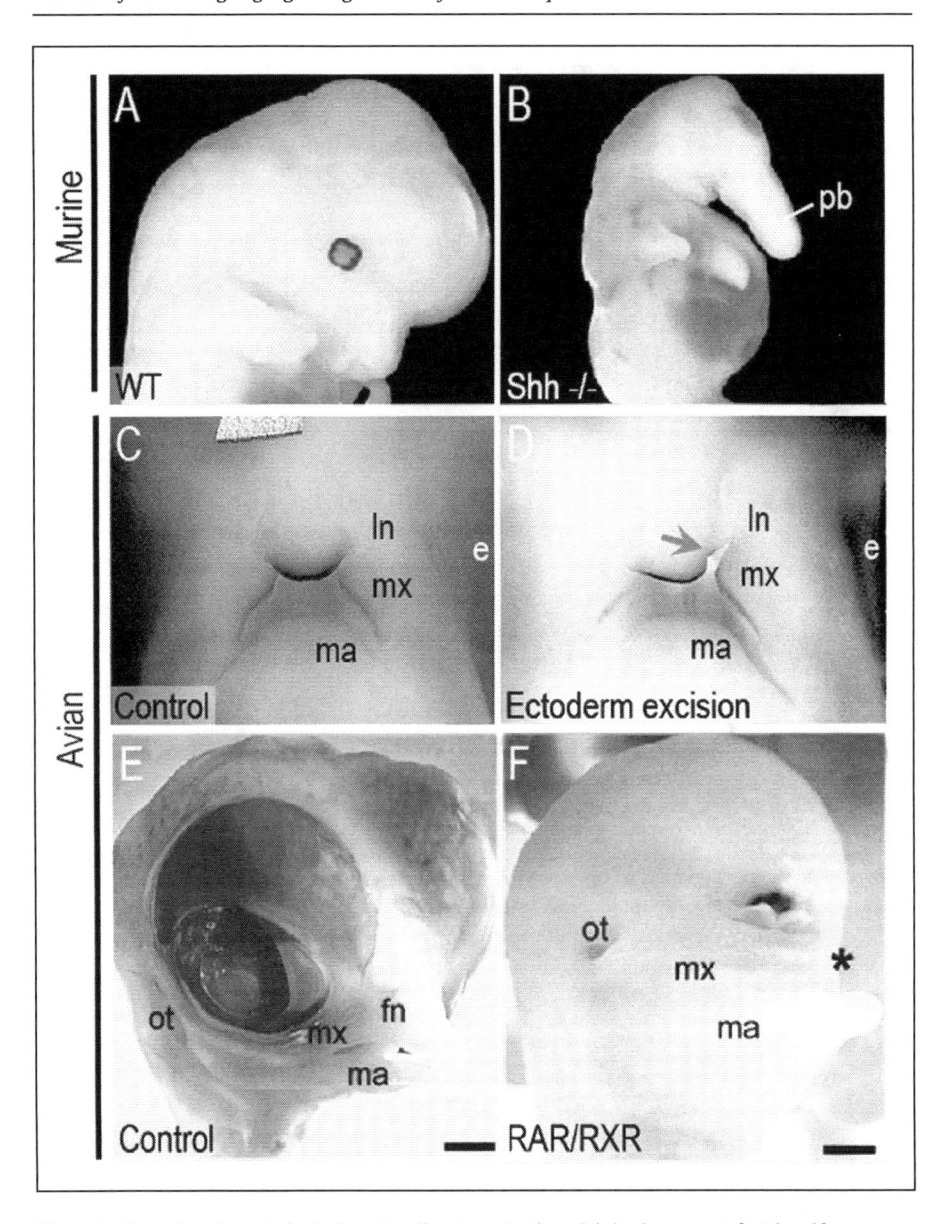

Figure 3. Alterations in sonic hedgehog signalling in animal models leads to craniofacial malformations. A,B) Oblique views of E15.5 murine embryos; C,D) frontal views of stage 30 chick embryos; E,F) oblique views of stage 36 chick embryos. A,B) Knockout of the *Shh* gene in the mouse leads to abnormal neural plate development and severe brain and facial malformations such as cyclopia, a proboscis (pb), and maxillary and mandibular hypoplasia as compared to wild type. C,D) Excision of *Shh* expressing facial ectoderm from stage 25 chick embryos leads to clefting of the upper beak (red arrow), which is equivalent to cleft lip and palate in humans. E,F) Inhibition of retinoic acid signalling in the face at stage 10 results in severe forebrain defects and facial malformations consisting of fused eyes and absence of derivatives of the frontonasal primordial (upper beak). Abbreviations: fn: frontonasal process; ln: lateral nasal process; ma: maxillary process; ot: otic process. A,B) Reprinted with permission from ref. 100. C,D) Reprinted courtesy of *Development*.[32] E, F) Reprinted courtesy of *Development*.[23]

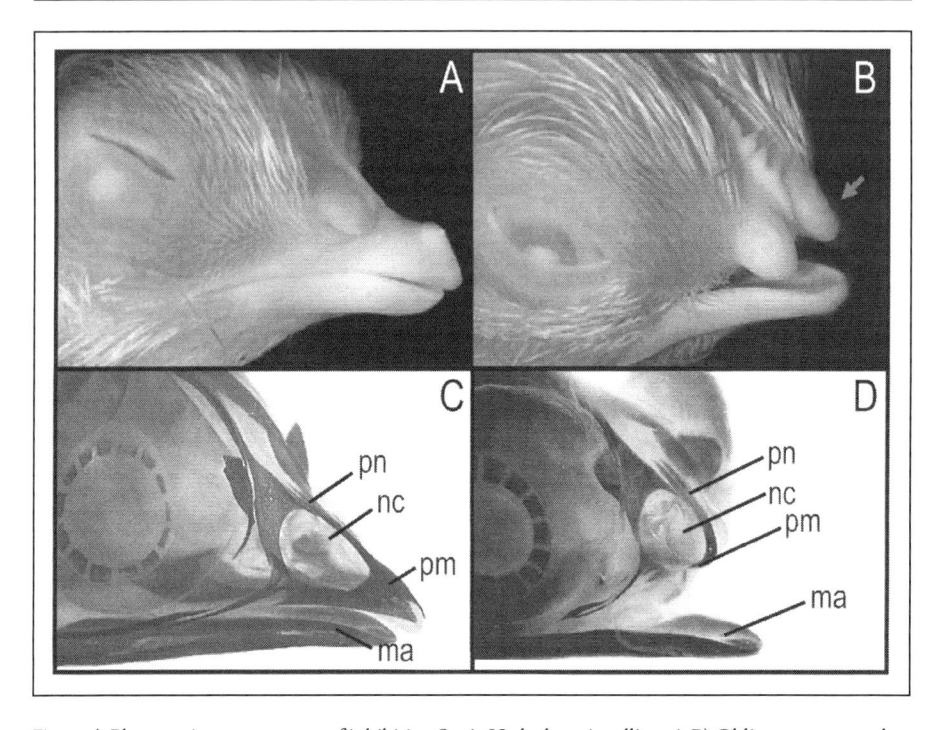

Figure 4. Phenotypic consequences of inhibiting Sonic Hedgehog signalling. A,B) Oblique gross morphological views of stage 41 chick embryos and (C,D) the corresponding alcian blue- and alizarin red-stained skeletal structures of these embryos. A,C) Control embryos show normal craniofacial features and structures. B,D). Embryos treated with cyclopamine at stage 17 exhibit mild microcephaly, hypotelorism, and truncation of the distal upper beak (red arrow). The body of the premaxillary bone (pm) is shortened (malformed and shifted ventrally, as compared to the pm in the control embryo). Abbreviations: nc: nasal capsule; pn: nasal process of premaxilla; ma: mandible. Reprinted courtesy of *J Clin Invest*.[24]

in embryos at HH stage 20 and later resulted in very perceptible facial anomalies or no discernable anomalies.[24]

Detailed skeletal analysis of cyclopamine treated embryos at HH stage 17 revealed that the observed distal upper beak truncation was secondary to a hypoplastic premaxilla, which was aberrantly positioned ventral to the nasal capsule (Fig. 4).[24] The palatine bones were medially located, indicating inhibition of their normal medial-lateral expansion (Fig. 4).[24] Molecular analysis revealed that the facial malformations were due to molecular mispatterning in the facial ectoderm and were not the consequence of CNCCs apoptosis within the FNP.[24] This highlights the importance of Shh during dorsal-ventral and medial-lateral patterning of the facial axes.

The loss of the *Shh* expression domain following exposure to cyclopamine at HH stages 15 and 17 was accompanied by an ectopic proximal expression of the Fgf8 domain from the Frontonasal Ectoderm Zone (FEZ).[24] The FEZ is a discrete region of facial ectoderm consisting of a ventral domain of *Shh* juxtaposed to, but not overlapping with, a dorsal domain of *Fgf8* expression (Fig. 1G).[19] It has organizer characteristics and regulates proximodistal growth and dorsoventral patterning within the FNP.[19] The loss of *Shh* and the shift in the *Fgf8* domain following cyclopamine treatment disrupted these organizing properties of the FEZ, affecting dorsal-ventral polarity and outgrowth of the FNP. These experiments suggest that Shh emanating from the forebrain is important for normal craniofacial development, and links forebrain and facial development. However, cyclopamine is capable of diffusing through facial ectoderm and mesenchyme to reach the neuroectoderm, and so selective inhibition of Shh signalling originating in the forebrain ectoderm was required to demonstrate the role(s) for brain-derived Shh on facial morphogenesis. Hybridoma cells, which produce the anti-Shh antibody 5E1, were injected into the brains of HH stage 10 chick embryos.[25] The resultant facial phenotypes were similar to those seen in HH stage 17 chicks following cyclopamine treatments,[25] and included truncation of the upper beak, ventralization of the premaxillary bone, decreased expansion of the medial-lateral axes and mispatterning of the FEZ.[25] The phenotype following 5E1 treatment was less severe than that observed after administration of cyclopamine at the same stage (Cordero and Helms, unpublished data). This may reflect the ability of cyclopamine to diffuse into multiple tissues and inhibit Shh signal transduction. These experiments did not address the consequences of blocking Shh signal transduction specifically within the facial mesenchyme. The effects of inhibiting Shh signalling in CNCCs have been studied using SMO conditional knockout mice to prevent the CNCCs from responding to hedgehog signalling.[37] Those embryos (Wnt-1-Cre, Smo$^{n/c}$) had extensive loss of craniofacial skeletal structures.[37] The authors suggest that *Fox* genes are involved in mediating Hh signalling during craniofacial development although specific role(s) have yet to be elucidated.[37]

Sonic Hedgehog in the Development of Lower Facial Structures

Shh in Tooth Development

Shh appears to play a number of critical roles in mediating the epithelial-mesenchymal interactions necessary for determining the spatial and structural information required for normal odontogenesis. In mice, *Shh* expression is localized in thickenings of oral epithelium that give rise to teeth and is absent from edentulous regions, the diastema mesenchyme. The relationship between the expression of *Shh* and the presence or absence of teeth reveals the importance of the spatial expression of *Shh* during odontogenesis. During tooth development, reciprocal interactions between the oral epithelium and mandibular mesenchyme appears to modulate Shh signalling.[56-59] Experiments where mandibular processes were cultured without their overlying mesenchyme, in which *Ptch1* and *Gli1* were up-regulated in conjunction with a down-regulation of *Gas1* expression in the underlying diastema mesenchyme,[59] suggested that mesenchymal Gas1 antagonizes the effects of Shh signalling in the epithelium. Such a relationship has been noted in other tissues[60] and in the initiation of tooth bud formation.[61]

Shh in Palatal Development

In humans and mice the definitive palate consists of the primary and secondary palate.[62] The primary palate arises from the fusion of the medial nasal prominences in the midline of the face.[62] The secondary palate is derived from the palatal shelves that grow out from the maxillary processes and form the majority of the hard and soft palate.[62] The palatal shelves consist of neural crest mesenchyme surrounded by epithelium. As mesenchymal cell proliferation progresses, the palatal shelves grow vertically downward with the tongue intervening between the two shelves before they elevate, take a horizontal position atop the tongue and subsequently fuse in the midline.

Development of the palate is dependent upon epithelial-mesenchymal interactions within the palatal shelves and involves Shh; both Shh and members of the signalling pathway are expressed during palatal development.[63] In mice, Shh, Fgf10 and the Fgf receptor 2b (Fgfr2b) appear to influence outgrowth of the palatal shelves.[64] The mesenchyme expresses *Fgf10*, a ligand for Fgfr2b, which is expressed in the epithelium and mesenchyme of the nasal aspect of the palatal shelves between E12-14.[64] Null mutations in *Fgf10* and *Fgfr2b* in mice result in cleft palate as a consequence of *Shh* down-regulation in the epithelium, decreased cell proliferation in the mesenchyme leading to inadequate outgrowth of the palatal shelves. In the palate, Shh appears to be a downstream target of Fgf10/Fgfr2b, and has been shown that recombinant Fgf proteins are capable of inducing *Shh* expression in the palatal epithelium in vitro.[64] Theoretically, this could have implications for human palatal clefting, since the genes involved in many instances of nonsyndromic facial clefting have not been delineated. However, to date there is no information regarding the presence of mutations in either the Shh or Fgf signalling pathways in nonsyndromic facial clefting.

Shh in Tongue Development

Development of the tongue requires complex epithelial-mesenchymal interactions in order to generate a structure capable of multiple biological functions. This is exemplified by the generation of the multiple types of papillae on the surface of the tongue. For example, the filiform papillae are involved in mechanical functions while the fungiform and circumvallate contain tastebuds with gustatory roles. The generation of papillae involves the localization of thickened regions of epithelial cells, the placodes, which evaginate into the mesenchyme creating raised papillae consisting of an epithelial surface and a mesenchymal core.[65,66]

As with the development of other facial structures, Shh appears to have key roles in epithelial-mesenchymal interactions, which are necessary for papillogenesis. During early stages of murine embryogenesis, *Shh*, *Ptch1* and *Gli1* are expressed diffusely in the tongue.[67] *Shh*, *Bmp2* and *Bmp4* expression is subsequently localized to regions of the anterior surface epithelium of the tongue where fungiform papillae will develop.[68] Shh may be involved in specifying the location where fungiform papillae form, as well as controlling the growth of and the spacing between individual papillae.[65] Inhibiting Shh signalling with either cyclopamine or 5E1 results in enlarged papillae in ectopic regions.[65,69]

Human Craniofacial Disorders

Genetic Etiologies

Mutations in *SHH* or components of the SHH signalling pathway have been shown to result in HPE and a spectrum of associated craniofacial phenotypes (Fig. 5).[43,44,70] However, no genotype-phenotype correlations have been found to explain the variability of phenotypes.[71,72] Molecular analysis of seven missense mutations in *SHH* that cosegregate with HPE appear to cause production of defective mature SHH, probably by destabilizing SHH or by altering the way in which it is processed.[73] Other disorders involving craniofacial dysmorphologies are due to mutations in components of the SHH pathway, and it is likely that the number of disorders involving aberrant SHH signalling will increase.

Gorlin's syndrome (Basal Cell Nevus syndrome) is caused by a gain of function mutation in *PTCH1*. Patients with this autosomal dominant disorder may present with craniofacial dysmorphologies including macrocephaly, frontal bossing, ocular malformations, cleft palate and odontogenic keratocysts of the jaws, in addition to multiple skin nevi.[74-76] The congenital malformations appear to be due to *PTCH1* haploinsufficiency.[77,78] The most serious complication of this disorder is the predisposition to cancers such as medulloblastomas, meningiomas, fibrosarcomas and basal cell carcinomas. A second hit model has been proposed[79] in which a somatic loss of function of the second allele, a tumor suppressor, leads to tumor formation.[80]

Greig Cephalopolysyndactyly syndrome (GCPS) and Pallister-Hall syndrome (PHS) are autosomal dominant disorders caused by mutations in *GLI3*. GCPS is characterized clinically by macrocephaly, hypertelorism and pre- or postaxial polysyndactyly.[81,82] Patients with PHS may have hypothalamic hamartoma, an imperforate anus and polydactyly.[83,84] Although elegant models have been proposed to explain the phenotypic differences in syndromes that result from mutations in *GLI3*,[85] further investigations concerning the genetics and cell biology involved are required.

Smith-Lemlli-Opitz (SLO) syndrome is an autosomal recessive disorder caused by impaired cholesterol biosynthesis resulting from a defect in 7-dehydrocholesterol-delta 7-reductase (DHCR7) activity.[86,87] Patients with SLO often present with microcephaly, a narrow frontal region, a broad-tipped nose, other face and limb abnormalities, and in some cases HPE. Defects in DHCR7 decrease embryonic/fetal de novo cholesterol biosynthesis which is critical for many processes during embryogenesis and fetal development. The manifestations may also be

The Role of Sonic Hedgehog Signalling in Craniofacial Development   53

Figure 5. Craniofacial dysmorphologies associated with holoprosencephaly. A-C) A spectrum of facial phenotypes is observed in holoprosencephaly (HPE). A) Prenatal ultrasound image of a second trimester human fetus with alobar HPE reveals a proboscis (red arrow) and cyclopia (yellow arrow). B) A child with semilobar HPE exhibits facial abnormalities including midline cleft lip and palate (red arrow), midface hypoplasia, and hypotelorism. C) Children with lobar HPE may exhibit normal facial features, as seen in this child. Photographs courtesy of (A) Dr. Ana Monteaguado; B,C) Dr. Jin Hahn. Reprinting of image provided by A) *Drug Discovery Today*;[99] B,C) Landes Bioscience/Eurekah.com and Springer Science+Business Media.[100]

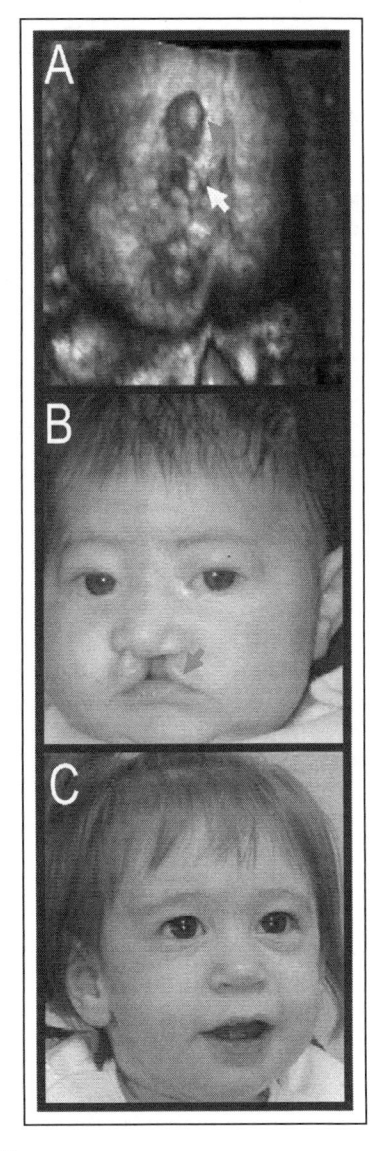

associated with perturbations in SHH signalling,[88] as cholesterol is required for the auto-processing and normal cellular transport of SHH.

Teratogenic Etiologies

Maternal exposure to vitamin A derivatives, ethanol and statins may adversely alter Shh signalling and lead to craniofacial malformations in embryos. The consequences of exposure to these agents depends on the gestational age at the time of exposure, the dose, the duration of exposure, and the genetic susceptibility of the embryo to the potential teratogen.

Craniofacial malformations have been reported in cases both of vitamin A deficiency and excess. An excess of vitamin A and its relationship to birth defects was highlighted after Accutane® (Isotretinoin) was introduced for the treatment of cystic acne and inadvertently taken during pregnancy.[89-91] Clinical manifestations included microcephaly, mandibulofacial dysplasia, microtia and conotruncal heart defects, and many of these features resemble the DiGeorge (Velocardiofacial) syndrome. The phenotype may be due to disruption in SHH signalling as administration of excess RA disrupts *Shh* expression in chick embryos.[26] The relationship between Shh and RA may also shed light on the molecular mechanisms leading to the Fetal Alcohol syndrome (FAS) in humans.

Ethanol use during pregnancy is the most common cause of preventable birth defects and mental retardation.[92] Classically, children with FAS present with microcephaly, a short nose, a smooth philtrum, a smooth and thin upper lip, and maxillary hypoplasia.[39,93,94] The molecular aberrations that lead to the craniofacial manifestations of FAS are unknown as yet but model systems are beginning to yield possible answers.[95] For example, chick embryos have revealed a link between a loss of Shh signalling and the craniofacial malformations associated with exposure to ethanol. Administration of ethanol to chick embryos led to a down-regulation of *Shh, Ptc, Gli1, Gli2* and *Gli3*, and neural crest cell death.[96]

Recently, a question has been raised regarding the teratogenic potential of statins.[97] These drugs are used clinically to treat hypercholesterolemia by inhibiting HMG-COA reductase, thereby lowering plasma levels of cholesterol. In humans, limb and central nervous system malformations such as HPE may be associated with their use in the first trimester of pregnancy.[98] The decreased availability of cholesterol in the developing embryo may adversely affect a number of developmental processes including auto-processing of SHH and its transport, thereby leading to the clinical phenotypes mentioned above. However, the teratogenic potential of this medication requires further rigorous studies.

Importance and Future Directions

Craniofacial malformations comprise approximately one third of all birth defects. This remarkable statistic underscores our need to determine the underlying genetic and environmental etiologies of these malformations. This should help in terms of providing effective preventative information, more sophisticated diagnosis and better treatment. The Shh pathway is required for a number of developmental events and has been implicated in the etiology of a number of disorders involving craniofacial dysmorphologics. We have begun a journey to uncover the many developmental complexities that lead to normal and perturbed craniofacial morphogenesis, and as we progress on this journey we will undoubtedly discover more roles for Shh in normal and abnormal craniofacial development.

References

1. Couly G, Le Douarin NM. Head morphogenesis in embryonic avian chimeras: Evidence for a segmental pattern in the ectoderm corresponding to the neuromeres. Development 1990; 108:543-558.
2. Wilson SW, Houart C. Early steps in the development of the forebrain. Dev Cell 2004; 6:167-181.
3. Schuurmans C, Guillemot F. Molecular mechanisms underlying cell fate specification in the developing telencephalon. Curr Opin Neurobiol 2002; 12:26-34.
4. Meulemans D, Bronner-Fraser M. Gene-regulatory interactions in neural crest evolution and development. Dev Cell 2004; 7:291-299.
5. Etchevers HC, Vincent C, Le Douarin NM et al. The cephalic neural crest provides pericytes and smooth muscle cells to all blood vessels of the face and forebrain. Development 2001; 128:1059-1068.
6. Noden DM. Origins and patterning of craniofacial mesenchymal tissues. J Craniofac Genet Dev Biol Suppl 1986; 2:15-31.
7. Helms JA, Schneider RA. Cranial skeletal biology. Nature 2003; 423:326-331.
8. Le Douarin NM, Dupin E. Multipotentiality of the neural crest. Curr Opin Genet Dev 2003; 13:529-536.
9. Trainor PA, Melton KR, Manzanares M. Origins and plasticity of neural crest cells and their roles in jaw and craniofacial evolution. Int J Dev Biol 2003; 47:541-553.
10. Kulesa P, Ellies DL, Trainor PA. Comparative analysis of neural crest cell death, migration, and function during vertebrate embryogenesis. Dev Dyn 2004; 229:14-29.
11. Le Douarin NM, Creuzet S, Couly G et al. Neural crest cell plasticity and its limits. Development 2004; 131:4637-4650.

12. Helms JA, Cordero D, Tapadia MD. New insights into craniofacial morphogenesis. Development 2005; 132:851-861.
13. Noden DM. The role of the neural crest in patterning of avian cranial skeletal, connective, and muscle tissues. Dev Biol 1983; 96:144-165.
14. Couly G, Grapin-Botton A, Coltey P et al. Determination of the identity of the derivatives of the cephalic neural crest: Incompatibility between Hox gene expression and lower jaw development. Development 1998; 125:3445-3459.
15. Schneider RA, Helms JA. The cellular and molecular origins of beak morphology. Science 2003; 299:565-568.
16. Trainor PA, Krumlauf R. Patterning the cranial neural crest: Hindbrain segmentation and Hox gene plasticity. Nat Rev Neurosci 2000; 1:116-124.
17. Couly G, Creuzet S, Bennaceur S et al. Interactions between Hox-negative cephalic neural crest cells and the foregut endoderm in patterning the facial skeleton in the vertebrate head. Development 2002; 129:1061-1073.
18. Trainor PA, Ariza-McNaughton L, Krumlauf R. Role of the isthmus and FGFs in resolving the paradox of neural crest plasticity and prepatterning. Science 2002; 295:1288-1291.
19. Hu D, Marcucio RS, Helms JA. A zone of frontonasal ectoderm regulates patterning and growth in the face. Development 2003; 130:1749-1758.
20. Trumpp A, Depew MJ, Rubenstein JL et al. Cre-mediated gene inactivation demonstrates that FGF8 is required for cell survival and patterning of the first branchial arch. Genes Dev 1999; 13:3136-3148.
21. Creuzet S, Schuler B, Couly G et al. Reciprocal relationships between Fgf8 and neural crest cells in facial and forebrain development. Proc Natl Acad Sci USA 2004; 101:4843-4847.
22. DeMyer W, Zeman W, Palmer CG. The face predicts the brain: Diagnostic significance of median facial anomalies for holoprosencephaly (arhinencephaly). Pediatrics 1964:256-263.
23. Schneider RA, Hu D, Rubenstein JL et al. Local retinoid signaling coordinates forebrain and facial morphogenesis by maintaining FGF8 and SHH. Development 2001; 128:2755-2767.
24. Cordero D, Marcucio R, Hu D et al. Temporal perturbations in sonic hedgehog signaling elicit the spectrum of holoprosencephaly phenotypes. J Clin Invest 2004; 114:485-494.
25. Marcucio RS, Cordero DR, Hu D et al. Molecular interactions coordinating the development of the forebrain and face. Dev Biol 2005; 284:48-61.
26. Helms JA, Kim CH, Hu D et al. Sonic hedgehog participates in craniofacial morphogenesis and is down-regulated by teratogenic doses of retinoic acid. Dev Biol 1997; 187:25-35.
27. Richman JM, Herbert M, Matovinovic E et al. Effect of fibroblast growth factors on outgrowth of facial mesenchyme. Dev Biol 1997; 189:135-147.
28. Francis-West PH, Tatla T, Brickell PM. Expression patterns of the bone morphogenetic protein genes Bmp-4 and Bmp-2 in the developing chick face suggest a role in outgrowth of the primordia. Dev Dyn 1994; 201:168-178.
29. Barlow AJ, Francis-West PH. Ectopic application of recombinant BMP-2 and BMP-4 can change patterning of developing chick facial primordia. Development 1997; 124:391-398.
30. Barlow AJ, Bogardi JP, Ladher R et al. Expression of chick Barx-1 and its differential regulation by FGF-8 and BMP signaling in the maxillary primordia. Dev Dyn 1999; 214:291-302.
31. Abzhanov A, Protas M, Grant BR et al. Bmp4 and morphological variation of beaks in Darwin's finches. Science 2004; 305:1462-1465.
32. Hu D, Helms JA. The role of sonic hedgehog in normal and abnormal craniofacial morphogenesis. Development 1999; 126:4873-4884.
33. Richman JM, Tickle C. Epithelia are interchangeable between facial primordia of chick embryos and morphogenesis is controlled by the mesenchyme. Dev Biol 1989; 136:201-210.
34. Gunhaga L, Jessell TM, Edlund T. Sonic hedgehog signaling at gastrula stages specifies ventral telencephalic cells in the chick embryo. Development 2000; 127:3283-3293.
35. Chiang C, Litingtung Y, Lee E et al. Cyclopia and defective axial patterning in mice lacking Sonic hedgehog gene function. Nature 1996; 383:407-413.
36. Hamburger V, Hamilton HL. A series of normal stages in the development of the chick embryo. Journal of Morphology 1951; 88:49-92.
37. Jeong J, Mao J, Tenzen T et al. Hedgehog signaling in the neural crest cells regulates the patterning and growth of facial primordia. Genes Dev 2004; 18:937-951.
38. Hu D, Helms JA. Unpublished data. 2006.
39. Young DL, Schneider RA, Hu D et al. Genetic and teratogenic approaches to craniofacial development. Crit Rev Oral Biol Med 2000; 11:304-317.
40. Machold R, Hayashi S, Rutlin M et al. Sonic hedgehog is required for progenitor cell maintenance in telencephalic stem cell niches. Neuron 2003; 39:937-950.

41. Kessaris N, Jamen F, Rubin LL et al. Cooperation between sonic hedgehog and fibroblast growth factor/MAPK signalling pathways in neocortical precursors. Development 2004; 131:1289-1298.
42. Palma V, Ruiz i Altaba A. Hedgehog-GLI signaling regulates the behavior of cells with stem cell properties in the developing neocortex. Development 2004; 131:337-345.
43. Belloni E, Muenke M, Roessler E et al. Identification of Sonic hedgehog as a candidate gene responsible for holoprosencephaly. Nat Genet 1996; 14:353-356.
44. Nanni L, Ming JE, Bocian M et al. The mutational spectrum of the sonic hedgehog gene in holoprosencephaly: SHH mutations cause a significant proportion of autosomal dominant holoprosencephaly. Hum Mol Genet 1999; 8:2479-2488.
45. Tamarin A, Crawley A, Lee J et al. Analysis of upper beak defects in chicken embryos following with retinoic acid. J Embryol Exp Morphol 1984; 84:105-123.
46. Abzhanov A, Cordero D. Unpublished data. 2006.
47. Hui CC, Joyner AL. A mouse model of greig cephalopolysyndactyly syndrome: The extra-toesJ mutation contains an intragenic deletion of the Gli3 gene. Nat Genet 1993; 3:241-246.
48. Mo R, Freer AM, Zinyk DL et al. Specific and redundant functions of Gli2 and Gli3 zinc finger genes in skeletal patterning and development. Development 1997; 124:113-123.
49. Duester G. Families of retinoid dehydrogenases regulating vitamin A function: Production of visual pigment and retinoic acid. Eur J Biochem 2000; 267:4315-4324.
50. Li H, Wagner E, McCaffery P et al. A retinoic acid synthesizing enzyme in ventral retina and telencephalon of the embryonic mouse. Mech Dev 2000; 95:283-289.
51. Mic FA, Molotkov A, Fan X et al. RALDH3, a retinaldehyde dehydrogenase that generates retinoic acid, is expressed in the ventral retina, otic vesicle and olfactory pit during mouse development. Mech Dev 2000; 97:227-230.
52. Keeler RF. Teratogenic compounds of Veratrum californicum (Durand) X. Cyclopia in rabbits produced by cyclopamine. Teratology 1970; 3:175-180.
53. Keeler RF. Livestock models of human birth defects, reviewed in relation to poisonous plants. J Anim Sci 1988; 66:2414-2427.
54. Incardona JP, Gaffield W, Kapur RP et al. The teratogenic Veratrum alkaloid cyclopamine inhibits sonic hedgehog signal transduction. Development 1998; 125:3553-3562.
55. Chen JK, Taipale J, Cooper MK et al. Inhibition of Hedgehog signaling by direct binding of cyclopamine to Smoothened. Genes Dev 2002; 16:2743-2748.
56. Peters H, Balling R. Teeth. Where and how to make them. Trends Genet 1999; 15:59-65.
57. Tucker AS, Sharpe PT. Molecular genetics of tooth morphogenesis and patterning: The right shape in the right place. J Dent Res 1999; 78:826-834.
58. Jernvall J, Thesleff I. Reiterative signaling and patterning during mammalian tooth morphogenesis. Mech Dev 2000; 92:19-29.
59. Cobourne MT, Miletich I, Sharpe PT. Restriction of sonic hedgehog signalling during early tooth development. Development 2004; 131:2875-2885.
60. Lee CS, Buttitta L, Fan CM. Evidence that the WNT-inducible growth arrest-specific gene 1 encodes an antagonist of sonic hedgehog signaling in the somite. Proc Natl Acad Sci USA 2001; 98:11347-11352.
61. Cobourne MT, Hardcastle Z, Sharpe PT. Sonic hedgehog regulates epithelial proliferation and cell survival in the developing tooth germ. J Dent Res 2001; 80:1974-1979.
62. Ferguson MW. Palate development. Development 1988; 103(Suppl):41-60.
63. Rice R, Connor E, Rice DP. Expression patterns of Hedgehog signalling pathway members during mouse palate development. Gene Expr Patterns 2006; 6:206-212.
64. Rice R, Spencer-Dene B, Connor EC et al. Disruption of Fgf10/Fgfr2b-coordinated epithelial-mesenchymal interactions causes cleft palate. J Clin Invest 2004; 113:1692-1700.
65. Hall JM, Bell ML, Finger TE. Disruption of sonic hedgehog signaling alters growth and patterning of lingual taste papillae. Dev Biol 2003; 255:263-277.
66. Farbman AI, Mbiene JP. Early development and innervation of taste bud-bearing papillae on the rat tongue. J Comp Neurol 1991; 304:172-186.
67. Hall JM, Hooper JE, Finger TE. Expression of sonic hedgehog, patched, and Gli1 in developing taste papillae of the mouse. J Comp Neurol 1999; 406:143-155.
68. Jung HS, Oropeza V, Thesleff I. Shh, Bmp-2, Bmp-4 and Fgf-8 are associated with initiation and patterning of mouse tongue papillae. Mech Dev 1999; 81:179-182.
69. Mistretta CM, Liu HX, Gaffield W et al. Cyclopamine and jervine in embryonic rat tongue cultures demonstrate a role for Shh signaling in taste papilla development and patterning: Fungiform papillae double in number and form in novel locations in dorsal lingual epithelium. Dev Biol 2003; 254:1-18.
70. Roessler E, Belloni E, Gaudenz K et al. Mutations in the C-terminal domain of Sonic Hedgehog cause holoprosencephaly. Hum Mol Genet 1997; 6:1847-1853.

71. Ming JE, Kaupas ME, Roessler E et al. Mutations in PATCHED-1, the receptor for SONIC HEDGEHOG, are associated with holoprosencephaly. Hum Genet 2002; 110:297-301.
72. Traiffort E, Dubourg C, Faure H et al. Functional characterization of sonic hedgehog mutations associated with holoprosencephaly. J Biol Chem 2004; 279:42889-42897.
73. Maity T, Fuse N, Beachy PA. Molecular mechanisms of Sonic hedgehog mutant effects in holoprosencephaly. Proc Natl Acad Sci USA 2005; 102:17026-17031.
74. Lacombe D, Chateil JF, Fontan D et al. Medulloblastoma in the nevoid basal-cell carcinoma syndrome: Case reports and review of the literature. Genet Couns 1990; 1:273-277.
75. Gorlin RJ. Gorlin (nevoid basal-cell carcinoma) syndrome. In: Gorlin RJ, Cohen MM, Hennekam RCM, eds. Syndromes of the Head and Neck. Oxford: Oxford Univ. Press, 2001.
76. Klein RD, Dykas DJ, Bale AE. Clinical testing for the nevoid basal cell carcinoma syndrome in a DNA diagnostic laboratory. Genet Med 2005; 7:611-619.
77. Bale AE. The nevoid basal cell carcinoma syndrome: Genetics and mechanism of carcinogenesis. Cancer Invest 1997; 15:180-186.
78. Bale AE, Yu KP. The hedgehog pathway and basal cell carcinomas. Hum Mol Genet 2001; 10:757-762.
79. Knudson Jr AG. Mutation and cancer: Statistical study of retinoblastoma. Proc Natl Acad Sci USA 1971; 68:820-823.
80. Bonifas JM, Bare JW, Kerschmann RL et al. Parental origin of chromosome 9q22.3-q31 lost in basal cell carcinomas from basal cell nevus syndrome patients. Hum Mol Genet 1994; 3:447-448.
81. Kalff-Suske M, Wild A, Topp J et al. Point mutations throughout the GLI3 gene cause Greig cephalopolysyndactyly syndrome. Hum Mol Genet 1999; 8:1769-1777.
82. Kang S, Graham Jr JM, Olney AH et al. GLI3 frameshift mutations cause autosomal dominant Pallister-Hall syndrome. Nat Genet 1997; 15:266-268.
83. Hall JG, Pallister PD, Clarren SK et al. Congenital hypothalamic hamartoblastoma, hypopituitarism, imperforate anus and postaxial polydactyly—A new syndrome? Part I: Clinical, causal, and pathogenetic considerations. Am J Med Genet 1980; 7:47-74.
84. Iafolla K, Fratkin JD, Spiegel PK et al. Case report and delineation of the congenital hypothalamic hamartoblastoma syndrome (Pallister-Hall syndrome). Am J Med Genet 1989; 33:489-499.
85. Shin SH, Kogerman P, Lindstrom E et al. GLI3 mutations in human disorders mimic Drosophila cubitus interruptus protein functions and localization. Proc Natl Acad Sci USA 1999; 96:2880-2884.
86. Honda A, Tint GS, Salen G et al. Defective conversion of 7-dehydrocholesterol to cholesterol in cultured skin fibroblasts from Smith-Lemli-Opitz syndrome homozygotes. J Lipid Res 1995; 36:1595-1601.
87. Shefer S, Salen G, Batta AK et al. Markedly inhibited 7-dehydrocholesterol-delta 7-reductase activity in liver microsomes from Smith-Lemli-Opitz homozygotes. J Clin Invest 1995; 96:1779-1785.
88. Roux C, Wolf C, Mulliez N et al. Role of cholesterol in embryonic development. Am J Clin Nutr 2000; 71:1270S-1279S.
89. Stern RS, Rosa F, Baum C. Isotretinoin and pregnancy. J Am Acad Dermatol 1984; 10:851-854.
90. Monga M. Vitamin A and its congeners. Semin Perinatol 1997; 21:135-142.
91. Nau H. Teratogenicity of isotretinoin revisited: Species variation and the role of all-trans-retinoic acid. J Am Acad Dermatol 2001; 45:S183-187.
92. CDC. Fetal Alcohol Information. 2004, (http://www.cdc.gov/ncbddd/fas/fasask.htm#how).
93. Jones KL, Smith DW, Ulleland CN et al. Pattern of malformation in offspring of chronic alcoholic mothers. Lancet 1973; 1:1267-1271.
94. Sampson PD, Streissguth AP, Bookstein FL et al. Incidence of fetal alcohol syndrome and prevalence of alcohol-related neurodevelopmental disorder. Teratology 1997; 56:317-326.
95. Sulik KK. Genesis of alcohol-induced craniofacial dysmorphism. Exp Biol Med (Maywood) 2005; 230:366-375.
96. Ahlgren SC, Thakur V, Bronner-Fraser M. Sonic hedgehog rescues cranial neural crest from cell death induced by ethanol exposure. Proc Natl Acad Sci USA 2002; 99:10476-10481.
97. Edison RJ, Muenke M. Mechanistic and epidemiologic considerations in the evaluation of adverse birth outcomes following gestational exposure to statins. Am J Med Genet A 2004; 131:287-298.
98. Edison RJ, Muenke M. Central nervous system and limb anomalies in case reports of first-trimester statin exposure. N Engl J Med 2004; 350:1579-1582.
99. Cordero DR, Tapadia MD, Helms JA. The etiopathologies of holoprosencephaly. Drug Discov Today: Disease Mech 2005; 2:529-537.
100. Cordero DR, Tapadia M, Helms JA. Sonic hedgehog signaling in craniofacial development. In: Ruiz i Altaba A, ed. Hedgehogh-Gli signaling in human disease. Georgetown: Eurekah.com; New York: Springer Science+Business Media, 2006:153-176.
101. Tapadia MD, Cordero D, Helms JA. It's all in your head: new insights into craniofacial development and deformation. J Anatomy 2005; 207:461-477.

CHAPTER 6

Multiple Roles for Hedgehog Signalling in Zebrafish Eye Development

Deborah L. Stenkamp*

Abstract

The Hedgehog signalling pathway is important in a large number of developmental contexts. In this review, several functions of this pathway for vertebrate eye formation and differentiation will be discussed, with an emphasis on information derived from the zebrafish model. The highlighted roles for Hedgehog signalling include those in photoreceptor development, ganglion cell development, retinal cell proliferation and cell death, and the initiation of retinal neurogenesis.

Introduction

A role for Hedgehog (Hh) signalling in development of the vertebrate eye was predicted based upon the known roles for Hh signalling in *Drosophila* eye development,[1,2] and the principle of evolutionary conservation of developmental function.[3] The surprise is that not only was this prediction correct, it turns out that the Hh signalling system is used for a variety of developmental purposes, and at several developmental stages during the formation and differentiation of the vertebrate eye. Many of the key experiments that uncovered these roles used the zebrafish as a genetic and organismal tool. The zebrafish has emerged as an outstanding model for the understanding of eye development, with large, rapidly-developing eyes, amenability to pharmacological and genetic manipulation, and suitability for a variety of imaging and gene expression studies.[4,5]

The vertebrate neural retina is derived from the embryonic neural tube, and is therefore an accessible model system for understanding central nervous system development. The optic vesicles that emerge from the diencephalon form (from proximal to distal) optic stalks, retinal pigmented epithelium (RPE), and neural retina. The neural retina further differentiates into a laminar structure containing photoreceptor cells in an outer nuclear layer adjacent to the RPE, processing neurons and Müller glia in an inner nuclear layer, and ganglion cells that project via the optic nerve to the brain. Proliferation of the multipotential retinal progenitor cells, as well as the generation and differentiation of these diverse cell types, is regulated by a combination of cell-intrinsic factors and cell-extrinsic (extracellular) cues;[6] the Hedgehog protein is now established as one of these very important extracellular cues.

The first role in eye development to be attributed to Hh signalling was the requirement for a midline source of Hh at late gastrulation/early neurulation for separating and patterning the eye fields along the proximo-distal axis.[7,8] Zebrafish with defects in midline tissues that serve as sources of Hh (notochord/prechordal plate and ventral diencephalon) develop a single (cyclopic)

---

*Deborah L. Stenkamp—Department of Biological Sciences, University of Idaho, Moscow, Idaho, U.S.A. Email: dstenkam@uidaho.edu

Shh and Gli Signalling and Development, edited by Carolyn E. Fisher and Sarah E.M. Howie. ©2006 Landes Bioscience and Springer Science+Business Media.

anterior eye rather than two lateral eyes. Zebrafish overexpressing the *sonic hedgehog* (*shh*) gene develop lateral optical structures with an excess of proximal tissues (optic stalk) at the expense of distal tissues (neural retina); this has been extensively reviewed.[9-13] However, these observations in zebrafish inspired the careful reconsideration of a similar, teratogenically-induced cyclopic phenomenon in other vertebrates, leading to the discovery that an alkaloid derived from the western cornlily plant (*Veratrum californicum*) interferes with Hh signal transduction.[14,15] Many of the experiments described in this chapter used this alkaloid, known as cyclopamine, as a pharmacological tool for investigating Hh signalling in eye development.

This chapter will consider those known roles for Hh signalling during eye development, which take place after the Hh signal separates the eye fields. The focus will be on experimental results from zebrafish, but with reference to supporting data from other vertebrates where appropriate. The various developmental functions of Hh signalling will be discussed in reverse developmental chronology, because the investigation and understanding of the later roles were necessary for the insights that allowed the unraveling of some earlier roles. Readers are also referred to the outstanding review by Amato et al[13] for a further discussion of Hedgehog signalling in eye development.

Hedgehog Signalling and Photoreceptor Differentiation

The Retinal Pigmented Epithelium Expresses Two hh Genes

Two of the three known zebrafish *hedgehog* genes, *sonic hedgehog* (*shh*), and *tiggy-winkle hedgehog* (*twhh*), are expressed in the RPE beginning at or near 45 hours post-fertilization (hpf),[16] corresponding to the time that the first photoreceptor cells exit the cell cycle.[17] When *hh* gene expression is evaluated on cryosections, a spatiotemporal pattern is revealed, such that expression is initiated in ventral RPE, then spreads nasally and finally dorsally and temporally (Fig. 1).[16] This pattern predicts the subsequent pattern of photoreceptor differentiation in the subjacent neural retina. Hh protein can be detected immunocytochemically in the RPE and in the subretinal space, suggesting that it is secreted toward the retina.[16] We have tentatively localized one of the Hh receptors, *patched-2* (*ptc-2*), to retinal neuroepithelial cells, beginning at 48 hpf.[16] Due to low expression, it has been very difficult to further evaluate expression of the *ptc* genes in the zebrafish eye.[16,18]

Hh Signalling from the RPE Propagates Photoreceptor Differentiation and Promotes Retinal Cell Survival

To test the hypothesis that Hh signalling from the RPE is necessary for photoreceptor differentiation, our laboratory has used several complementary methods that knock down Hh signalling during the time of photoreceptor development. Microinjection of antisense oligonucleotides, either phosphorothioate- or morpholino-conjugated, at 51-54 hpf results in measurable reduction in Hh expression in the RPE, and in significant attenuation of photoreceptor differentiation.[16,19] In most cases a rudimentary outer nuclear layer still forms, but these cells do not express the photoreceptor-specific opsin genes. Knockdown of both *shh* and *twhh* expression was needed to result in significant effects on photoreceptor development, indicating some degree of redundancy in the function of these two genes.[16] Treatment of zebrafish embryos with cyclopamine during the same developmental period, also results in this phenotype.[19] Finally, embryos genetically deficient in Hh signalling display the photoreceptor differentiation defect.[19,20] We have observed the failed spread of opsin expression in both the *sonic-you* (*syu*) mutant, which is a deletion spanning the entire *shh* gene,[21] and in the slow muscle-omitted (*smu*) mutant, which is a functional null mutation in the *smoothened* gene.[22] Smoothened encodes a critical component of the Hh signal transduction pathway. When Hh binds to the Patched receptor, inhibition from Patched on Smoothened is relaxed, and intracellular signals are generated.[23] Therefore, the null mutation in *smoothened* is predicted to completely disable Hh signal transduction.

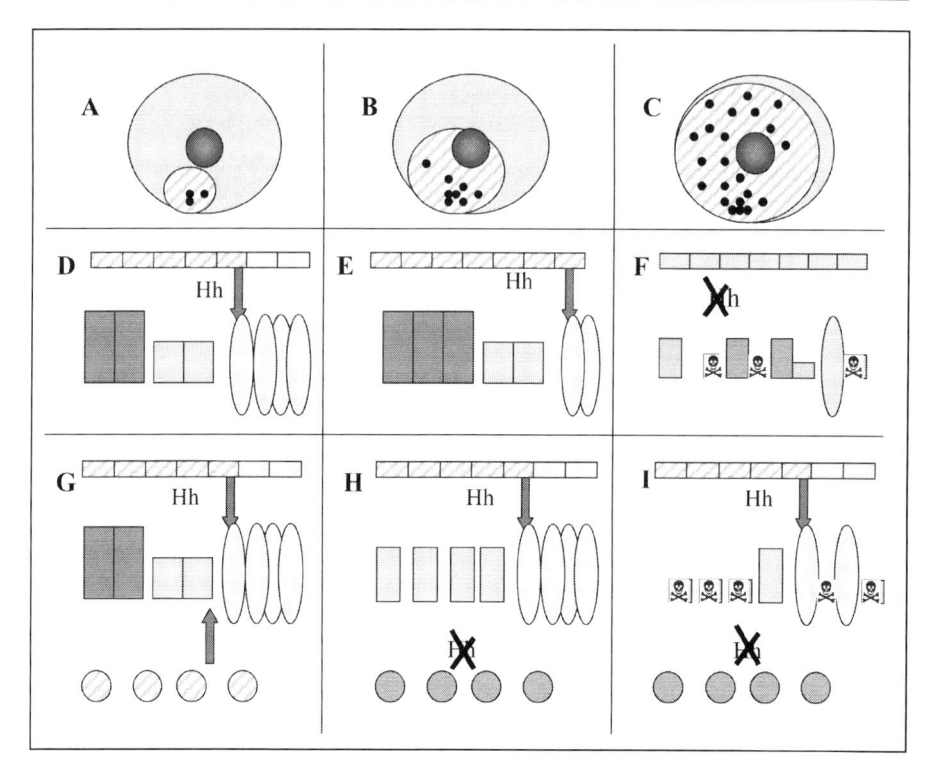

Figure 1. Hedgehog signalling is necessary for photoreceptor differentiation and survival. A-C) Spread of *hh* gene expression through the RPE predicts the spread of photoreceptor opsin expression in the retina.[16] Illustrations represent transparent lateral views of the zebrafish eye; ventral is on the bottom, nasal to the left. Hh expression in the RPE is represented by stripes; opsin expression in a photoreceptor is represented by a small dark profile. A) 48 hpf. B) 54 hpf. C) 60 hpf. D-F) Hh signalling from RPE may propagate photoreceptor differentiation by influencing expression of *rx1*.[19,20] D and E) represent steps in propagation; opsin-expression photoreceptors are dark rectangles and *rx1*-expressing photoreceptors are grey rectangles. F) represents outcome following genetic or pharmacological knockdown of Hh signalling from the RPE; this outcome includes substantial cell death. G-I) Hh signalling from amacrine cells may also be important for photoreceptor development. I) Represents failed photoreceptor differentiation in the absence of Hh signalling from amacrine cells (though RPE still expresses *shh*).[18] F) represents one alternative interpretation, in which disrupting the Hh signal from amacrine cells results in cell death and therefore there are no photoreceptors to differentiate.

We further evaluated the photoreceptor differentiation defect phenotype by using in situ hybridization for several additional photoreceptor-specific genes. The *cone-rod homeobox* (*crx*) gene is expressed normally in the rudimentary outer nuclear layer of *syu* mutants and morpholino- or cyclopamine-treated embryos.[19,20] *NeuroD* also shows a normal photoreceptor expression pattern in the *syu* mutants.[20] However, the retinal homeobox gene, *rx1*,[24] is not expressed in photoreceptors in the *syu* mutant, the *smu* mutant, or following other treatments designed to reduce Hh signalling.[19,20] Because *rx* genes can regulate the expression of other photoreceptor-specific genes in vitro[25] and in vivo,[26] we consider *rx1* a candidate for mediating the effects of Hh signalling on photoreceptor differentiation (Fig. 1). Finally, reduced Hh signalling results in cell death within the developing retina.[20] The timing of this cell death is consistent with an important role for the Hh signal, specifically from the RPE, in promoting retinal cell survival. Cell death is initially highest in neuroepithelial cells and photoreceptors, and then spreads to other retinal layers.

Hh Signalling from Amacrine Cells Is Also Involved in Photoreceptor Differentiation

Transgenic zebrafish, developed by Carl Neumann, expressing green fluorescent protein under control of the *shh* promoter (*shh*-GFP), revealed that this *hh* gene is also expressed in amacrine cells.[18] Expression has been verified by in situ hybridization, and the spatiotemporal pattern of expression mimics that of the slightly earlier wave of *hh* gene expression in ganglion cells (see below). The onset and spread of the amacrine cell Hh signal is independent of that from ganglion cells, and of expression of the retinal transcription factor *ath5*. The wave of Hh in amacrine cells still takes place in the *lakritz/ath5* mutant, which lacks ganglion cells.[18,27]

What is the functional role of the amacrine cell Hh signal in the retina? It is possible that this signal is needed for cell differentiation in the retina's inner nuclear layer. Indeed, when examined at 64-96 hpf, a time when inner nuclear layer neurons normally show cell-specific markers, the *syu* mutant expresses none.[18] However, during this developmental time, cell death in the *syu*-/- embryo is widespread, with the majority of dying neurons localized to the inner nuclear layer by 75 hpf.[20] Therefore, it is likely that specific cell markers do not appear in the inner nuclear layer of the *syu* mutants because these cells are dead or dying. The Hh signal from amacrine cells may be important for promoting retinal cell survival, although this function has not been specifically tested.

It is also possible that the amacrine cell Hh signal is needed for photoreceptor differentiation, a function we attributed to the Hh protein originating from the RPE.[16] To address this question, Shkumatava et al[18] created mosaic embryos, consisting of a combination of *syu*-/- cells and wild-type, *shh*-GFP cells. The goal was to determine whether failed photoreceptor differentiation was associated with the absence of *shh* expression in nearby RPE or in nearby amacrine cells. In their experiments, the retinal regions displaying normal photoreceptor differentiation (as assessed by the expression of the specific marker, zpr-1), were radially contiguous with the regions containing wild-type, *shh*-GFP amacrine cells.[18] However, in regions where wild-type, *shh*-GFP RPE cells were located, zpr-1-expressing photoreceptors were not found unless wild-type, *shh*-GFP amacrine cells were also present.[18] These findings suggest that the amacrine cell Hh signal, rather than the RPE Hh signal, may act to promote photoreceptor differentiation (Fig. 1). There are several alternative explanations that are consistent with these and other data. One possibility is that the amacrine cell Hh signal is required for retinal cell survival, while the RPE signal is needed for photoreceptor differentiation (Fig. 1). Another is that Hh signals from a basal (amacrine) as well as an apical (RPE) source are needed for photoreceptor differentiation. Finally, it may be that the total amount of Hh signal available to the photoreceptor layer must exceed a certain threshold before differentiation can take place. The significance of gradients and thresholds for Hh signalling is exemplified during cell determination and differentiation events within the embryonic spinal cord.[28] These latter possibilities would be difficult to address using mosaic embryos, because the *twhh* gene is expressed in both wild-type and *syu*-/- RPE, and its functions overlap with those of *shh*.[16,29] To investigate these issues as well as many others, we are developing lines of transgenic zebrafish that express *shh* under the control of a heat shock promoter. Our goal is to apply local heat shock by using a laser,[30] to achieve spatiotemporally-selective expression of *shh*. While these lines have not yet been established, we have confirmed the feasibility of this approach through mosaic expression studies (data not shown).

Discussion and Significance

In the zebrafish, Hh signalling is clearly required for photoreceptor differentiation; however, the cellular source of this important signal remains unclear. Possible sources include the RPE, amacrine cells, or a combination of the two. In contrast to studies in other vertebrates, cyclopamine treatment in live *Xenopus* during the time-frame in which photoreceptor development ensues had no effect on this cell type. Instead, treatments caused defects in RPE differentiation.[31] One explanation for this apparent difference is that Hh signalling also regulates RPE differentiation in the zebrafish, and that the photoreceptors are affected indirectly, via subtle

RPE functional deficiencies. Arguing against this interpretation are results from a rodent model in which purified Hh protein added to cultured retinal progenitors stimulates the differentiation of photoreceptor cells,[32] suggesting a direct influence of Hh on cells of the neural retina.

One highly significant finding that perhaps deserves more attention is the requirement for Hh signalling for retinal cell survival.[20] Although we do not know whether this is a direct or indirect effect, it suggests Hh as a candidate survival factor to be considered for treating retinal degenerative disorders. In addition, any roles for Hh signalling in adult retina remain unexplored.

Hedgehog Signalling and Ganglion Cell Differentiation

Ganglion Cells Express Two hh Genes

Both *shh* and *twhh* are also expressed in ganglion cells (GCs); this pattern was first appreciated through the use of a transgenic, *shh*-GFP reporter line, and was subsequently verified by in situ hybridization.[29] Like expression in the RPE, there is a pronounced spatiotemporal gradient of expression that resembles the pattern of retinal cellular differentiation. However, GCs express *hh* genes at a much earlier time, 28 hpf, shortly after the first GCs withdraw from the cell cycle.[17] Immunocytochemical techniques reveal Hh protein expression in GCs at this time.[19]

Hh Signalling from GCs Promotes Retinal Cell Proliferation

At the time of GC differentiation, and of Hh signalling from the GCs, the neural retina remains proliferative, as other cell types have not yet become postmitotic.[17] This fact, along with the microphthalmic phenotype of the *syu* mutant, prompted us to evaluate the extent of cell proliferation in wild-type vs. *syu-/-* animals at 34 hpf, when GCs are differentiating. Compared to their wild-type siblings, mutants possessed significantly fewer cells that could be labeled with a marker for M-phase, indicating that one of the roles of the Hh signal from GCs is to promote continued retinal cell proliferation.[20] However, since amacrine cells may also express *shh* during this time,[18] they must also be considered a potential source for this proliferative signal.

Hh Signalling from GCs Propagates GC Differentiation and Further Hh Signalling

Treatment of the transgenic, *shh*-GFP zebrafish with cyclopamine at 26 hpf blocks the spread of transgene expression in GCs, as well as the appearance of a marker for GC differentiation, zn5.[29] Furthermore, when the transgenic line is crossed onto a *syu* background (with genetically reduced Hh signalling), the wave of GC differentiation and transgene expression is not fully propagated.[29] This defect can be partially rescued by supplemental expression of *shh* in ganglion cells.[29] These data strongly suggest that the GC Hh signal propagates itself, by stimulating production of additional GCs, similar to the situation in the *Drosophila* retina, where Hh secretion by newly-generated photoreceptors promotes the subsequent generation of additional photoreceptors.[33] Knockdown of Hh signalling at 27 hpf with cyclopamine or with antisense morpholinos also reduces the spread of expression of the transcription factor *ath5*.[19] Zebrafish *ath5* is an ortholog of *Drosophila atonal*, which is regulated by Hh signalling in developing fly photoreceptors.[34] In zebrafish, *ath5* is necessary for GC differentiation.[27] Collectively, these data suggest that Hh signalling from GCs stimulates *ath5* expression in nearby cells, promoting the generation of additional GCs that will then express the Hh signal to further propagate this process (Fig. 2).

Discussion and Significance

A slightly different role for GC-derived Hh signalling has been described in the chick, although this role is not inconsistent with the functions discussed above. In the chick, high levels of exogenous Hh actually inhibit the formation of additional GCs, as part of a negative feedback system for regulating GC production.[35] The similarities to the *Drosophila* system are

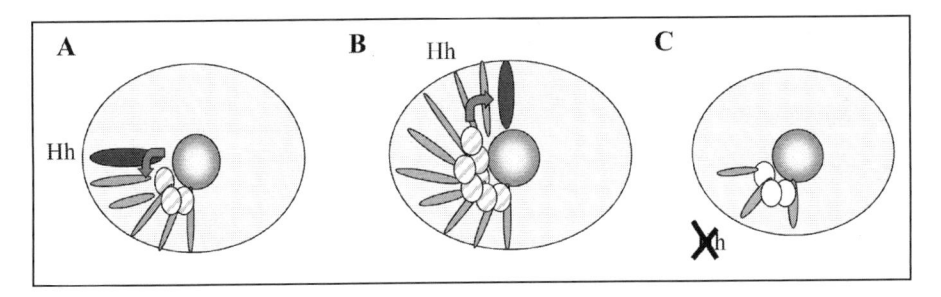

Figure 2. Hedgehog signalling is necessary for GC differentiation, and for progenitor cell proliferation. A) and B) represent steps in propagation of GC differentiation, mediated by propagation of expression of the transcription factor *ath5* (dark profiles).[19,29] C) represents failed propagation of GCs following genetic or pharmacological knockdown of Hh signalling, as well as reduced progenitor cell proliferation, resulting in microphthalmia.[19,20,29]

startling. In the fly eye, Hh signalling is needed both to drive *atonal* expression (though in this case, in nascent photoreceptors), and to inhibit *atonal* expression so that the proper number and pattern of photoreceptors are formed.[34] Interestingly, the rhabdomeric photoreceptors of the insect eye may be evolutionarily related to the melanopsin-containing (photosensitive) GCs of the vertebrate eye.[36] Perhaps the subset of GCs that expresses *hh* genes is also the subset that is directly photosensitive.

In the mammalian model, Hh signal derived from ganglion cells has also been shown to be involved in a number of additional developmental processes related to the visual system. For example, selective genetic knockdown of Hh in GCs results in disorganization of Müller glial processes and a poorly organized retina.[37] Hedgehog protein may also signal from GC axons to regulate gliogenesis in the developing optic nerve.[38]

The apparent proliferation-promoting activity for Hh protein has been confirmed in vitro, in two different rodent models; exogenous Hh protein stimulates cell proliferation in cultures of rat or mouse embryonic retinal cells.[32,39] Mice heterozygous for a null mutation in the *patched* gene, and hence predicted to have constitutively higher levels of Hh signalling, display prolonged, post-embryonic retinal proliferation.[40] Collectively these observations justify the ongoing interest in manipulation of the Hh signalling system as part of stem cell-based strategies for treatment of retinal disorders.

Hedgehog Signalling and Retinal Neurogenesis

The Sonic Hedgehog Mutant, Syu, Can Fail to Initiate Retinal Neurogenesis

In our evaluation of retinal gene expression in the *syu-/-* embryo, we consistently observed that these mutants, identified by a curved body axis, pericardial swelling and microphthalmia,[21] could be further separated into two phenotypic categories based upon retinal histology at 58 hpf, when retinal differentiation is well under way. Half of the *syu* mutants displayed recognizable retinal layers, while the other half consisted of undifferentiated retinal progenitor cells, which could be labeled by neuroepithelial markers.[20] Interestingly, Shkumakava et al[18] also observed that *syu* mutants show lamination defects, but did not pursue this phenotype with markers for retinal progenitor cells. In our experiments, wild-type siblings never displayed anything resembling this phenotype, and so the *shh* gene must be one of the genetic factors involved. Therefore, Hh signalling must be required not only for photoreceptor and ganglion cell differentiation, but also for the *initiation* of retinal neurogenesis. What is the source and timing of the Hh signal for this activity? Expression of *hh* genes in the RPE, amacrine cells, and ganglion cells occurs too late to influence the onset of neurogenesis in the retina, with the additional problem that neurogenesis must be initiated in order for amacrine and ganglion cells to form. Therefore, the relevant Hh signalling event must commence earlier in development, and originate outside of the eye.

Failed Retinal Neurogenesis Is Associated with Prechordal Plate Abnormalities

A likely source for this early Hh signal was revealed by the elegant studies of Masai et al,[41] which identified the prechordal plate as a source of an unknown signal that promoted optic stalk development and indirectly initiated retinal neurogenesis. These experiments utilized a series of zebrafish mutants with defects in prechordal plate development or migration. These mutants are all cyclopic, because the Hh signal from the prechordal plate is necessary during late gastrulation for separating the developing eye fields.[7,8] Mutants in which the prechordal plate defect persisted also failed to initiate retinal neurogenesis in the single eye that formed, while in those in which the prechordal plate formed by the time of neurulation (10 hpf), retinal differentiation commenced.[41] Because the timing of this unknown signal (10 hpf) was far earlier than when retinal neurogenesis is initiated (27 hpf), the authors pursued the optic stalk as the tissue that relayed the prechordal plate signal to the retina. A series of tissue transplant experiments led to the conclusion that the prechordal plate induces retinal neurogenesis via the optic stalk.[41]

Reduced Hh Signalling from the Prechordal Plate Results in Failed Retinal Neurogenesis

The experiments of Masai et al,[41] along with our own results using the *syu* mutant,[20] inspired us to pursue Hh as the unidentified prechordal plate signal required for the initiation of retinal neurogenesis. We performed a series of Hh knockdown experiments designed to reduce Hh signalling beginning at 10 hpf (the time of the unidentified signal from the prechordal plate), at 27 hpf (when GCs begin to express *hh* genes), or at 51 hpf (when the RPE is expressing *hh* genes). Retinal phenotypes were evaluated at 34 and 58 hpf for signs that retinal neurogenesis had commenced. At 34 hpf this was revealed by normal expression of *ath5*, the first specific marker for the onset of retinal neurogenesis,[41] and at 58 hpf this was revealed by normal retinal lamination. We used a combination of antisense (morpholino) and other pharmacological (cyclopamine) approaches. The morpholino antisense approach was at best 40% effective at reducing expression of the target protein, Hh,[19] most likely because of our need to inject the morpholinos well after the stage when cellular uptake of these compounds is efficient.[42] Therefore we monitored the proportions at which specific phenotypes were observed, in order to verify that this frequency was consistent with the success rate of the experimental treatment. Our findings supported our original hypothesis: early Hh knockdown treatments consistently resulted in abnormal or failed *ath5* expression, and resulted in failed retinal lamination.[19] These phenotypes were not replicated in the later Hh knockdown experiments, and therefore could not be the result of reduced Hh signalling from ganglion cells, amacrine cells, or RPE. We conclude that a Hh signal from a developmentally early source(s) outside the eye is required for the initiation of retinal neurogenesis (Fig. 3).[19]

The Role of Hh Signalling for Retinal Neurogenesis Is Independent of Its Role in Optic Stalk Development

In our temporally-selective Hh knockdown experiments, we also monitored the status of optic stalk development by in situ hybridization for the transcription factor *pax2*.[19] In the early (10 hpf) knockdown experiments, we rarely observed minor optic stalk abnormalities, even in experiments where retinal neurogenesis defects were abundant.[19] Because these results were at odds with those of Masai et al,[41] we pursued this issue further by using the *smu* mutant. The *smu* mutant is a functional null for *smoothened* expression, but is not cyclopic because of significant maternal expression of *smu*, which typically persists until early somitogenesis (10-15 hpf).[22] We predicted that the eyes of the *smu-/-* embryo should therefore phenocopy those of our earliest Hh knockdown experiments. Furthermore, optic stalk defects had been reported for the *smu-/-* embryo;[22] this consequently appeared to be the ideal genetic experiment to test the hypothesis that the optic stalk mediated the effects of prechordal plate Hh signalling on

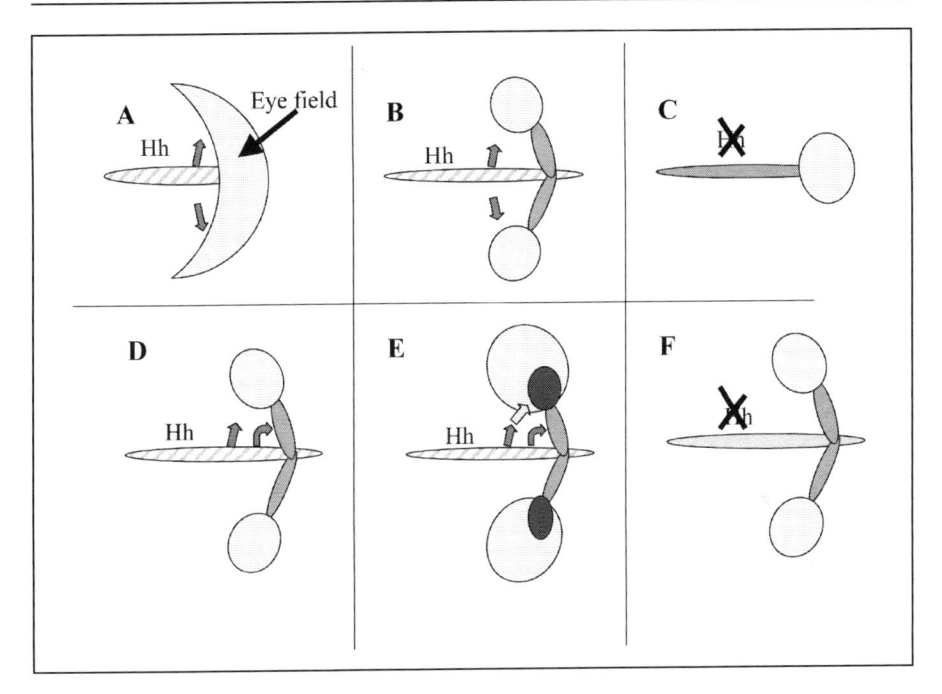

Figure 3. Persistent Hedgehog signalling is needed for proximo-distal patterning of the developing eye fields, including those that may specify ventral retina as the site of initiation of retinal neurogenesis. A-C) Hedgehog signalling early in neurulation is required for separating the eye fields. A and B) represent steps in this process; C) represents a cyclopic outcome when Hh signalling is eliminated; anterior is to the right. D-E) Hedgehog signalling slightly later in neurulation is required for maintenance of optic stalk identity and for initiation of retinal neurogenesis (indicated by dark-colored ventral regions of retina). These effects are independent, and the latter is likely indirect.[19,41] D and E) represent steps in this process; F) depicts failed retinal neurogenesis and microphthalmia resulting from genetic or pharmacological knockdown of Hh signalling.

retinal neurogenesis. We analyzed *smu* mutants and their wild-type siblings with an optic stalk marker (either *pax2*[43] or *fgf8*[44]), in combination with a marker for retinal neurogenesis (*ath5* or *fgf8*, which is expressed in newly-generated ganglion cells as well as in the optic stalk[44]). At 34 hpf, wild-type embryos always showed normal optic stalks and retinal gene expression consistent with the normal initiation of neurogenesis.[19] However, *smu-/-* embryos displayed a variety of phenotypes: in some there was no initiation of retinal neurogenesis but they had normal optic stalks, in some retinal neurogenesis was initiated but they lacked optic stalks, and in others both defects were apparent.[19] The optic stalk phenotype could therefore be uncoupled from the retinal neurogenesis phenotype, perhaps due to variability in when and where maternal wild-type *smu* is depleted. We conclude that Hh signalling from the prechordal plate at the time of neurulation (10 hpf), is important both for optic stalk development and for the initiation of retinal neurogenesis, but that these effects are independent.[19] The large temporal interval between the signalling event (10 hpf), and the effect of the signal (27 hpf), still requires that this effect be mediated by some other tissue (Fig. 3).

Discussion and Significance

The importance of early Hh signalling for allowing the later initiation of retinal neurogenesis may be yet another example of the influence of Hh on proximo-distal patterning in the developing eye fields (see Introduction). For example, the developing chick optic cup can be divided into dorsal vs. ventral compartments based upon morphology and gene expression; features of these compartments are related to their proximity to a midline source of Hh signalling, and can be manipulated by manipulating Hh signalling.[45] In *Xenopus* treated so as to increase Hh signalling, there is an expansion of proximal (ventral) optic cup fates.[31] Therefore, in the zebrafish, failure to initiate retinal neurogenesis may be a manifestation of a loss of proximal tissue identity, as ventral retina is the site at which neurogenesis is initiated.[17] An additional proximo-distal effect of midline Hh signalling is the regulation of the transcription factors *vax1* and *vax2* in optic stalk and ventral retina.[46] These transcription factors in turn regulate several key events in the development of the optic cup.[46]

In the zebrafish, microphthalmia is associated with failed retinal neurogenesis due to Hh knockdown.[19,20] This observation lends some insight into the possible etiology of microphthalmia as part of the suite of abnormalities that may occur in human holoprosencephaly. Holoprosencephaly is a developmental syndrome resulting from impaired midline separation of the embryonic forebrain; genetic causes include mutations in the *shh* gene,[47,48] and teratogenic causes (in livestock) include maternal ingestion of cyclopamine.[49] Resultant defects can range from complete cyclopia and severe craniofacial abnormalities, to a virtually normal appearance with the exception of a single median maxillary central incisor. Microphthalmia and/or anophthalmia are commonly associated with other holoprosencephalic features, including those resulting from defects in the *shh* gene,[47] and from cyclopamine ingestion,[50] but the mechanistic connection between this abnormality and midline signalling has not been established. We suggest that in mammals as well as other vertebrates, the midline Hh signal may be needed for an ocular proximo-distal patterning event that is required for normal eye development.

Summary

Hedgehog genes were identified in vertebrates in 1993; the first demonstration that Hh signalling was needed for proper eye development was reported in 1995. Since then, there has been an explosion of information and interest in this signalling pathway, and in the key roles it plays during development of the eye. Several important questions remain regarding the cellular sources of Hh signals, the significance of Hh signalling for retinal cell survival, the involvement of Hh signalling defects in developmental disorders, and the evolutionary relationships among systems that rely upon the Hh signal. The next experimental challenges will also include those designed to unravel interactions of Hh signalling with other pathways, such as FGFs, wnts, BMPs and retinoic acid.[51] Studies in the zebrafish will continue to illuminate the important roles for Hh signalling in the vertebrate eye.

Acknowledgements

I am grateful to Dr. Brian Link and Ms. Ruth Frey for helpful comments on the manuscript. This work was supported by NIH EY012146.

References

1. Heberlein U, Singh CM, Luk AY et al. Growth and differentiation in the Drosophila eye coordinated by hedgehog. Nature 1995; 373(6516):709-711.
2. Cho KO, Chern J, Izzadoost S et al. Novel signaling from the peripodial membrane is essential for eye disc patterning in Drosophila. Cell 2000; 103(2):331-342.
3. Bagunà J, Garcia-Fernàndez J. Evo-Devo: The long and winding road. Int J Dev Biol 2003; 47(7-8):705-713.
4. Neumann CJ. Pattern formation in the zebrafish retina. Sem Cell Dev Biol 2001; 12:485-490.
5. Easter Jr SS, Malicki JJ. The zebrafish eye: Developmental and genetic analysis. Results Probl Cell Differ 2002; 40:346-370.
6. Livesey FJ, Cepko CL. Vertebrate neural cell-fate determination: Lessons from the retina. Nat Rev Neurosci 2001; 2(2):109-118.
7. Ekker SC, Ungar AR, Greenstein P et al. Patterning activities of vertebrate hedgehog proteins in the developing eye and brain. Curr Biol 1995; 5(8):944-955.
8. Macdonald R, Barth KA, Xu Q et al. Midline signalling is required for Pax gene regulation and patterning of the eyes. Development 1995; 121(10):3267-3278.

9. Incardona JP, Roelink H. The role of cholesterol in Shh signaling and teratogen-induced holoprosencephaly. Cell Mol Life Sci 2000; 57:1709-1719.
10. Cohen MM, Shiota K. Teratogenesis of holoprosencephaly. Am J Med Gen 2002; 109:1-15.
11. Roessler E, Muenke M. Midline and laterality defects: Left and right meet in the middle. Bioessays 2001; 23(10):888-900.
12. Ingham PW, McMahon AP. Hedgehog signaling in animal development: Paradigms and principles. Genes Dev 2001; 15(23):3059-3087.
13. Amato MA, Boy S, Perron M. Hedgehog signaling in vertebrate eye development: A growing puzzle. Cell Mol Life Sci 2004; 61:899-910.
14. Incardona JP, Gaffield W, Kapur RP et al. The teratogenic Veratrum alkaloid cyclopamine inhibits sonic hedgehog signal transduction. Development 1998; 125(18):3553-3562.
15. Cooper MK, Porter JA, Young KE et al. Teratogen-mediated inhibition of target tissue response to Shh signaling. Science 1998; 280(5369):1603-1607.
16. Stenkamp DL, Frey RA, Prabhudesai SN et al. Function for Hedgehog genes in zebrafish retinal development. Dev Biol 2000; 220(2):238-252.
17. Hu M, Easter SS. Retinal neurogenesis: The formation of the initial central patch of postmitotic cells. Dev Biol 1999; 207(2):309-321.
18. Shkumatava A, Fischer S, Muller F et al. Sonic hedgehog, secreted by amacrine cells, acts as a short-range signal to direct differentiation and lamination in the zebrafish retina. Development 2004; 131(16):3849-3858.
19. Stenkamp DL, Frey RA. Extraretinal and retinal hedgehog signaling sequentially regulate retinal differentiation in zebrafish. Dev Biol 2003; 258(2):349-363.
20. Stenkamp DL, Frey RA, Mallory DE et al. Embryonic retinal gene expression in sonic-you mutant zebrafish. Dev Dyn 2002; 225(3):344-350.
21. Schauerte HE, van Eeden FJ, Fricke C et al. Sonic hedgehog is not required for the induction of medial floor plate cells in the zebrafish. Development 1998; 125(15):2983-2993.
22. Varga ZM, Amores A, Lewis KE et al. Zebrafish smoothened functions in ventral neural tube specification and axon tract formation. Development 2001; 128(18):3497-3509.
23. Lum L, Beachy PA. The Hedgehog response network: Sensors, switches, and routers. Science 2004; 304(5678):1755-1759.
24. Chuang JC, Mathers PH, Raymond PA. Expression of three Rx homeobox genes in embryonic and adult zebrafish. Mech Dev 1999; 84(1-2):195-198.
25. Kimura A, Singh D, Wawrousek EF et al. Both PCE-1/RX and OTX/CRX interactions are necessary for photoreceptor-specific gene expression. J Biol Chem 2000; 275(2):1152-1160.
26. Chen CM, Cepko CL. The chicken RaxL gene plays a role in the initiation of photoreceptor differentiation. Development 2002; 129(23):5363-5375.
27. Kay JN, Finger-Baier KC, Roeser T et al. Retinal ganglion cell genesis requires lakritz, a Zebrafish atonal homolog. Neuron 2001; 30(3):725-736.
28. Ericson J, Briscoe J, Rashbass P et al. Graded sonic hedgehog signaling and the specification of cell fate in the ventral neural tube. Cold Spring Harb Symp Quant Biol 1997; 62:451-466.
29. Neumann CJ, Nuesslein-Volhard C. Patterning of the zebrafish retina by a wave of sonic hedgehog activity. Science 2000; 289(5487):2137-2139.
30. Halloran MC, Sato-Maeda M, Warren JT et al. Laser-induced gene expression in specific cells of transgenic zebrafish. Development 2000; 127(9):1953-1960.
31. Perron M, Boy S, Amato MA et al. A novel function for Hedgehog signaling in retinal pigment epithelium differentiation. Development 2003; 130:1565-1577.
32. Levine EM, Roelink H, Turner J et al. Sonic hedgehog promotes rod photoreceptor differentiation in mammalian retinal cells in vitro. J Neurosci 1997; 17(16):6277-6288.
33. Ma C, Zhou Y, Beachy PA et al. The segment polarity gene hedgehog is required for progression of the morphogenetic furrow in the developing Drosophila eye. Cell 1993; 75(5):927-938.
34. Dominguez M. Dual role for Hedgehog in the regulation of the proneural gene atonal during ommatidia development. Development 1999; 126(11):2345-2353.
35. Zhang XM, Yang XJ. Regulation of retinal ganglion cell production by Sonic hedgehog. Development 2001; 128(6):943-957.
36. Arendt D, Tessmar-Raible K, Snyman H et al. Ciliary photoreceptors with a vertebrate-type opsin in an invertebrate brain. Science 2004; 306:869-871.
37. Wang YP, Dakubo G, Howley P et al. Development of normal retinal organization depends on Sonic hedgehog signaling from ganglion cells. Nat Neurosci 2002; 5(9):831-2.
38. Wallace VA, Raff MC. A role for Sonic hedgehog in axon-to-astrocyte signalling in the rodent optic nerve. Development 1999; 126(13):2901-2909.

39. Jensen AM, Wallace VA. Expression of Sonic hedgehog and its putative role as a precursor cell mitogen in the developing mouse retina. Development 1997; 124(2):363-371.
40. Moshiri A, Reh TA. Persistent progenitors at the retinal margin of ptc+/- mice. J Neurosci 2004; 24(1):229-237.
41. Masai I, Stemple DL, Okamoto H et al. Midline signals regulate retinal neurogenesis in zebrafish. Neuron 2000; 27(2):251-263.
42. Nasevicius A, Ekker SC. Effective targeted gene 'knockdown' in zebrafish. Nat Genet 2000; 26(2):216-220.
43. Krauss S, Johansen T, Korzh V et al. Expression of the zebrafish paired box gene pax[zf-b] during early neurogenesis. Development 1991; 113(4):1193-1206.
44. Reifers F, Adams J, Mason IJ et al. Overlapping and distinct functions provided by fgf17, a new zebrafish member of the Fgf8/17/18 subgroup of Fgfs. Mech Dev 2000; 99(1-2):39-49.
45. Zhang XM, Yang XJ. Temporal and spatial effects of Sonic hedgehog signaling in chick eye morphogenesis. Dev Biol 2001; 233(2):271-290.
46. Take-uchi M, Clarke JDW, Wilson SW. Hedgehog signaling maintains the optic stalk-retinal interface through regulation of Vax gene activity. Development 2003; 130:955-968.
47. Hehr U, Gross C, Diebold U et al. Wide phenotypic variability in families with holoprosencephaly and a sonic hedgehog mutation. Eur J Pediatr 2004; 163:347-352.
48. Lazaro L, Dubourg C, Pasquier L et al. Phenotypic and molecular variability of the holoprosencephalic spectrum. Am J Med Gen 2004; 129A:21-24.
49. Keeler RF, Binns W. Teratogenic compounds of Veratrum californicum (Durand). V. Comparison of cyclopian effects of steroidal alkaloids from the plant and structurally related compounds from other sources. Teratology 1968; 1(1):5-10.
50. Keeler RF. Cyclopamine and related steroidal alkaloid teratogens: Their occurrence, structural relationship, and biologic effects. Lipids 1978; 13:708-715.
51. Yang XJ. Roles of cell-extrinsic growth factors in vertebrate eye pattern formation and retinogenesis. Semin Cell Dev Biol 2004; 15(1):91-103.

CHAPTER 7

Sonic Hedgehog Signalling during Tooth Morphogenesis

Martyn T. Cobourne, Isabelle Miletich and Paul T. Sharpe*

Abstract

The Sonic hedgehog (Shh) peptide belongs to a small family of signalling molecules that have a complex mode of action and wide range of function during normal vertebrate development. In common with many regions of the embryo, *Shh* is expressed in the developing tooth in a regionally restricted manner. Specifically, *Shh* expression is localised to the epithelial component of the tooth germ at various stages during the odontogenic process; however, both tooth-forming epithelium and mesenchyme are responsive to the signal. A number of studies have analysed the role of Shh during tooth development, utilising both culture based and genetic systems, and it is clear that this signalling pathway is essential for normal development of the tooth. During the initiation of odontogenesis, localised signalling is important for growth and development of the tooth bud, whilst later during morphogenesis, Shh plays a role in cellular differentiation and polarization in the epithelial component of the tooth germ. These complex interactions are mediated by intra-epithelial and epithelial-mesenchymal signalling by Shh throughout these stages of tooth development.

Introduction

Since their characterisation in the early part of the last decade, members of the Hedgehog family of signalling peptides have been shown to play a fundamental role during the development of both invertebrate and vertebrate species.[1,2] In vertebrates, the Sonic hedgehog (Shh) protein has proved to be the most versatile, demonstrating a wide range of influence upon cellular behaviour in both embryonic and postnatal tissues. Not surprisingly, such diversity of function in association with a single molecule is reflected in complex biochemical mediation of the signalling pathway. At almost every stage, from generation and modification of the protein, movement through fields of competent cells, reception, transduction, and ultimately interpretation of the signal in the cellular response, Shh signalling has demonstrated both novel and versatile mechanisms of action.[1-7] This chapter will review current knowledge with respect to the action of Shh signalling during tooth development. Work in this area is of interest to a number of laboratories around the world and both in vitro and more latterly, in vivo genetic manipulations using transgenic mice, have provided considerable insight into the mode of action of this signalling molecule during various stages of odontogenesis.

*Corresponding Author: Paul T. Sharpe—Department of Craniofacial Development, GKT Dental Institute, King's College London, Floor 28, Guy's Hospital, London SE 19RT, U.K. Email: paul.sharpe@kcl.ac.uk

Shh and Gli Signalling and Development, edited by Carolyn E. Fisher and Sarah E.M. Howie. ©2006 Landes Bioscience and Springer Science+Business Media.

Early Generation of the Tooth

The developing tooth has proved to be a useful model system for understanding the principles of organogenesis. Whilst the origins of teeth are relatively simple, being derived from the ectoderm and cranial neural crest of the fronto-nasal process and first branchial arch, the subsequent tissue interactions required to produce a mature tooth of a particular class are complex. These interactions are mediated by molecular mechanisms under rigid genetic control and are orchestrated by a host of signalling cascades and developmentally active molecules.[8-11]

The first morphological evidence of early odontogenesis occurs with the formation of a localised thickening in the oral epithelium. This thickening or primary epithelial band undergoes localised proliferation in discreet regions of the dental axis to form the tooth buds. Simultaneously, cranial neural crest-derived ectomesenchyme condenses around each bud and collectively these two tissue populations constitute the tooth germ. The epithelial component or enamel organ will ultimately form enamel of the tooth crown, whilst the remainder of the tooth and its periodontal attachment forms from neural crest-derived dental papilla and follicle, respectively.[12] Following the bud stage, the enamel organ assumes a cap shape, mediated by the formation of a distinct, nonproliferating and biologically active region of epithelial cells called the primary enamel knot.[13-16] This enamel knot rapidly disappears, but secondary enamel knots, situated at the sites of the future coronal cusp tips continue to signal and subtly modify cuspal architecture according to the class of tooth. These changes in three-dimensional shape of the enamel organ produce the bell stage. During the transition from cap to bell stage, cellular differentiation within compartments of the tooth germ produces distinct cell populations responsible for generating all the specialised tissues of the mature tooth. These include enamel-forming ameloblasts, dentine-forming odontoblasts and the surrounding dental follicle that generates the periodontal attatchment of the tooth. Ultimately, the mature tooth will erupt into the oral cavity and assume its occlusal position.

The mature adult dentition is ultimately composed of serially homologous groups of teeth organised within symmetrical classes within each quadrant of the upper and lower jaws. In human populations, the primary or deciduous dentition is replaced by the permanent, consisting of four tooth classes; incisors, canines, premolars and molars. In contrast, the rodent dentition is composed of a single set of teeth, characterised by an absence of both canines and premolars. In their place is an edentulous region or diastema, separating the incisor and molar fields. In addition, the rodent incisor is specially adapted to its function; lacking enamel on the lingual surface it continually erupts throughout life to provide a constant occlusal surface. The mouse currently provides the best available model for the study of odontogenesis, however, as fundamental differences do exist between the murine and human dentition; care has to be taken in extrapolating data between the two species.

The Shh Pathway Is Active in the Developing Tooth

Shh transcripts and downstream components of the pathway are active in the developing tooth germ (Fig. 1). *Shh* has been demonstrated at several specific stages of development in the mouse tooth germ, but expression is always restricted to the epithelial component.[14,17-21] This expression is first seen in a highly restricted region within the epithelial thickening during initiation and later, in cells situated at the tip of the tooth bud. These cells are presumed to be precursors of the enamel knot and by the early cap stage, *Shh* is only expressed in this cell population.[14,21] The late cap and early bell stages are characterised by *Shh* transcripts in the internal enamel epithelium, stratum intermedium and stellate reticulum, with progressive upregulation in ameloblasts of the bell stage tooth germ; following their terminal differentiation this expression declines.[17,20,22-24]

Ptc1 encodes the Shh receptor and the developing tooth shows regional expression of *Ptc1* in both epithelium and mesenchyme (Fig. 1). During initiation, *Ptc1* is expressed strongly in odontogenic mesenchyme underlying the thickening and weakly in epithelial cells at the tip of the invagination.[21,25] At later stages, expression is widespread in the epithelial and

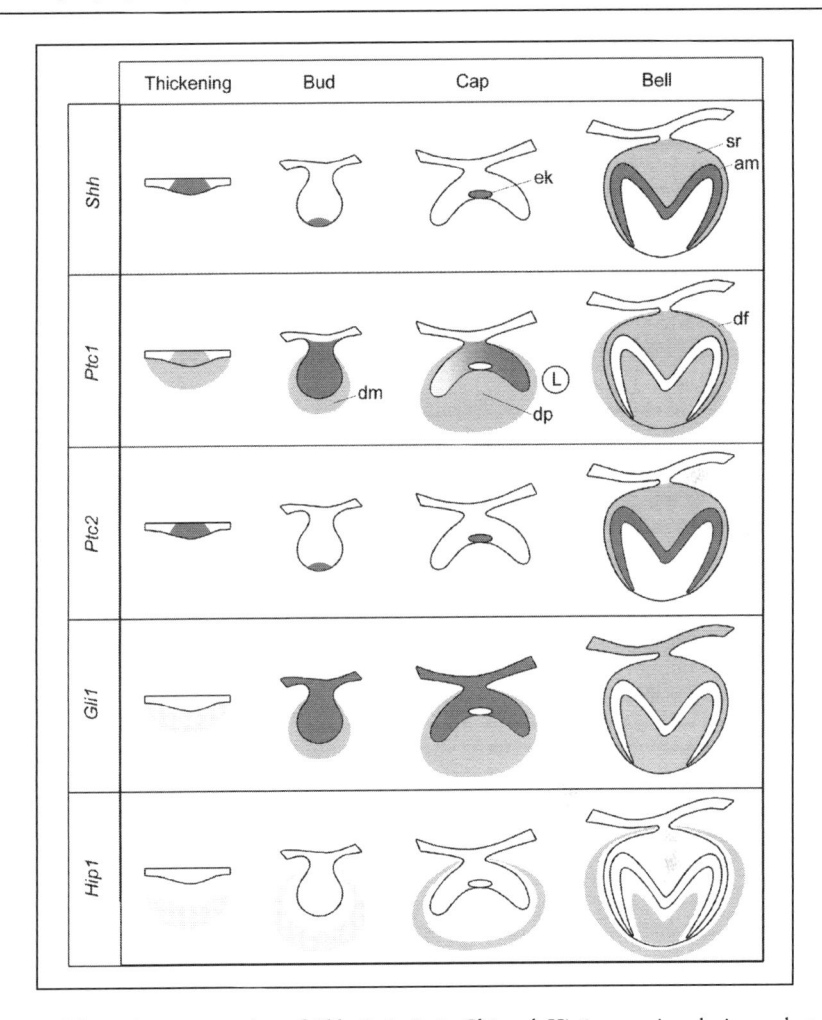

Figure 1. Schematic representation of *Shh*, *Ptc1*, *Ptc2*, *Gli1* and *Hip1* expression during early tooth development. Gene expression in epithelial and mesenchymal compartment of the tooth is indicated in red and blue, respectively. am: ameloblasts; df: dental follicle; dm: dental mesenchyme; dp: dental papilla; ek: enamel knot; L: lingual side; sr: stellate reticulum. A color version of this figure is available online at http://www.Eurekah.com.

mesenchymal components of the tooth germ, but is specifically absent from epithelial cells of the enamel knot.[21] During hard tissue formation, *Ptc1* is upregulated in differentiating odontoblasts and secretory ameloblasts, indicating both these cell populations are responsive to signalling, and consistent with Shh protein distribution observed in these regions.[26] Interestingly, *Ptc1* demonstrates increased activity in the lingual region of the cap stage molar enamel organ, implying a degree of asymmetry in the epithelial response to Shh.[21,23] An additional *Ptc2* gene is present in vertebrates[27,28] and transcripts seem to be localised within epithelial compartments of the tooth, being coexpressed with *Shh* in the early thickening, bud tip, enamel knot and stratum intermedium.[21,22,24,27,29,30] This expression pattern suggests that Ptc2 may well be an additional receptor for Shh during odontogenesis and a target of Shh signalling. Gli family members are ultimately responsible for the interpretation of vertebrate Hedgehog signalling and Gli genes have also been identified in the developing tooth. *Gli1* is expressed in both cell compartments, particularly the mesenchyme, throughout development.[21,22] *Gli2* and *Gli3* expression is fairly ubiquitous early in development, but from the bud stage progressive localisation to the mesenchymal component occurs.[21]

More recently, further novel components in the Hedgehog signalling pathway have been isolated in vertebrate species and they too are active during odontogenesis. *Hip1* (Hedgehog-interacting protein) encodes a membrane glycoprotein capable of binding all mammalian Hedgehog proteins and attenuating signalling,[31] whilst *Gas1* (Growth arrest-specific gene) encodes a glycosylphosphatidylinositol-linked membrane glycoprotein demonstrated to have an antagonistic effect on Shh signalling in the somites.[32] Both of these genes are upregulated in peripheral mesenchyme surrounding the tooth germ, particularly the dental follicle from the cap stage of development, suggesting a role in pathway inhibition around the outer limits of odontogenic mesenchyme.[22,33] In addition Rab23, encoded by the mouse *open brain* (*opb*) gene, belongs to the Rab family of GTPases involved in vesicle transport. Rab23 has been shown to antagonise Shh signalling in the mouse spinal cord by acting intracellularly, downstream of Shh.[34] Interestingly, *Rab23* demonstrates contrasting expression domains in the incisor and molar dentition, being restricted to the mesenchymal compartment of molar teeth and the epithelium of the enamel knot in incisor teeth. These findings provide some evidence of distinct regulatory pathways for Shh in teeth of different classes.[35]

Long and Short Range Shh Signalling in the Tooth

The active signalling form of Hedgehog proteins are notable in having a cholesterol moiety covalently placed at the carboxyl-terminal end[36,37] and an amino-terminal palmitoyl group.[38] The presence of dual hydrophobic groups might be expected to restrict movement of the secreted protein by tethering it to the surface of producing cells, as demonstrated for *Drosophila* hedgehog in the wing imaginal disc.[39] However, in genetically modified mice these hydrophobic moieties seem to be essential for the normal distribution of Shh protein. In the mouse limb, cholesterol modification actually seems to allow enhanced movement of Shh from its site of production,[40] possibly in the form of a multimeric complex.[41] Similarly, gene-targeted mice producing a nonpalmitoylated form of Shh exhibit deficiencies in protein distribution and associated defects in the limbs and neural tube, known sites of long range Shh signalling.[42]

Within the developing tooth, Shh appears to act as both a short and long range signal on the basis of strong immunohistochemical staining being observed in the epithelial cells (sites of strong *Shh* transcription) and graded reactivity in the mesenchymal components, including the dental papilla and follicle, at later stages.[24,26,43] Importantly, whilst Shh protein is strongly detected in the basement membrane separating epithelial and mesenchymal components of the tooth germ, it is also present within differentiating odontoblasts and extracellularly in their predentine product.[26] What are the relative contributions of short and long range signalling in the developing tooth? In mice producing a functional Shh protein lacking the cholesterol modification, the teeth are essentially normal; in contrast to a number of other regions, including the brain, face and limb.[23] Certainly in the limb bud, these mice demonstrate that cholesterol modification of Shh is dispensable for signalling over a limited range, but essential for long range signalling (up to thirty cell diameters). The absence of a significant dental phenotype in these animals invites speculation as to the nature or significance of true long range signalling in the developing tooth. The craniofacial phenotype of these mice was not described in detail[23] but mice lacking a palmitoylated form of Shh have marked holoprosencephaly, a characteristic feature of deficient Shh signalling.[25] Preliminary analysis suggests that the severity of this holoprosencephaly means this mouse line will not be informative with regard to tooth development (unpublished data).

Shh Interacts with Multiple Gene Families in the First Branchial Arch

A number of signalling pathways are active in the first branchial arch during odontogenesis and reciprocal interactions between these molecules within the epithelium and mesenchyme

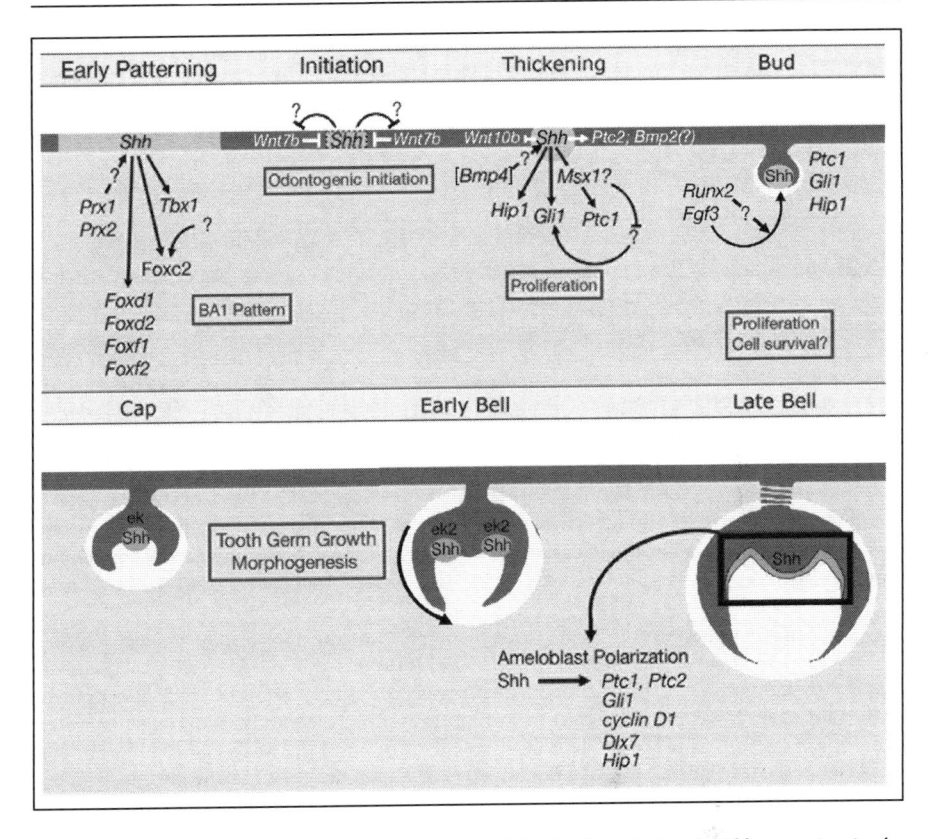

Figure 2. Shh interactions during early development of the first branchial arch. *Shh* expression in the epithelium is required for normal growth, development and survival of neural crest during early patterning of the maxillary and mandibular first arch derivatives. Localised *Shh* in the epithelium demarcates the sites of tooth development along the dental axis, mediating proliferation and possibly cell survival as the tooth buds invaginate into the underlying mesenchyme. During the later stages of odontogenesis, Shh signalling in the primary and secondary enamel knots is important for morphogenesis during the transition from cap to bell stages and the establishment of crown shape; in particular during growth of the lingual side of the tooth germ. From early to late bell stages, Shh signalling in the stratum intermedium ensures correct growth and polarization of the enamel-secreting ameloblasts.

mediate the complex tissue interactions that are required to generate a tooth.[8-11] The Shh pathway is active during tooth development and interactions have been demonstrated with downstream Hedgehog components and members of other signalling families (Fig. 2).[44]

In common with a variety of regions in the developing embryo, *Ptc1* and *Gli1* are transcriptional targets of Shh in the tooth germ. Both genes are upregulated in odontogenic mesenchyme cultured in the presence of Shh-loaded beads.[21,25,33,45] However, *Ptc1* induction by Shh appears to require *Msx1* in dental mesenchyme, whereas *Gli1* does not.[45] In addition, with the exception of the enamel knot, *Ptc2* transcription is also Shh-dependent in the tooth germ.[22] Shh can also induce *Hip1* in isolated mandibular arch mesenchyme[33] but in contrast to *Ptc1* and *Gli1*, at E11.5 during initiation, *Hip1* expression is completely lost over a 24 hour culture period in isolated mandibular mesenchyme.[43] It is clear that subtle differences exist in the regulation of downstream Shh targets in odontogenic tissues. A general principle of Hedgehog signalling is that in the resting state Ptc1 inhibits the pathway, with this inhibition only being relieved by binding of ligand.[46-48] However, as *Ptc1* is also a direct transcriptional target of

Hedgehog proteins, activation of the pathway immediately puts in place a mechanism to ultimately shut it down, via Ptc-mediated sequestration of ligand.[49] Hip1 can also bind Hedgehog proteins and attenuate signalling.[31] Thus in odontogenic regions of the first arch, in common with *Ptc1* induction, a further mechanism of Hip1-mediated autoregulation exists to control this signalling pathway.[18]

Interactions between Shh and members of other genetic pathways have also been reported in the first arch. Shh signalling interacts with *Prx* genes; in *Prx1/Prx2* double knockout mice *Shh* is downregulated in the oral epithelium from around E9.5, whilst these genes are not required to induce *Shh*, it would appear that they do regulate the production of a mesenchymal signal that maintains *Shh* transcription in the overlying oral epithelium.[50] *Tbx1* encodes a member of the T-Box family of transcription factors, being expressed in the mesodermal core and endoderm of the first branchial arch, and is dependent upon Shh signalling in first branchial arch ectoderm based upon the analysis of *Shh -/-* embryos.[51] Interactions also exist between Shh and members of the Wnt signalling pathway; in odontogenic epithelium around E11.5, Wnt7b can repress *Shh* when ectopically expressed in early epithelial thickenings,[52] whilst Shh can repress *Wnt10a* in molar epithelium.[25] Genetically mediated loss of Shh signalling in the developing tooth has also revealed further targets of this pathway in the epithelial component of the tooth germ, including *cyclin D1* in preameloblasts and stratum intermedium and *Dlx7* in differentiating ameloblasts.[22] Interestingly, a yeast two-hybrid screen has demonstrated that Ptc1 is able to interact with cyclin B1,[53] although no evidence for an interaction exists in the tooth.

Runx2 encodes one member of the three-member runt-domain transcription factor family and is a key regulator of osteoblast function. *Runx2* is expressed in the mesenchymal compartment of the developing tooth and *Runx2 -/-* mutant mice exhibit arrest of tooth development at the bud stage. *Shh* expression is absent from the lower molar tooth buds of these mice, even though a putative enamel knot does seem to form. In contrast, weak expression of *Shh* is observed in the upper molars. An identified target of *Runx2* in dental mesenchyme is *Fgf3*, but exogenous Fgf signalling is unable to rescue *Shh* expression in cultured mutant tooth germs. Thus, additional mesenchymal targets, under the regulation of *Runx2*, would appear to be able to regulate *Shh* transcription in the epithelium of the enamel knot during a critical stage of odontogenesis.[54]

Members of the Bone Morphogenetic Protein (BMP) family are also involved in signalling interactions during early tooth development and Shh interacts with several members. *Bmp2* is coexpressed with *Shh* in early dental epithelium[25] and in vitro inhibition of Shh using 5E1 (a function-blocking antibody) suggests that *Bmp2* is a downstream target of Shh in the developing tooth germ.[55] In addition, several findings suggest that mesenchymal Bmp4 regulates *Shh* expression in dental epithelium in a concentration-dependant manner; *Shh* is downregulated in the epithelium of *Msx1* mutant mice and this can be rescued by exogenous Bmp4, inhibition of Bmp4 by Noggin represses *Shh* in wild type dental epithelium and ectopic expression of human *Bmp4* in murine dental mesenchyme either restores *Shh* in *Msx1* mutants or represses it in wild type tissue.[55,56] However, not all of these observations are consistent with those from mice generated with conditional inactivation of *Shh* in odontogenic epithelium; *Bmp2* does not downregulate in the epithelium under these conditions.[23] In addition, both *Ptc1* and *Gli1* downregulate in odontogenic mesenchyme of these conditional mice; only *Ptc1* is attenuated in *Msx1* mutants. Can *Msx1* repress a gene that is capable of inducing *Gli1* expression independently of Shh?[23,45,55,56] In chick mandibular explants, ectopic Bmp4 and Fgf4 can promote artificial development of cap stage tooth rudiments and *Shh* is expressed in these tooth germs in a distribution similar to that seen in the enamel knot of murine tooth germs. Under normal circumstances no *Shh* expression is observed in the epithelium of chick mandibles cultured in the absence of Bmp4 and Fgf4.[57] In addition, a novel BMP-inducible BMP inhibitor, Ectodin, has been found to be repressed by Shh in developing tooth buds, prompting the suggestion that Shh in the enamel knot can oppose the effect of BMP's inducing their own antagonist.[58]

The Functional Significance of Shh during Development of the Tooth

During the initiation stage of tooth development, the appearance of highly restricted areas of *Shh* expression in the oral epithelium precisely demarcates the position of future teeth along the dental arch primordia. Biological activity of Shh alone appears to be sufficient to induce the primary events at the origin of tooth formation. Although definitive genetic evidence for a critical role in odontogenic initiation is lacking, in vitro manipulation of Shh signalling in the first arch derivatives using a variety of techniques suggests that signalling from this pathway is required to initiate tooth development. Gain-of-function experiments, such as implantation of agarose beads soaked with recombinant Shh protein[21] or electroporation of ectopic *Shh* (unpublished data) into the oral epithelium can mimic tooth initiation by inducing ectopic invagination of oral epithelial cells. Conversely, loss-of-function experiments result in tooth development arresting at the epithelial thickening stage when Shh biological activity is inactivated with 5E1 or signal transduction is inhibited with the PKA-activator forskolin during the very earliest stages of odontogenesis.[59] Therefore Shh mitogenic activity plays a key role in early proliferation and budding of the oral epithelium into the underlying mesenchyme at the sites of tooth formation.

Given this important role in initiation, a key question is how discrete expression domains of *Shh* are established along the proximo-distal axis of the future jaws. Whilst inducers of *Shh* in the oral epithelium at the initiation stage have yet to be identified, restriction mechanisms have been discovered that spatially limit either *Shh* transcription or Shh protein activity. At the initiation stage, *Wnt7b* is expressed in nondental epithelium in a pattern complementary to *Shh*. Ectopic expression of *Wnt7b* in early dental (*Shh*-expressing) epithelium leads to loss of *Shh* expression and arrest of tooth development, suggesting that *Wnt7b* can repress *Shh* expression in oral epithelium.[52] A further example of Shh restriction has been observed at the post-transcriptional level during the bud stage of development.[43] The symmetrical diastema regions of the mouse jaw remain edentulous, although located between two sources of Shh protein in the adjacent incisor and molar tooth buds. Shh protein is able to diffuse into diastema mesenchyme but is rendered inactive and nonfunctional in this area through a poorly understood mechanism that depends on the presence of the overlying epithelium. Gas1, an inhibitor of Shh signalling in somitic mesoderm[32] is also expressed in diastema mesenchyme, in an epithelial-dependent manner. Further, ectopic expression of *Gas1* into this mesenchyme can downregulate the Shh target *Ptc1*, suggesting that Gas1 can also negatively regulate Shh signalling in discreet regions of the developing jaws.[43]

The important role of Shh in modulating growth of the dental tissues has been established not only at the initiation stage, but also during the later phases of tooth development. *Shh* function has been selectively removed from the dental epithelium and its derivatives shortly after the initiation stage of tooth development using Cre/loxP site-specific recombination. Mice expressing a *Shh* conditional allele, flanked with loxP sites, crossed with mice expressing Cre recombinase under control of the keratin-14 (K14) promoter lack Shh signalling activity in odontogenic epithelium from the bud stage of development.[23] Both molars and incisors are severely reduced in size, indicating that Shh acts as a growth factor during tooth development. In addition, these teeth occupy abnormal positions within the jaw, being fused to the oral epithelium with an absence of dental cords and alveolar bone on the oral side. Importantly, *Shh* conditional mutants also suggest a role for *Shh* in patterning and morphogenesis of the developing tooth. Mutant molars are abnormally shaped, displaying broad and underdeveloped cusps. Interestingly, *Shh* may be involved in patterning molars along the buccolingual axis, since the lingual side of *Shh* conditional mutant molars appears to be more severely affected;[23] this is consistent with higher levels of *Ptc1* expression on the lingual side of the cap stage enamel organ. Shh does not seem to be required for terminal differentiation of ameloblasts and odontoblasts, as both dentine and enamel are deposited in conditional mutant tooth germs cultured under the kidney capsule. However, *Shh* clearly plays a role in polarisation and growth of both these cell populations.[23] Since the Shh receptor *Ptc1* is expressed in both the epithelial and mesenchymal compartment of the developing tooth, Shh signalling would appear to mediate epithelial-mesenchymal as well as intra-epithelial cell interactions. Genetic removal of *Smo* expression from the oral epithelium using K14 driven Cre recombination of a conditional *Smo* allele has therefore been used to abrogate intra-epithelial transduction of the Shh signal.[22] Morphological defects of the teeth were essentially similar to those of *Shh* conditional mutants, except for molar size which was normal overall, and odontoblast polarization, which occurred normally along the whole epithelial-mesenchymal interface. Ameloblasts failed to grow in size and polarize, suggesting that although signalling from preodontoblasts is necessary for ameloblast cytodifferentiation, *Shh*-dependent intra-epithelial signalling is also required.[22]

The analyses of mice engineered with targeted disruption of varying combinations of Gli genes also implicate a requirement for Shh transduction during normal development of the dentition. *Gli2 -/-* embryos either have absent or fused maxillary incisors, whilst *Gli2 -/-; Gli3 -/-* double mutants exhibit single small incisor buds that do not develop beyond that stage.[21] However, it is not clear if these midline defects are caused by a mild holoprosencephaly or more localised disruption of incisor initiation. The molar dentition is normal in *Gli2 -/-* mutants, whereas *Gli2 -/-;Gli3 -/-* mutants show no sign of any molar development. Functional redundancy between *Gli2* and *Gli3*, which are largely coexpressed during tooth development, can explain the more severe phenotype in the double mutant.

A localised role for the Shh signal during odontogenesis needs to be appreciated within the wider context of development of the whole craniofacial region. The first branchial arch fails to form in *Shh* null mutants due to severe holoprosencephaly.[60] However, genetic removal of responsiveness to Hedgehog family members within cranial neural crest cells results in a less severe craniofacial phenotype.[61] These mice have a severe, but incomplete loss of branchial arch-derived skeletal elements associated with marked apoptosis and decreased proliferation within the neural crest. These defects are associated with loss of several Hedgehog targets, including members of the Forkhead transcription factor family (*Foxc2, Foxd1, Foxd2, Foxf1* and *Foxf2*) and *Pax9* in the distal region of the mandibular arch. Within the dentition of these conditional mutants the lower incisors are absent, whilst only a single maxillary incisor and variable numbers of molars develop. Interestingly, those teeth that are present are malformed and arrested, emphasising the important role of normal Shh signal transduction to the underlying ectomesenchyme during odontogenesis.[61]

Conclusions

Secreted Hedgehog proteins mediate their effects upon competent cell populations in the developing embryo via a complex and currently poorly understood biochemical pathway. In common with many regions of embryogenesis, Shh signalling is active in the developing tooth germ. Specifically, *Shh* transcripts are localised to epithelial compartments; the early epithelial thickening, enamel knot, stratum intermedium, stellate reticulum and ameloblast cell populations; whilst the secreted protein is able to travel considerable distance within the enamel organ and underlying mesenchyme. Available evidence suggests a crucial role for Shh in the growth, morphogenesis and cytodifferentiation of cells within the epithelial component of the developing tooth, mediated by epithelial-epithelial and epithelial-mesenchymal transduction. The tooth has proved to be a useful model for the study of this pathway; both in vitro, and in vivo using genetically modified mice. This system will undoubtedly continue to provide insight into activity of this signalling peptide, information that will be of relevance to other regions of the developing embryo. In particular, the importance of signalling during odontogenic initiation, how pathway activity is restricted and the relative contributions of downstream components to the signalling process.

References

1. Ingham PW, McMahon AP. Hedgehog signaling in animal development: Paradigms and principles. Genes Dev 2001; 15:3059-3087.
2. McMahon AP, Ingham P, Tabin C. Developmental roles and clinical significance of hedgehog signalling. Curr Top Dev Biol 2003; 53:1-114.

3. Bijlesma MF, Speck CA, Peppelenbosch MP. Hedgehog: An unusual signal transducer. Bioessays 2004; 26:387-394.
4. Goodrich LV, Scott MP. Hedgehog and patched in neural development and disease. Neuron 1998; 21:1243-1257.
5. Lum L, Beachy PA. The hedgehog response network: Sensors, switches and routers. Science 2004; 304:1755-1759.
6. Nybakken K, Perrimon N. Hedgehog signal transduction: Recent findings. Curr Opin Genet Dev 2002; 12:503-511.
7. Ogden SK, Ascano Jr M, Stegman MA et al. Regulation of hedgehog signalling: A complex story. Biochem Pharm 2004; 67:805-814.
8. Tucker AS, Sharpe P. The cutting edge of mammalian development; how the embryo makes teeth. Nat Rev Genet 2004; 5:499-508.
9. Jernvall J, Thesleff I. Reiterative signaling and patterning during mammalian tooth morphogenesis. Mech Dev 2000; 92:19-29.
10. Tucker AS, Sharpe PT. Molecular genetics of tooth morphogenesis and patterning: The right shape in the right place. J Dent Res 1999; 78:826-834.
11. Peters H, Balling R. Teeth. Where and how to make them. Trends Genet 1999; 15:59-65.
12. Chai Y, Jiang X, Ito Y et al. Fate of the mammalian cranial neural crest during tooth and mandibular morphogenesis. Development 2000; 127:1671-1679.
13. Jernvall J, Kettunen P, Karavanova I et al. Evidence for the role of the enamel knot as a control center in mammalian tooth cusp formation: Nondividing cells express growth stimulating Fgf-4 gene. Int J Dev Biol 1994; 38:463-469.
14. Vaahtokari A, Åberg T, Jernvall J et al. The enamel knot as a signaling center in the developing mouse tooth. Mech Dev 1996; 54:39-43.
15. Vaahtokari A, Åberg T, Thesleff I. Apoptosis in the developing tooth: Association with an embryonic signaling center and suppression by EGF and FGF-4. Development 1996; 122:121-129.
16. Jernvall J, Åberg T, Kettunen P et al. The life history of an embryonic signaling center: BMP-4 induces p21 and is associated with apoptosis in the mouse tooth enamel knot. Development 1998; 125:161-169.
17. Bitgood MJ, McMahon AP. Hedgehog and Bmp genes are coexpressed at many diverse sites of cell- cell interaction in the mouse embryo. Dev Biol 1995; 172:126-138.
18. Kronmiller JE, Nguyen T, Berndt W et al. Spatial and temporal distribution of sonic hedgehog mRNA in the embryonic mouse mandible by reverse transcription/polymerase chain reaction and in situ hybridization analysis. Arch Oral Biol 1995; 40:831-838.
19. Iseki S, Araga A, Ohuchi H et al. Sonic hedgehog is expressed in epithelial cells during development of whisker, hair, and tooth. Biochem Biophys Res Commun 1996; 218:688-693.
20. Koyama E, Yamaai T, Iseki S et al. Polarizing activity, sonic hedgehog, and tooth development in embryonic and postnatal mouse. Dev Dyn 1996; 206:59-72.
21. Hardcastle Z, Mo R, Hui CC et al. The Shh signalling pathway in tooth development: Defects in Gli2 and Gli3 mutants. Development 1998; 125:2803-2811.
22. Gritli-Linde A, Bei M, Maas R et al. Shh signalling within the dental epithelium is necessary for cell proliferation, growth and polarization. Development 2002; 129:5323-5337.
23. Dassule HR, Lewis P, Bei M et al. Sonic hedgehog regulates growth and morphogenesis of the tooth. Development 2000; 127:4775-4785.
24. Koyama E, Wu C, Shimo T et al. Development of stratum intermedium and its role as a sonic hedgehog-signalling structure during odontogenesis. Dev Dyn 2001; 222:178-191.
25. Dassule HR, McMahon AP. Analysis of epithelial-mesenchymal interactions in the initial morphogenesis of the mammalian tooth. Dev Biol 1998; 202:215-227.
26. Gritli-Linde A, Lewis P, McMahon AP et al. The whereabouts of a morphogen: Direct evidence for short- and graded long-range activity of hedgehog signaling peptides. Dev Biol 2001; 236:364-386.
27. Motoyama J, Takabatake T, Takeshima K et al. Ptch2, a second mouse patched gene is coexpressed with Sonic hedgehog. Nat Genet 1998; 18:104-106.
28. Takabatake T, Ogawa M, Takahashi TC et al. Hedgehog and patched gene expression in adult ocular tissues. FEBS Lett 1997; 410:485-489.
29. Motoyama J, Heng H, Crackower MA et al. Overlapping and nonoverlapping Ptch2 expression with Shh during mouse embryogenesis. Mech Dev 1998; 78:81-84.
30. Wu C, Shimo T, Liu M et al. Sonic hedgehog functions as a mitogen during bell stage of odontogenesis. Conn Tiss Res 2003; 44:92-96.
31. Chuang PT, McMahon AP. Vertebrate hedgehog signalling modulated by induction of a Hedgehog- binding protein. Nature 1999; 397:617-621.
32. Lee CS, Buttitta L, Fan CM. Evidence that the WNT-inducible growth arrest-specific gene 1 encodes an antagonist of sonic hedgehog signaling in the somite. Proc Natl Acad Sci USA 2001; 98:11347-11352.

33. Cobourne MT, Sharpe PT. Expression and regulation of hedgehog-interacting protein during early tooth development. Conn Tiss Res 2002; 43:143-147.
34. Eggenschwiler JT, Espinoza E, Anderson KV. Rab23 is an essential negative regulator of the mouse sonic hedgehog signalling pathway. Nature 2001; 412:194-198.
35. Miletich I, Cobourne MT, Abdeen M et al. Expression of Rab23 and Slimb/B-TrCP, two antagonists of Shh signalling, during mouse tooth development. Arch Oral Biol 2005; 50(2):147-51.
36. Porter JA, von Kessler DP, Ekker SC et al. The product of hedgehog autoproteolytic cleavage active in local and long-range signalling. Nature 1995; 374:363-366.
37. Porter JA, Ekker SC, Park WJ et al. Hedgehog patterning activity: Role of a lipophilic modification mediated by the carboxy-terminal autoprocessing domain. Cell 1996; 86:21-34.
38. Pepinsky RB, Zeng C, Wen D et al. Identification of a palmitic acid-modified form of human sonic hedgehog. J Biol Chem 1998; 273:14037-14045.
39. Burke R, Nellen D, Bellotto M et al. Dispatched, a novel sterol-sensing domain protein dedicated to the release of cholesterol-modified hedgehog from signaling cells. Cell 1999; 99:803-815.
40. Lewis PM, Dunn MP, McMahon JA et al. Cholesterol modification of sonic hedgehog is required for long-range signaling activity and effective modulation of signaling by Ptc1. Cell 2001; 105:599-612.
41. Zeng X, Goetz JA, Suber LM et al. A freely diffusible form of sonic hedgehog mediates long-range signalling. Nature 2001; 411:716-720.
42. Chen M-H, Li Y-L, Kawakami T et al. Palmitoylation is required for the production of a soluble multimeric hedgehog protein complex and long-range signaling in vertebrates. Genes Dev 2004; 18:641-659.
43. Cobourne MT, Miletich I, Sharpe PT. Restriction of sonic hedgehog signalling during early tooth development. Development 2004; 131:2875-2885.
44. Cobourne MT, Sharpe PT. Sonic hedgehog signalling and the developing tooth. Curr Top Dev Biol 2004; 65:255-287.
45. Zhang Y, Zhao X, Hu Y et al. Msx1 is required for the induction of patched by sonic hedgehog in the mammalian tooth germ. Dev Dyn 1999; 215:45-53.
46. Marigo V, Davey RA, Zuo Y et al. Biochemical evidence that patched is the Hedgehog receptor. Nature 1996; 384:176-179.
47. Stone DM, Hynes M, Armanini M et al. The tumour-suppressor gene patched encodes a candidate receptor for Sonic hedgehog. Nature 1996; 384:129-134.
48. Briscoe J, Chen Y, Jessell TM et al. A hedgehog-insensitive form of patched provides evidence for direct long-range morphogen activity of sonic hedgehog in the neural tube. Mol Cell 2001; 7:1279-1291.
49. Chen Y, Struhl G. Dual roles for patched in sequestering and transducing Hedgehog. Cell 1996; 87:553-563.
50. Ten Berge D, Brouwer A, Korving J et al. Prx1 and Prx2 are upstream regulators of sonic hedgehog and control cell proliferation during mandibular arch morphogenesis. Development 2001; 128:2929-2938.
51. Garg V, Yamagishi C, Hu T et al. Tbx1, a DiGeorge syndrome candidate gene, is regulated by sonic hedgehog during pharyngeal arch development. Dev Biol 2001; 235:62-73.
52. Sarkar L, Cobourne M, Naylor S et al. Wnt/Shh interactions regulate ectodermal boundary formation during mammalian tooth development. Proc Natl Acad Sci USA 2000; 97:4520-4524.
53. Barnes EA, Kong M, Ollendorff V et al. Patched1 interacts with cyclin B1 to regulate cell cycle progression. EMBO J 2001; 20:2214-2223.
54. Åberg T, Wang X-B, Kim J-H et al. Runx2 mediates FGF signalling from epithelium to mesenchyme during tooth morphogenesis. Dev Biol 2004; 270:76-93.
55. Zhang Y, Zhang Z, Zhao X et al. A new function of BMP4: Dual role for BMP4 in regulation of sonic hedgehog expression in the mouse tooth germ. Development 2000; 127:1431-1443.
56. Zhao X, Zhang Z, Song Y et al. Transgenically ectopic expression of Bmp4 to the Msx1 mutant dental mesenchyme restores downstream gene expression but represses Shh and Bmp2 in the enamel knot of wild type tooth germ. Mech Dev 2000; 99:29-38.
57. Chen Y, Zhang Y, Jiang T-X et al. Conservation of early odontogenic signaling pathways in Aves. Proc Natl Acad Sci USA 2000; 97:10044-10049.
58. Laurikkala J, Kassai Y, Pakkasjarvi L et al. Identification of a secreted BMP antagonist, ectodin, integrating BMP, FGF, and SHH signals from the tooth enamel knot. Dev Biol 2003; 264:91-105.
59. Cobourne MT, Hardcastle Z, Sharpe PT. Sonic hedgehog regulates epithelial proliferation and cell survival in the developing tooth germ. J Dent Res 2001; 80:1974-1979.
60. Chiang C, Litingtung Y, Lee E et al. Cyclopia and defective axial patterning in mice lacking Sonic hedgehog gene function. Nature 1996; 383:407-413.
61. Jeong J, Mao J, Tenzen T et al. Hedgehog signaling in the neural crest cells regulates the patterning and growth of facial primordia. Genes Dev 2004; 18:937-951.

CHAPTER 8

Limb Pattern Formation:
Upstream and Downstream of Shh Signalling

Aimée Zuniga* and Antonella Galli

Abstract

The vertebrate limb is an attractive model system for studying the interplay of signalling molecules that coordinate growth and patterning during organogenesis. *Sonic Hedgehog* (Shh) plays a key regulatory role during vertebrate limb development as a mediator of the zone of polarizing activity, which directs antero-posterior patterning and ensures that a thumb develops anteriorly and a little finger at the posterior edge of the hand.

The purpose of this chapter is to discuss the different aspects of Shh signalling function during vertebrate limb development. In particular, we will describe the sequence of events leading to the induction and formation of the *Shh* expression domain at the posterior limb bud margin. These events are critical to define the role of Shh in subsequent patterning of the distal limb bud and to establish the initial antero-posterior polarity. We then focus mainly on describing the molecular mechanisms supporting the potential role of Shh as a morphogen during digit patterning. Furthermore, we review the role of Gli family members in mediating Shh signal transduction with special emphasis on Shh-Gli3 interactions. Finally we will report on recent work that challenges the relevance of Shh as a spatial morphogen.

Introduction

The vertebrate embryonic limb is an excellent experimental model to study fundamental developmental processes including cell-cell signalling and pattern formation. The limb is not a vital organ and the easy accessibility allows researchers to combine experimental manipulation with genetic analysis. Moreover, the key signalling pathways regulating limb development are also involved in the development of other organs and their deregulation play roles in pathological conditions. Vertebrate limb development starts with proliferation of mesenchymal cells located in the lateral plate mesoderm (LPM) at the presumptive limb level. These mesenchymal cells can be considered as limb stem cell-like progenitors as they will give rise to all the skeletal elements and connective tissue of the future limb. In contrast, the muscle elements derive from migratory progenitors originating from the somites, a different cell lineage. At the time of limb induction, the mesenchymal cells of the LPM proliferate and accumulate under the ectoderm creating a bulge, the limb bud. Those small protrusions arising from the body wall of the embryo are composed of a core of mesenchymal cells and an outer layer of ectodermal epithelial cells. During progression of limb bud outgrowth the mesenchymal cells differentiate to give rise to the skeletal elements. Cell proliferation and reciprocal interaction between mesoderm and ectoderm are the driving force for outgrowth of the limb bud that gradually extends to form the typical tetrapod limb (Fig. 1A). Three axes characterize the fundamental

---

*Corresponding Author: Aimée Zuniga—Developmental Genetics, Centre for Biomedicine, University of Basel, Mattenstrasse 28, 4058 Basel, Switzerland. Email: Aimee.Zuniga@unibas.ch

Shh and Gli Signalling and Development, edited by Carolyn E. Fisher and Sarah E.M. Howie. ©2006 Landes Bioscience and Springer Science+Business Media.

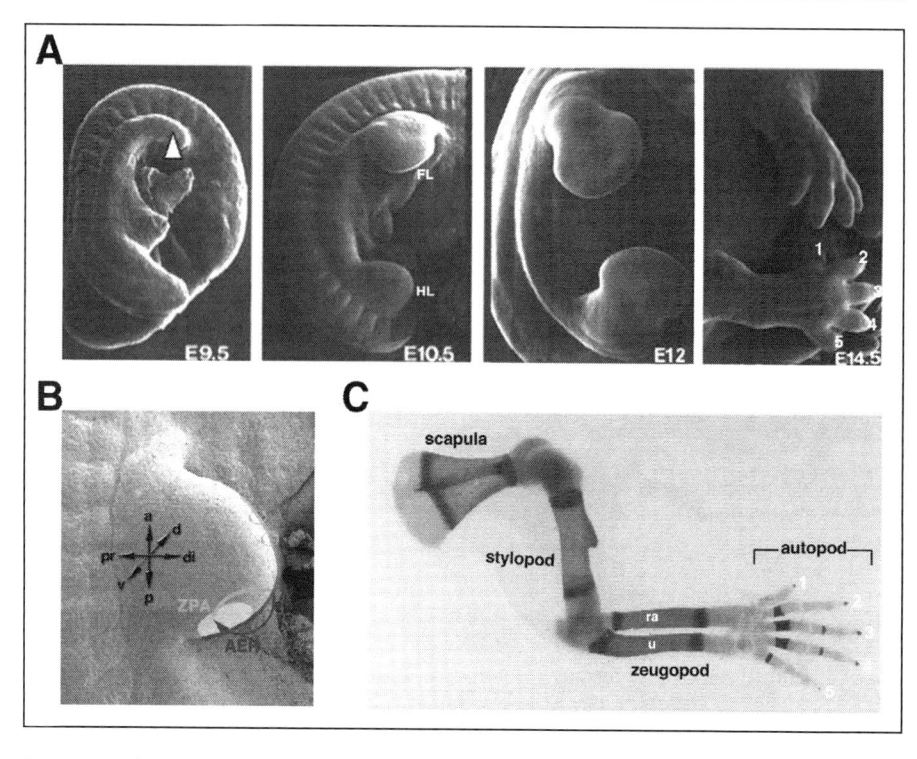

Figure 1. Embryonic development of vertebrate limb. A) Scanning EM of a mouse embryo showing a temporal overview of fore and hindlimb development. At embryonic day 9.5 (E9.5, 21-29 somites) the forelimb bud (white arrowhead) starts to protrude as a bulge from the flank of the embryo, while the hindlimb is not yet visible. Forelimb development is initiated about half a day ahead of the hindlimb. One day later (E10.5, 35-39 somites) the forelimb bud has completely emerged from the body wall. At E12 hand- and footplate are distinct and differentiation of skeletal elements is proceeding. At E14.5, 5 digits (1 to 5) are clearly visible. Reprinted with permission from: Martin P. Int J Dev Biol 1990; 34:323-336. © UBC Press.[73] B) Outgrowth and patterning of the vertebrate limb are controlled by reciprocal interactions (green and blue arrows) between two signalling centers: the ZPA (zone of polarizing activity) depicted in green and located in the posterior limb bud mesenchyme and the AER (apical ectodermal ridge) marked in blue. The three embryonic axes are also represented. Abbreviations: a, anterior; p, posterior; d, dorsal; v, ventral; pr, proximal; di, distal; FL, forelimb; HL, hindlimb. C) Limb skeletal staining with alizarin red (for bones) and alcian blue (for cartilage) of a mouse forelimb at E17.5. Different skeletal elements are shown: the most proximal one is the scapula followed by the stylopod (humerus), the zeugopod (radius, ra and ulna, u) and the more distal elements represented by the autopod (metacarpals and digits). A color version of this figure is available online at http://www.Eurekah.com.

organization of the vertebrate limb (Fig. 1B). The proximal-distal axis goes from shoulder to digit, the antero-posterior axis from thumb (digit 1) to little finger (digit 5) and the dorsal-ventral axis is represented by knuckles and palm. The proximal-distal axis is divided in three main regions (Fig. 1C): stylopod (humerus for the forelimb and femur for the hindlimb), zeugopod (radius-ulna for forelimb and tibia-fibula for hindlimb) and autopod (metacarpal and digits for forelimb and metatarsal and toes for hindlimb).

Experiments in the last sixty years have shown that growth and patterning of the vertebrate limb depends on reciprocal interactions between the ectoderm and the mesoderm and has led to the discovery of two signalling centres, the zone of polarizing activity (ZPA or polarizing region) and the apical ectodermal ridge (AER; Fig. 1B).[1] The ZPA consist of a

Limb Pattern Formation 81

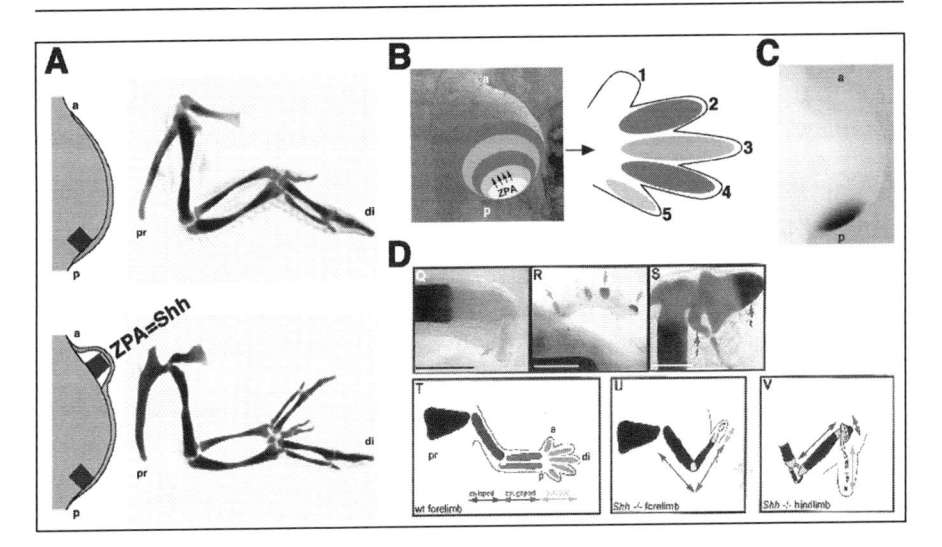

Figure 2. The ZPA is an organizer and Shh acts as a morphogen. A) Classical transplantation experiment of ZPA cells in the anterior mesenchyme of an early chick wing bud leads to a complete mirror image digit duplication (compare the top panel to the lower one). Anterior grafts of *Shh*-expressing cells achieve the same effect. B) Interpretation of Wolpert's morphogen model. Cells of the ZPA (marked in yellow) produce a diffusible molecule called morphogen (arrows) in a gradient manner from posterior to anterior (represented by green gradient). Mesenchymal cells respond differentially to various morphogen thresholds to specify digits. For example, specification of the most posterior digit (digit 5) requires higher level of morphogen activity in comparison to a more anterior digit (digit 4). Digits are numbered from anterior (a, digit 1) to posterior (p, digit 5). C) Whole mount *in situ hybridization* showing the expression of *Shh* in the ZPA in a mouse forelimb bud at E10. D) Loss of posterior-distal limb elements in *Shh* deficient mouse embryos. High power view showing skeletal preparation of developing $Shh^{-/-}$ distal forelimb (Q), distal hindlimb (R) and intermediate hindlimb (S). Schematic representation of wild-type forelimb at E18.5 in comparison to $Shh^{-/-}$ forelimb (U) and hindlimb (V). In Q and R, green arrows point to the presumptive phalanges and carpals or tarsal, fore and hindlimb respectively. Red arrows in S indicate the short zeugopod consisting by the fibula (f) and the tibia (t). In T, U, V the coloured bars identify the three elements: stylopod, zeugopod and autopod and gray areas represent cartilage. Reprinted with permission from: Kraus P, Fraidenraich D, Loomis CA. Mech Dev. 2001; 100: 48-58. © 2001 Elsevier.[9] In all the panels anterior (a) is at the top and posterior (p) at the bottom, proximal (pr) at the left and distal (di) at the right side. A color version of this figure is available online at http://www.Eurekah.com.

group of proliferating and undifferentiated mesenchymal cells located in the posterior limb bud while the AER is a ridge of columnar epithelial cells running along the distal margin of the limb bud ectoderm. Manipulation of the chick limb bud has shown that these signalling centres are essential for limb morphogenesis. The AER is required for proximal-distal limb development. Removal of the AER at an early stage of development leads to apoptosis of the distal limb bud mesenchyme and formation of only proximal limb structures. In contrast, removal of the AER at later developmental stages leads to progressively more distally restricted truncations depending on the time of the surgical ablation.[2,3] AER functions are mediated by members of the family of fibroblast growth factors (FGF), which are expressed in the AER. FGFs keep the mesenchymal cells directly underlying the AER in a proliferating and undifferentiated state.[4,5] The ZPA drives cell proliferation and cell fate specification of the limb bud mesenchyme and is essential for outgrowth and patterning. Transplantation of ZPA cells to the anterior mesenchyme induces complete mirror image duplications of the digits (Fig. 2A).[1] For this reason, the ZPA is often referred to as the limb organiser.[6] Chick limb bud recombination experiments led Lewis Wolpert to propose the so-called "French flag" model in 1969.[7] To explain the role of the ZPA in digit patterning Wolpert proposed that it produces a diffusible molecule, termed morphogen. The morphogen would be distributed in a gradient manner from posterior to anterior along the limb bud axis. Mesenchymal cells would respond to the morphogen gradient in a threshold dependent manner and would specify the mesenchyme with respect to its identity (proximo-distal and antero-posterior). The model predicts that specification of the most posterior digit requires the highest level of morphogen activity, while specification of more anterior digits would require lower thresholds of morphogen activity (Fig. 2B). In agreement with Wolpert's model, the vertebrate homolog of the *Drosophila Hedgehog (Hh)* gene, *Sonic hedgehog (Shh)*, was identified as the instructive diffusible signal expressed by all cells of the ZPA (Fig. 2C).[8] Shh is sufficient to mimic ZPA activity as ectopic grafts of *Shh* expressing cells into the anterior limb bud mesenchyme cause complete mirror image digits duplication like ZPA grafts (Fig. 2A).[1] Moreover, inactivation of the *Shh* gene in mouse and equivalent mutations in human and chick embryos disrupt patterning and formation of posterior-distal elements such as the ulna and fibula and loss of all the digits except digit one (the most anterior digit) (Fig. 2D).[9-11] This demonstrates that Shh signalling is required for antero-posterior patterning during distal limb bud outgrowth.

Limb Bud Initiation

Limb bud initiation is an example of interplay between different growth factor signalling pathways. Members of the FGF and WNT families play key regulatory roles during limb bud initiation.[12,13] Prior to limb bud initiation *Fgf8* is expressed in the intermediate mesoderm (IM) underlying the presumptive fore and hindlimb areas. *Fgf10* is expressed in a wider region including the IM, the segmental plate and the LPM.[12] The *Wnt2b* gene is expressed in the IM and in the LPM at the forelimb level, while *Wnt8c* is expressed in the LPM at the hindlimb level.[14] During onset of limb bud initiation, axial tissues such as the IM and the somites adjacent to the limb forming areas are capable of producing factors that confine *Fgf10* expression to the LPM of the presumptive limb bud. Gain-of-function experiments in chick embryos suggest that Wnt2b and Wnt8c, for fore and hindlimb respectively, contribute to restrict and/or maintain *Fgf10* expression at the appropriate levels in the LPM (Fig. 3A).[14] Upon restriction of *Fgf10* expression to the LPM of the presumptive limb areas, Fgf10 signals to the overlying ectoderm to induce expression of *Wnt3a*, which is required for the establishment and the maintenance of the early AER (Fig. 3B). Indeed, Wnt3a in the ectoderm signals through β-catenin to activate *Fgf8* expression, the earliest marker for early AER formation.[15,16] Subsequently, a positive signalling feedback loop between *Fgf8* and *Fgf10* is initiated to maintain *Fgf10* expression in the nascent limb mesenchyme (Fig. 3B)[17] and to maintain *Shh* expression to the posterior margin of the limb bud (Fig. 3D).[18,19]

Early Limb Bud Polarisation and Establishment of the ZPA

Polarization of the limb field along the antero-posterior axis of the embryo leads to positioning of the ZPA. Genetic analysis in the mouse in combination with chick embryo manipulations has led to the identification of some of the essential components implicated in the establishment of the ZPA. Posteriorly restricted genes of the *Hox* class have been proposed to regulate *Shh* induction. *Hoxb8* is transiently expressed at the posterior side of the nascent limb bud shortly before the appearance of *Shh* expression. Transgenic mice ectopically expressing *Hoxb8* in the anterior part of the limb bud display polydactyly with mirror image digit duplications reminiscent of ZPA grafts in chick.[20] These data, together with the temporal and spatial expression patterns of *Hoxb8* during normal limb bud development, suggested that *Hoxb8* could be the ZPA inducer. However, targeted disruption of *Hoxb8* in the mouse and the deletion of all *Hox8* paralogues, does not result in inhibition of limb bud outgrowth or changes in antero-posterior patterning.[21,22] More recent work suggests that 5'*Hoxd* (*Hoxd11, -12 and -13*) genes play a role in positioning the *Shh* expression domain in the posterior limb bud

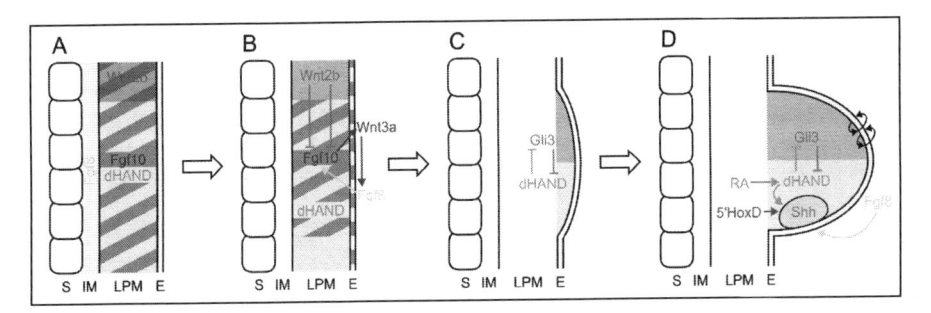

Figure 3. Induction and prepatterning mechanisms acting upstream of Shh signalling. Regulatory interactions involved in forelimb bud induction (panels A, B) and polarization of the early limb bud (panel C, D). A) Prior to limb bud induction *Fgf10* and *dHAND* are expressed throughout the LPM while *Fgf8* is expressed in the IM at the level of the presumptive forelimb. During the onset of limb bud induction, *Wnt2b* is expressed in the LPM and contributes to restrict and maintain *Fgf10* expression at the forelimb level in the LPM. B) Fgf10 signals to the overlying ectoderm to induce *Wnt3a*, which in turn activates *Fgf8* in the AER. Subsequently, a feedback loop between *Fgf8* and *Fgf10* is established. C) Early polarization of limb bud requires a reciprocal genetic repression between *Gli3* and *dHAND*. *Gli3* is expressed in the anterior mesenchyme and restricts *dHAND* in the posterior mesenchyme, while *dHAND* keeps *Gli3* anteriorly restricted. This mutual antagonism interaction prepatterns the limb bud mesenchyme prior to Shh signalling. D) *Gli3/dHAND* reciprocal antagonism leads to the establishment of the ZPA and *Shh* expression in the posterior limb bud mesenchyme. Other molecules such as RA, acting via dHAND, and *5'Hoxd* genes play roles in the activation of the *Shh*-expressing domain in the ZPA. In addition, *Fgf8* from the AER is important to maintain *Shh* expression. Abbreviation: S: somites; IM: intermediate mesoderm; LPM: lateral plate mesoderm; E: ectoderm. Anterior (a) is at the top; posterior (p) is at the bottom.

mesenchyme (Fig. 3D). Anterior misexpression of *5'Hoxd* locus induces ectopic expression of *Shh* in the anterior limb bud mesenchyme, thereby generating a double posterior limb with a loss of antero-posterior asymmetry rather than a mirror image duplication.[23]

Retinoic acid (RA), the active derivative of vitamin A is a potent activator of limb bud polarization as RA-soaked beads implanted in the anterior side of the developing limb bud induce digit duplications similar to a ZPA graft or implantation of Shh coated beads.[24] Furthermore, limb bud development in chick embryos is inhibited by RA receptor antagonists or by synthetic inhibitors of RA synthesis.[25,26] The role of RA has also been addressed genetically. Gene inactivation in mouse of retinaldehyde dehydrogenase-2 (*Raldh2*), an enzyme necessary for RA synthesis, completely disrupts limb bud formation and Shh activation.[27,28] Interestingly, RA can rescue forelimb bud development in *Raldh2* deficient mouse embryos in a dose-dependent manner. These studies establish that RA is required for Shh activation during initiation of forelimb bud outgrowth (Fig. 3D). Furthermore, molecular analysis performed on *Raldh2* mutant limb buds also suggests that RA may interact upstream and/or with the basic helix-loop-helix (bHLH) transcription factor *dHAND* (also called *Hand2*) to activate Shh signalling.[27,29] Indeed, *dHAND* expression can be activated anteriorly by ectopic application of RA in chick limb buds or in transgenic mice; this ectopic expression can activate the Shh signalling pathway and induce digit duplications. Conversely, no Shh expression is detected in the limb buds of *dHAND* deficient mouse embryos.[30] In chicken, mouse and zebrafish embryos, *dHAND* is initially expressed throughout the LPM (Fig. 3A,B) and rapidly restricted to the posterior limb bud mesenchyme during onset of outgrowth (Fig. 3C).[29-31] This restriction appears crucial in establishing the ZPA in the posterior limb bud mesenchyme.

Until recently, only transcriptional activation had been considered to play a role in the polarisation of the limb field and establishment of the ZPA. However, it has emerged that transcriptional repression is also a crucial component of this process. *Gli3* expression is activated in the anterior mesenchyme in a manner complementary to *dHAND*.[32] *Gli3* is required for *dHAND* restriction in the posterior mesenchyme during initiation of limb bud outgrowth. In particular, in limb buds deficient for Gli3 function, *dHAND* remains expressed throughout the entire early limb bud mesenchyme. In turn, *dHAND* is required to restrict *Gli3* expression in the anterior limb bud mesenchyme, as *Gli3* transcripts are expressed throughout the entire limb bud mesenchyme in early *dHAND* deficient embryos.[32] Genetic analysis of limbs buds deficient for both *Gli3* and *Gremlin* (*Grem1*, see below) also implicates *Gli3* in positioning the ZPA.[33,34] Moreover, *dHAND* and *Gli3* are normally activated in *Shh* deficient limb buds, demonstrating that this mutual genetic antagonism occurs upstream of and is independent of Shh. The *dHAND/Gli3* genetic interaction also controls posterior restriction of other key regulators such as *5'Hoxd* genes and *Grem1*.[31,32] In summary, the mutual antagonistic interaction between *Gli3* and *dHAND* prepatterns the limb bud mesenchyme upstream of Shh signalling and participates in *Shh* activation in the posterior limb bud mesenchyme (Fig. 3C,D).[30,33] Unfortunately, *dHAND* deficient mouse embryos die shortly after embryonic day 9.5 (E9.5), which has precluded a molecular and phenotypic analysis of later limb pattern formation.[35] In contrast, *Gli3* deficient embryos die shortly before birth and mutations in *Gli3* result in polydactyly associated with severe loss of digit identity in both mouse and humans.[36-38] During the onset of limb bud outgrowth, patterning genes such as the *5'Hoxd* genes are activated in the posterior part of the emerging limb bud and remain posteriorly expressed until about E11 in mouse embryo. Shortly before the forming handplate becomes visible, their expression domains expand distally and anteriorly under the control of Shh signalling. In *Gli3* deficient embryos, *5'Hoxd* genes are expressed throughout the entire limb bud mesenchyme without posterior restriction until advanced limb bud stages. Expression of other posterior genes such as *Fgf4* in the AER and *Grem1* in the mesenchyme is also expanded anteriorly, demonstrating a general role for *Gli3* as a transcriptional repressor during progression of limb patterning (Fig. 5A).[32] Analysis of *Shh* and *Shh;Gli3* compound mutant embryos has shed further light on the roles of *Gli3* and *Shh* in patterning of the limb bud. In the absence of *Shh*, posterior genes of both the mesenchyme (i.e., *5'Hoxd* genes, *Grem1*) and AER (i.e., *Fgf4, Fgf8, Fgf9* and *Fgf17*) are activated normally, except for *Gli1* and *Ptc*, whose transcriptional activation and upregulation are absolutely dependant on Shh signalling.[39,40] Shh-independent activation of both mensechyme and AER genes in a spatially restricted manner provides further evidence that the limb is prepatterned. Subsequent regulation of all these genes is Shh-dependent; in $Shh^{-/-}$ mutant embryos their expression is rapidly lost and cells undergo apoptosis, resulting in the loss of the handplate. Disruption of one or both *Gli3* alleles in a $Shh^{-/-}$ mutant background improves limb bud development and digit formation in a dose dependent manner.[39] The limbs of *Gli3; Shh* double homozygous mutant embryos are indistinguishable from *Gli3* single mutant embryos, both phenotypically and molecularly. Taken together, these studies showed that *Gli3* initially acts upstream of *Shh* and subsequently downstream to repress anterior ectopic expression of distal limb patterning genes throughout the entire limb bud.[39] The temporarily and spatially restricted activation and propagation of these posterior genes arises from the ability of dHAND and Shh signalling (see below) to overcome *Gli3*-mediated repression activity of progression of limb bud morphogenesis. In summary, genes involved in distal limb patterning are activated prior to establishment of the *Shh* organiser and one of the essential roles of Shh signalling is to upregulate and propagate their expression. Accordingly, the polydactyly and associated loss of digit identity in *Gli3* mutant limb buds is a likely consequence of losing unequal distribution of patterning genes along the antero-posterior axis of the limb bud.

Molecular Mechanisms of the Shh Response

Conservation of the components of the Hh signalling pathway between flies and vertebrates is extremely high. Briefly, both in fly and vertebrate, in the absence of secreted Shh, the transmembrane receptor Patched (Ptc) suppresses the activity of Smoothened (Smo), a transmembrane protein that transduces the Shh signal inside the responding cell. The pathway is activated by the stoechiometric binding of active Shh to Ptc at the surface of the responding

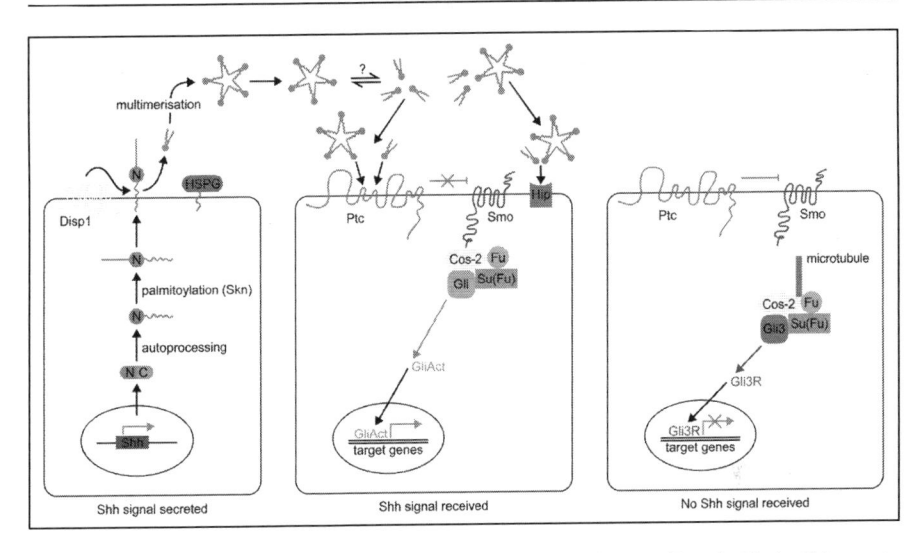

Figure 4. A simplified view of the Shh signalling network. "Shh signal secreted": as the Native Shh protein enters the secretory pathway, it undergoes autoprocessing which generates an N-terminal signalling domain with a C-terminal cholesterol moiety. Palmitoylation follows, resulting in an N-terminal palmitate, a process mediated by Skinny hedgehog. As modified Shh is secreted from the cell, it becomes tethered to the membrane due to the cholesterol-modification and perhaps also due to the palmitoylation. The transmembrane protein *Disp1* functions to release Shh into the extracellular space while the GPI anchored HSPGs are involved in the regulation of Shh transport. Shh becomes multimeric and diffuses to the responding cell. "Shh signal received": it is not clear whether multimeric or monomeric Shh ligand binds preferentially to the transmembrane receptor Ptc. The Shh/Ptc interaction lifts the repression that Ptc normally exerts on the transmembrane protein Smo. Smo transduces the Shh signal through its cytoplasmic tail which recruits the Cos-2/Fu/Su(Fu)/Gli complex. Gli proteins (it is not know whether it is Gli1, Gli2 or Gli3) are turned into activator forms and are translocated to the nucleus where they activate Shh target genes. Shh binds to Ptc and to the transmembrane protein Hip and excess of these receptors can sequester the Shh ligand, leading to a limitation of Shh diffusion. "No Shh signal received": in the absence of Shh, Ptc represses Smo. Inside the cell, Cos-2 may as in *Drosophila* anchor a Cos-2/Fu/Su(Fu)/Gli3 complex to the microtubule, although such a complex has not been identified in vertebrates. Gli3 is however processed into the putative repressor form, Gli3R. Gli3R translocates to the nucleus where it represses Shh target genes.

cells, releasing Smo from Ptc repression. In turn, Smo regulates downstream cytoplasmic targets such as Gli family members (vertebrate orthologues of Cubitus interruptus (Ci) in *Drosophila*) (Fig. 4).[41]

Long and Short Range Signalling: The Molecular Basis for Shh Versatility

The morphogen model predicts that a diffusible molecule secreted by the ZPA is distributed in a gradient along the antero-posterior axis of the limb bud resulting in highest levels of morphogen posteriorly and lowest levels anteriorly. Consequently, mesenchymal cells respond differentially to the morphogen according to their distance from the ZPA to specify digit identity. However, unlike small molecules such as RA, proteins do not freely diffuse in the extracellular space over long distances, raising the question of how the Shh signal is propagated. Recent advances have shed light on the molecular mechanisms mediating the Shh response, and explain how Shh can act directly, both locally and at a distance (Fig. 4). The native Shh protein is initially composed of an N-terminal signalling domain and a C-terminal catalytic domain. Shh enters the secretory pathway and undergoes autocleavage, resulting in the release of the N-terminal signalling domain, which is modified at its C-terminal by an ester-linked cholesterol adduct (Fig. 4; "Shh signal secreted").[42] Genetic studies have demonstrated the biological importance of this cholesterol modification for limb development. A mutant form of Shh protein that cannot be cholesterol-modified was generated by gene targeting in mice. In embryos carrying this mutation, formation of the anterior digits 2 and 3 is disrupted and the expression domains of transcriptional targets of Shh do not extend as anteriorly as in wild type limb buds, suggesting that long range signalling across the antero-posterior axis is reduced.[43] Following cholesterol modification, the N-terminal part of the peptide is palmitoylated, a process mediated by an acyl-transferase encoded by the *Skinny hedgehog* gene (*Skn*). Limbs of mice deficient for *Skn* lack digit 2 and display fusions of digits 3 and 4, demonstrating that palmitoylation is also critical for Shh long range activity.[44] The resulting active Shh ligand is a doubly lipid-modified peptide, whose tight association with the cell membrane triggers local and high level of signalling response. Membrane tethering precludes direct effects on distant cells, however several mechanisms contribute to ligand release and subsequent transport. At least four different mechanisms are likely to contribute to Shh signalling versatility in embryonic tissues during patterning and offer ways to regulate Shh activity (Fig. 4).

1. As in *Drosophila*, membrane-tethered Shh is released from the secreting cells by the action of the transmembrane transporter-like protein encoded by the *Dispatched* gene (*Disp1*).[45,46] Mice deficient for *Disp1* have defects characteristic of loss of Hedgehog signalling, and evidence from several groups shows that Disp1 is essential to permit movement of the Shh ligand to its target tissues.[46-48] Lethality of *Disp1* deficient mice at E9.5 has so far precluded analysis of the limb phenotype, but tissue-specific inactivation will allow researchers to address its function during limb bud morphogenesis.

2. Multimerisation of Shh can account for aspects of direct long-range activity. Multimerisation, possibly in lipid rafts, results in the Shh lipid modifications being trapped inside the multimer. Multimerised Shh is soluble, freely diffusible and it seems to form a gradient across the antero-posterior axis of the chick limb, which points to a role in mediating Shh activity.[49] Furthermore, palmitoylation is required for producing a soluble multimeric protein complex, suggesting that Shh signalling in distant cells is triggered by multimeric forms of the Shh ligand.[44]

3. Components of the extracellular matrix such as heparan sulfate proteoglycans (HSPG) play a role in modulating growth factor signals and in *Drosophila*, mutations in the GPI-anchored HSPG encoded by the *Dally* gene, and in enzymes required for HSPG biosynthesis such as *Toutvelu* affect Hh distribution and signalling.[50] These studies show that HSPG play a role in transferring the Hh ligand along the cell membrane. Analysis of the role of the vertebrate orthologues of these genes (*Glypican* genes for *Dally* and *Ext* genes for *Toutvelu* respectively) should further our understanding of how Shh transport and activity is regulated in embryos.

4. Finally, the distance over which Shh is able to diffuse can be restricted by ligand quenching, which involves a self-regulatory negative feedback loop mechanism: the Shh receptor *Ptc* and the membrane-bound glycoprotein *Hedgehog interacting protein* (*Hip*) genes are positively regulated by cells responding to Shh signalling in the developing limb bud.[51,52] Both Ptc and Hip function to down-regulate Shh signalling activity by binding the Shh ligand. Therefore, Shh sequestration at the cell surface by Hip and Ptc will limit its signalling range and activity.

Inside the Responding Cell: The Duality of the Gli Family of Transcriptional Activators and Repressors

In flies, the zinc finger transcription factor Ci is the only and essential mediator of Hh signalling and transcriptional regulation expression of Hh targets depends on Ci processing and its subcellular localisation. Full-length Ci functions as an activator of gene expression while a truncated form, CiR is a transcriptional repressor.[53] Full-length Ci forms a tetrameric protein complex together with the serine-threonine kinase Fused (*Fu*), the PEST domain-containing protein Suppressor of Fused (*Su(Fu)*) and the kinesin-like protein Costal-2 (Cos-2). In the absence of Hh, Cos-2 binds to microtubules and sequesters the protein

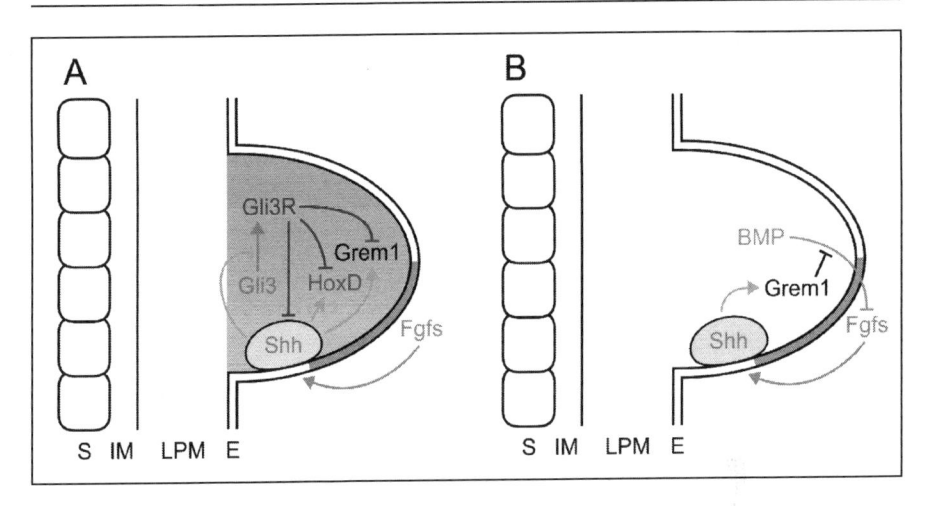

Figure 5. Shh signalling interactions during limb bud outgrowth. A) Antagonistic interaction between Shh and Gli3 signals is represented. Shh signals from the ZPA to block processing of full length Gli3 to Gli3R. In turn, Gli3R keeps *Shh* expression posteriorly restricted. Gli3R also restricts posterior factors such as *Grem1* and *HoxD* genes. Gradient of Gli3-Gli3R is represented: high Gli3R anterior in blue and high Gli3 posterior in red. B) Schematic representation of the Shh/Grem1/Fgf feedback loop. Shh up-regulates and maintains the expression of the BMP antagonist *Grem1* in the posterior mesenchyme. In turn, this enables the expression of *Fgf4*, *Fgf9* and *Fgf17* in the posterior AER. Fgf signalling by the AER is necessary to propagate Shh signalling in the posterior distal mesenchyme. Establishment and maintenance of this feedback loop controls distal limb bud outgrowth. A color version of this figure is available online at http://www.Eurekah.com.

complex containing full-length Ci in the cytoplasm. In turn, Ci is proteolytically cleaved into the CiR form, which lacks the nuclear export signal, its cytoplasmic anchoring and transcriptional activation domains. Thus CiR accumulates in the nucleus and thereby represses Hh transcriptional targets. Upon Hh stimulation, Smo dissociates the tetrameric complex from the microtubule by recruiting Cos-2, which results in blocking CiR formation and in the accumulation of full-length Ci in the nucleus.[54] In vertebrates, much less is known about the circuitry molecules acting downstream of Shh. Nevertheless, the role of three vertebrate *Ci* orthologues, *Gli1*, *Gli2* and *Gli3* has been extensively studied during limb bud development. All three *Gli* transcriptional regulators are expressed by the limb bud mesenchyme with the exception of the ZPA. *Gli1* is restricted to the posterior part of the limb bud mesenchyme and marks all Shh-responding cells, while *Gli2* and *Gli3* are expressed through the entire limb bud mesenchyme and partially overlap with *Gli1* expression.[55] It has been hypothesised that in vertebrates, the activator and repressor functions of Ci are performed by different Gli homologues. The best evidence to date stems from genetic studies of Gli family members in the neural tube and indicate that Gli1 acts solely as an activator while Gli2 and Gli3 can function both as repressors and activators.[56,57] However, genetic studies of the different Gli members have established that only *Gli3* loss-of-function mutations cause semi-dominant limb malformations in mice and humans,[37,58,59] while neither *Gli1* nor *Gli2*, or *Gli1;Gli2* mouse mutants display any significant limb phenotypes.[60] Furthermore, *Gli2;Gli3* double mutants display the same limb phenotype as *Gli3* single mutants.[61] Recent studies also indicate that digit identities in response to Shh signalling are not mediated by Gli1 and Gli2 transcriptional activator functions but rather by graded Gli3 repressor functions.[61] To date, Gli3 is the only Ci orthologue demonstrated to have an essential function during limb bud patterning as it regulates correct digit numbers and identities. Just like Ci in flies, full-length Gli3 is processed into the repressor form Gli3R, in chick limb buds (Fig. 5A).[62]

Levels of Gli3R decrease from anterior to posterior, most probably because the conversion of full-length Gli3 is antagonised by Shh signalling from the posterior mesenchyme (Fig. 5A).[62] Furthermore, limbs of mice overexpressing Shh in the entire limb bud mesenchyme exhibit polydactylies with loss of digit identities that phenocopy *Gli3* deficient limbs, in line with evidence that Shh signalling blocks Gli3R activity, a process that is crucial to limb bud patterning.[44] While there is no evidence for a role for full-length Gli3 in the limb bud, recent data show that direct interaction of Gli3R with Hoxd proteins converts Gli3R into a transcriptional activator of Shh targets.[63]

Maintenance of the ZPA by Signal Relay

Epithelial-mesenchymal interactions are essential for distal limb outgrowth and patterning (see "Introduction"). These interactions are mediated by Shh in the mesenchyme and FGFs in the AER, and possibly other AER growth factors. Removing the AER leads to rapid loss of *Shh* expression and subsequent loss of distal structures by apoptosis, while addition of Fgf-soaked beads following AER removal are sufficient to maintain expression and outgrowth. Conversely, anterior grafts of Shh-expressing cells cause expansion of the expression domains of posterior *Fgf* in the AER. This positive feedback loop between Shh in the mesenchyme and Fgf in the AER is essential for maintenance and propagation of the Shh organiser.[64,65] Furthermore, it ensures that the ZPA stays in close proximity to the most distal part of the limb bud where digits will form as limb bud morphogenesis progresses. In spite of its ability to act at long range, Shh signals to the AER by a signal relay mechanism. Both gain and loss-of-function studies have demonstrated that Grem1-mediated BMP antagonism is essential to establish positive feedback regulation.[66-70] *Grem1* is activated upstream of Shh and is essential to activate expression of *Fgf4*, *-9* and *-17* in the AER. Fgfs in the AER propagate *Shh* expression by the ZPA and in turn Shh propagates *Grem1* expression in the mesenchyme, most likely by blocking Gli3R production. A self-propagating *Shh/Grem1/Fgf* positive feedback loop is thus established to control progression of distal limb bud outgrowth and patterning (Fig. 5B).[66,70]

Digit Patterning by Shh: The End of the Spatial Gradient Model?

The data reviewed here points to the role of Shh as a classical morphogen instructing digit number and identity.
1. Ectopic Shh grafts cause mirror-image digit duplications.
2. Abrogation of a potential spatial gradient by Shh overexpression throughout the limb bud mesenchyme causes polydactyly with associated loss of digit identities.
3. The loss of digits 2 to 5 in limbs of *Shh* deficient embryos points to an essential role in digit formation.
4. Mutations altering Shh diffusion particularly affect more anterior digits.

However, the proposal that digit identity is simply specified by cells responding to a spatial gradient of diffusible Shh ligand needs to be reconsidered in light of recent studies (Fig. 6).[61,71] In mice, genetic marking of cells expressing *Shh* and their descendants in the limb bud, reveals that the descendants of Shh-expressing cells themselves give rise to all of the most posterior skeletal elements, digits 5 and 4, the ulna/fibula, and contribute significantly to digit 3. Interestingly, the descendants of Shh-expressing cells do not contribute to digit 2, in agreement with previous studies that digit 2 is formed in response to Shh long-range signalling.[43] The phenotypic analysis of *Shh* deficient limbs has previously shown that digit 1 is specified in a Shh-independant manner.[9] One of Wolpert's morphogen gradient predictions states that the cells expressing the morphogen would not contribute themselves significantly to the structures they pattern, and that cells responding to the highest levels of the morphogen would give rise to the most posterior skeletal elements. However, the levels of *Gli1* transcriptional activation in Shh-responding cells do not reflect the antero-posterior gradient as predicted.[61] Instead, the cells that have expressed *Shh* the longest form themselves into the most posterior digit 5, and descendants contribute to digits 4 and 3.

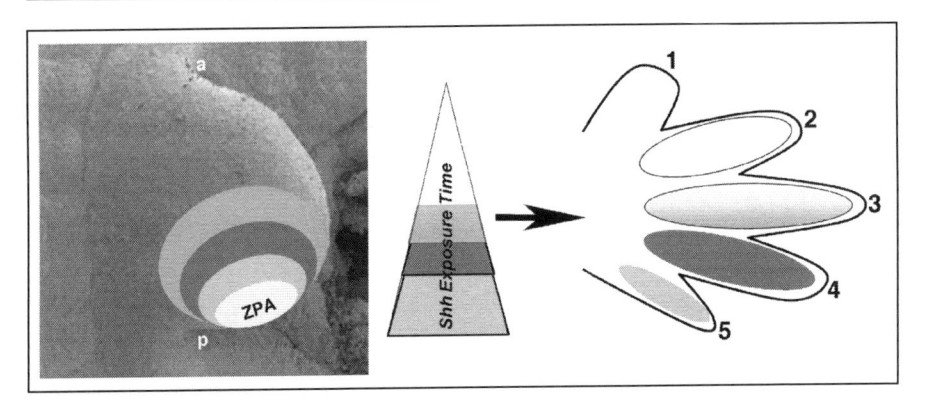

Figure 6. The temporal Shh gradient model. Schematic representation of the expansion-based temporal Shh gradient model. Descendant of Shh-expressing cells (green gradient) give rise to digits 5, 4 and part of digit 3. The identity of these three digits is specified by the length of time that the cells giving rise to these digits have been exposed to the Shh signal. Digit 5, the most posterior one, derives from cells that have been exposed to Shh signalling for a longer time in comparison to cells that will form digit 4 and digit 3. In contrast, specification of digit 2 depends entirely on Shh diffusion (long-range signal) while digit 1 is Shh-independent. The ZPA is indicated in yellow and anterior (a) is at the top while posterior (p) is at the bottom. A color version of this figure is available online at http://www.Eurekah.com.

During the stages when the handplate becomes morphologically distinct and differentiation proceeds (in mouse, E12 onwards, Fig. 1A), *Shh* becomes gradually down-regulated, *Fgf* expression ceases and *Grem1* becomes restricted to the interdigit area. This points to a breakdown of the *Shh/Grem/Fgf* feedback loop and recent work in chick has investigated the underlying mechanism resulting in its termination and the relevance for limb bud development.[72] It appears that cells expressing *Shh* and their descendants cannot express *Grem1*. As limb bud outgrowth proceeds, this cell population becomes larger and the gap between the source of Shh signal and cells competent to respond to it by expressing *Grem1*, widens over time. The authors provide experimental evidence in support of the idea that the gap becomes too wide for Shh signalling to continue controlling *Grem1* expression. As a consequence, loss of *Grem1* expression in the mesenchyme causes down-regulation of *Fgf* transcription in the AER and thereby shuts down the feedback loop. Experimental maintenance of the feedback loop for longer than normal results in formation of an extra phalange, which provides evidence that timely shutdown of the feedback loop is necessary for proper limb bud morphogenesis.[72] However, it remains unclear why Shh descendant cells cannot express *Grem1* and whether long range Shh signalling normally regulates *Grem1* expression levels. Further understanding of the mechanism that terminates the feedback loop is required to gain insight into the way organs self-regulate their final size and shape.

Ten years after its discovery as the signal produced by the ZPA, Shh continues to fascinate. While currents efforts in the field are focusing on understanding the mechanisms establishing the antero-posterior prepattern prior to Shh and Gli signalling, more research is needed to understand all the roles attributed to Shh. Much insight has been gained from investigating the role of Gli proteins as the downstream effectors of Shh signalling. However, less is known about the role of many of the target genes, and understanding the effects of these target genes (belonging to all the major signalling pathways) will shed light on the mechanisms coordinating differentiation of the various cell types contributing to the vertebrate limb.

Acknowledgements

We thank R. Zeller for helpful discussions and suggestions, our group members and L. Panman for their critical input for this manuscript. We are grateful to A. Roulier from the Art Department at the Pharmazentrum in Basel for making Figures 3, 4 and 5. We also thank J. Brown for providing the Shh graft shown in Figure 2A and O. Michos for the scanning EM of the limb bud shown in Figure 2B and Figure 6. The authors are also grateful to the publishers Elsevier and UBC Press as well as to P. Martin for giving their permission to use data already published.

References

1. Panman L, Zeller R. Patterning the limb before and after SHH signalling. J Anat 2003; 202(1):3-12.
2. Saunders JWJ. The proximo-distal sequence of origin of limb parts of the chick wing and the role of the ectoderm. J Exp Zoology 1948; (108):363-404.
3. Summerbell D. A quantitative analysis of the effect of excision of the AER from the chick limb-bud. J Embryol Exp Morphol 1974; 32(3):651-660.
4. Niswander L, Tickle C, Vogel A et al. FGF-4 replaces the apical ectodermal ridge and directs outgrowth and patterning of the limb. Cell 1993; 75(3):579-587.
5. Vogel A, Tickle C. FGF-4 maintains polarizing activity of posterior limb bud cells in vivo and in vitro. Development 1993; 119(1):199-206.
6. Spemann HaMH. Induction of embryonic primordia by implantation of organizers from a different species. Int J Dev Biol 2001; (reprinted)45:13-38.
7. Wolpert L. Positional information and the spatial pattern of cellular differentiation. J Theor Biol 1969; 25:1-47.
8. Riddle RD, Johnson RL, Laufer E et al. Sonic hedgehog mediates the polarizing activity of the ZPA. Cell 1993; 75:1401-1416.
9. Kraus P, Fraidenraich D, Loomis CA. Some distal limb structures develop in mice lacking Sonic hedgehog signaling. Mech Dev 2001; 100(1):45-58.
10. Chiang C, Litingtung Y, Lee E et al. Cyclopia and defective axial patterning in mice lacking Sonic hedgehog gene function. Nature 1996; 383:407-413.
11. Chiang C, Litingtung Y, Harris MP et al. Manifestation of the limb prepattern: Limb development in the absence of sonic hedgehog function. Dev Biol 2001; 236:421-435.
12. Martin GR. The roles of FGFs in the early development of vertebrate limbs. Genes Dev 1998; 12:1571-1586.
13. Tickle C, Munsterberg A. Vertebrate limb development—the early stages in chick and mouse. Curr Opin Genet Dev 2001; 11(4):476-481.
14. Kawakami Y, Capdevila J, Buscher D et al. WNT signals control FGF-dependent limb initiation and AER induction in the chick embryo. Cell 2001; 104(6):891-900.
15. Kengaku M, Capdevila J, Rodriguez-Esteban C et al. Distinct WNT pathways regulating AER formation and dorsoventral polarity in the chick limb bud. Science 1998; 280(5367):1274-1277.
16. Barrow JR, Thomas KR, Boussadia-Zahui O et al. Ectodermal Wnt3/beta-catenin signaling is required for the establishment and maintenance of the apical ectodermal ridge. Genes Dev 2003; 17(3):394-409.
17. Isaac A, Cohn MJ, Ashby P et al. FGF and genes encoding transcription factors in early limb specification. Mech Dev 2000; 93:41-48.
18. Lewandoski M, Sun X, Martin GR. Fgf8 signalling from the AER is essential for normal limb development. Nat Genet 2000; 26(4):460-463.
19. Moon AM, Capecchi MR. Fgf8 is required for outgrowth and patterning of the limbs. Nat Genet 2000; 26(4):455-459.
20. Charite J, de Graaff W, Shen S et al. Ectopic expression of Hoxb-8 causes duplication of the ZPA in the forelimb and homeotic transformation of axial structures. Cell 1994; 78(4):589-601.
21. van den Akker E, Reijnen M, Korving J et al. Targeted inactivation of Hoxb8 affects survival of a spinal ganglion and causes aberrant limb reflexes. Mech Dev 1999; 89(1-2):103-114.
22. van den Akker E, Fromental-Ramain C, de Graaff W et al. Axial skeletal patterning in mice lacking all paralogous group 8 Hox genes. Development 2001; 128(10):1911-1921.
23. Zakany J, Kmita M, Duboule D. A dual role for Hox genes in limb anterior-posterior asymmetry. Science 2004; 304(5677):1669-1672.
24. Tickle C, Alberts BM, Wolpert L et al. Local application of retinoic acid in the limb bud mimics the action of the polarizing region. Nature 1982; 296:564-565.

25. Helms JA, Kim CH, Eichele G et al. Retinoic acid signaling is required during early chick limb development. Development 1996; 122(5):1385-1394.
26. Stratford T, Horton C, Maden M. Retinoic acid is required for the initiation of outgrowth in the chick limb bud. Curr Biol 1996; 6(9):1124-1133.
27. Niederreither K, Vermot J, Schuhbaur B et al. Embryonic retinoic acid synthesis is required for forelimb growth and anteroposterior patterning in the mouse. Development 2002; 129(15):3563-3574.
28. Mic FA, Sirbu IO, Duester G. Retinoic acid synthesis controlled by Raldh2 is required early for limb bud initiation and then later as a proximodistal signal during apical ectodermal ridge formation. J Biol Chem 2004; 279(25):26698-26706.
29. Fernandez-Teran M, Piedra ME, Kathiriya IS et al. Role of dHAND in the anterior-posterior polarization of the limb bud: Implications for the Sonic hedgehog pathway. Development 2000; 127:2133-2142.
30. Charite J, McFadden DG, Olson EN. The bHLH transcription factor dHAND controls Sonic hedgehog expression and establishment of the zone of polarizing activity during limb development. Development 2000; 127(11):2461-2470.
31. Yelon D, Baruch T, Halpern ME et al. The bHLH transcription factor Hand2 plays parallel roles in zebrafish heart and pectoral fin development. Development 2000; 127:2573-2582.
32. te Welscher P, Fernandez-Teran M, Ros MA et al. Mutual genetic antagonism involving GLI3 and dHAND prepatterns the vertebrate limb bud mesenchyme prior to SHH signaling. Genes Dev 2002; 16(4):421-426.
33. Zuniga A, Zeller R. Gli3 (Xt) and formin (ld) participate in the positioning of the polarising region and control of posterior limb-bud identity. Development 1999; 126(1):13-21.
34. Zuniga A, Michos O, Spitz F et al. Mouse limb deformity mutations disrupt a global control region within the large regulatory landscape required for Gremlin expression. Genes Dev 2004; 18(13):1553-1564.
35. Srivastava D, Thomas T, Lin Q et al. Regulation of cardiac mesodermal and neural crest development by the bHLH transcription factor dHAND. Nat Genet 1997; 16:154-160.
36. Schimmang T, Lemaistre M, Vortkamp A et al. Expression of the zinc finger gene Gli3 is affected in the morphogenetic mouse mutant extra-toes (Xt). Development 1992; 116:799-804.
37. Hui C, Joyner A. A mouse model of greig cephalopolysyndactyly syndrome: The extra-toesJ mutation contains an intragenic deletion of the Gli3 gene. Nat Genet 1993; 3(3):241-246.
38. Theil T, Kaesler S, Grotewold L et al. Gli genes and limb development. Cell Tissue Res 1999; 296(1):75-83.
39. te Welscher P, Zuniga A, Kuijper S et al. Progression of vertebrate limb development through shh-mediated counteraction of GLI3. Science 2002; 298:827-830.
40. Litingtung Y, Dahn RD, Li Y et al. Shh and Gli3 are dispensable for limb skeleton formation but regulate digit number and identity. Nature 2002; 418(6901):979-983.
41. Cohen Jr MM. The hedgehog signaling network. Am J Med Genet A 2003; 123(1):5-28.
42. Lee JJ, Ekker SC, von Kessler DP et al. Autoproteolysis in hedgehog protein biogenesis. Science 1994; 266(5190):1528-1537.
43. Lewis PM, Dunn MP, McMahon JA et al. Cholesterol modification of sonic hedgehog is required for long-range signaling activity and effective modulation of signaling by Ptc1. Cell 2001; 105(5):599-612.
44. Chen MH, Li YJ, Kawakami T et al. Palmitoylation is required for the production of a soluble multimeric Hedgehog protein complex and long-range signaling in vertebrates. Genes Dev 2004; 18(6):641-659.
45. Burke R, Nellen D, Bellotto M et al. Dispatched, a novel sterol sensing domain protein dedicated to the release of cholesterol-modified hedgehog from signaling cells. Cell 1999; 99:803-815.
46. Ma Y, Erkner A, Gong R et al. Hedgehog-mediated patterning of the mammalian embryo requires transporter-like function of dispatched. Cell 2002; 111(1):63-75.
47. Caspary T, Garcia-Garcia MJ, Huangfu D et al. Mouse dispatched homolog1 is required for long-range, but not juxtacrine, Hh signaling. Curr Biol 2002; 12(18):1628-1632.
48. Kawakami T, Kawcak T, Li YJ et al. Mouse dispatched mutants fail to distribute hedgehog proteins and are defective in hedgehog signaling. Development 2002; 129(24):5753-5765.
49. Zeng X, Goetz JA, Suber LM et al. A freely diffusible form of Sonic hedgehog mediates long-range signalling. Nature 2001; 411(6838):716-720.
50. Lin X. Functions of heparan sulfate proteoglycans in cell signaling during development. Development 2004; 131(24):6009-6021.
51. Goodrich LV, Milenkovic L, Higgins KM et al. Altered neural cell fates and medulloblastoma in mouse patched mutants. Science 1997; 277(5329):1109-1113.

52. Chuang P-T, McMahon AP. Vertebrate hedgehog signalling modulated by induction of a hedgehog-binding protein. Nature 1999; 987:617-621.
53. Ingham PW, McMahon AP. Hedgehog signaling in animal development: Paradigms and principles. Genes Dev 2001; 15(23):3059-3087.
54. Lum L, Beachy PA. The Hedgehog response network: Sensors, switches, and routers. Science 2004; 304(5678):1755-1759.
55. Buscher D, Ruther U. Expression profile of Gli family members and Shh in normal and mutant mouse limb development. Dev Dyn 1998; 211(1):88-96.
56. Lee J, Platt KA, Censullo P et al. Gli1 is a target of Sonic hedgehog that induces ventral neural tube development. Development 1997; 124(13):2537-2552.
57. Bai CB, Stephen D, Joyner AL. All mouse ventral spinal cord patterning by hedgehog is Gli dependent and involves an activator function of Gli3. Dev Cell 2004; 6(1):103-115.
58. Johnson DR. Extra-toes: A new mutant gene causing multiple abnormalities in the mouse. J Embryol Exp Morph 1967; 17(3):543-581.
59. Vortkamp A, Gessler M, Le Paslier D et al. Isolation of a yeast artificial chromosome contig spanning the Greig cephalopolysyndactyly syndrome (GCPS) gene region. Genomics 1994; 22(3):563-568.
60. Park HL, Bai C, Platt KA et al. Mouse Gli1 mutants are viable but have defects in SHH signaling in combination with a Gli2 mutation. Development 2000; 127:1593-1605.
61. Ahn S, Joyner AL. Dynamic changes in the response of cells to positive hedgehog signaling during mouse limb patterning. Cell 2004; 118(4):505-516.
62. Wang B, Fallon JF, Beachy PA. Hedgehog-Regulated processing of Gli3 produces an anterior/posterior repressor gradient in the developing vertebrate limb. Cell 2000; 100:423-434.
63. Chen Y, Knezevic V, Ervin V et al. Direct interaction with Hoxd proteins reverses Gli3-repressor function to promote digit formation downstream of Shh. Development 2004; 131(10):2339-2347.
64. Niswander L, Tickle C, Vogel A et al. Function of FGF-4 in limb development. Mol Reprod Dev 1994; 39(1):83-88.
65. Laufer E, Nelson CE, Johnson RL et al. Sonic hedgehog and Fgf-4 act through a signalling cascade and feedback loop to integrate growth and patterning of the development limb bud. Cell 1994; 79:993-1003.
66. Zuniga A, Haramis AP, McMahon AP et al. Signal relay by BMP antagonism controls the SHH/FGF4 feedback loop in vertebrate limb buds. Nature 1999; 401(6753):598-602.
67. Merino R, Rodriguez-Leon J, Macias D et al. The BMP antagonist Gremlin regulates outgrowth, chondrogenesis and programmed cell death in the developing limb. Development 1999; 126(23):5515-5522.
68. Capdevila J, Tsukui T, Rodriquez Esteban C et al. Control of vertebrate limb outgrowth by the proximal factor Meis2 and distal antagonism of BMPs by Gremlin. Mol Cell 1999; 4(5):839-849.
69. Khokha MK, Hsu D, Brunet LJ et al. Gremlin is the BMP antagonist required for maintenance of Shh and Fgf signals during limb patterning. Nat Genet 2003; 34(3):303-307.
70. Michos O, Panman L, Vintersten K et al. Gremlin-mediated BMP antagonism induces the epithelial-mesenchymal feedback signaling controlling metanephric kidney and limb organogenesis. Development 2004; 131(14):3401-3410.
71. Harfe BD, Scherz PJ, Nissim S et al. Evidence for an expansion-based temporal Shh gradient in specifying vertebrate digit identities. Cell 2004; 118(4):517-528.
72. Scherz PJ, Harfe BD, McMahon AP et al. The limb bud Shh-Fgf feedback loop is terminated by expansion of former ZPA cells. Science 2004; 305(5682):396-399.
73. Martin P. Tissue patterning in the developing mouse limb. Int J Dev Biol 1990; 34:323-336.

CHAPTER 9

Sonic Hedgehog Signalling in the Developing and Regenerating Fins of Zebrafish

Fabien Avaron, Amanda Smith and Marie-Andrée Akimenko*

Abstract

Zebrafish is now a well established model for the study of developmental and regenerative processes. Indeed, the genetic cascades that control the early development of the structure that will form the paired fins (the fin bud) present similarities with the early formation of the tetrapod fore and hindlimb buds. One of these conserved molecular pathways involves secreted factors of the Hedgehog family [sonic hedgehog (shh) and tiggywinkle hedgehog (twhh)]. As in the tetrapod limbs, hedgehog proteins are initially expressed in the posterior region of the early fin bud where they contribute to the patterning of the antero-posterior axis, then are involved in cell proliferation and the formation of various skeletal elements. The hedgehog pathway is reactivated in adult fish following fin amputation, an event that triggers the regeneration program. During this process, the hedgehog signal is involved in various processes such as the growth and maintenance of the blastema and patterning of the fin ray.

The Zebrafish Hedgehog Genes

Vertebrate *Hedgehog* (*Hh*) genes are classified into three classes: *sonic* (*shh*), *indian* (*ihh*[a]) and *desert* (*dhh**) *hedgehog* class. Most vertebrate species possess one member from each gene family. However, the teleost *danio rerio* (zebrafish) possesses at least five *hedgehog* genes: two *sonic*-class genes: *shh*[1] and *tiggywinkle* (*twhh*),[2] two *Indian*-class genes: *echidna* (*ehh*)[3] and *ihh*[4] and one *desert*-class gene (Fig. 1).[4] Despite the high number of *Hh* genes, we will see that only *shh* and *twhh* are expressed during fin bud development and only *shh* seems to be required for their proper development. During fin ray regeneration, both *shh* and the newly identified *ihh* are expressed, and functional data indicates that the Hh signalling pathway is involved in blastema formation and maintenance, and later in fin ray patterning.

Overview of the Zebrafish Pectoral Fin Bud Development

Zebrafish possesses five sets of fins divided into two types (Fig. 2A,B): the paired fins (pectoral and pelvic) and the median fins (dorsal, caudal and anal). The development of the two types of fins are somehow different: the median fins develop directly from the epidermal fold

---

[a] in *Xenopus laevis*, *ihh* and *dhh* have respectively been named *banded hedgehog* (*bhh*) and *cephalic hedgehog* (*chh*).[2]

*Corresponding Author: Marie-Andrée Akimenko—Department of Medicine and Cellular and Molecular Medicine, Ottawa Health Research Institute, University of Ottawa, 725 Parkdale Avenue, Ottawa ON K1Y4E9, Canada. Email: makimenko@ohri.ca

Shh and Gli Signalling and Development, edited by Carolyn E. Fisher and Sarah E.M. Howie. ©2006 Landes Bioscience and Springer Science+Business Media.

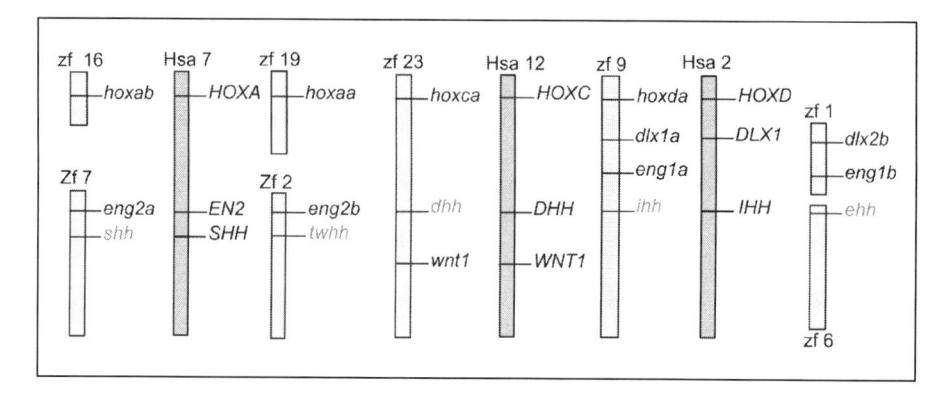

Figure 1. Chromosomal location of human (Hsa) and zebrafish (zf) *Hedgehog* genes.

surrounding the caudal half of the young larvae (the median fin fold[5]), whereas the paired fins first arise from a local proliferation of the lateral plate mesoderm to form the fin bud.[6] However, the visible part of the pectoral fins which contains the exoskeleton (the fin rays) eventually develops inside of an epidermal fin fold in a process that resembles the development of the median fins from the median fin fold.[5] Early fish fin buds and tetrapod limb buds show striking morphological resemblances and they both contain equivalent signalling centers: The ZPA (zone of polarizing activity) in the posterior mesenchyme,[1,7,8] the apical ectoderm[6,9,10] (equivalent to the apical ectodermal ridge, or AER, in tetrapods), and the ventral ectoderm,[11,12] which are responsible for the specification of the antero-posterior, proximo-distal and dorso-ventral axes, respectively. However, the AER of the tetrapod limb progressively degenerates during development, whereas the zebrafish apical ectoderm will form an elongated fin fold in which the external part of the fin, including the fin rays will eventually develop.[6] The divergence of the outcome of the apical epidermis between tetrapod limb and larval fin is thought to be a major component of the initial morphological differences between the two types of appendages.

One of the molecular pathways involved in both limb and fin development is the hedgehog (*Hh*) signalling pathway. This pathway has been extensively studied in zebrafish, in particular in the pectoral fin bud which constitutes a practical and accessible model for developmental and functional studies of early limb development.

Fin Bud and Early Larval Fin

The early pectoral fin buds arise by 24 hours-post-fertilization (hpf) from the limb fields which consist of a pair of small aggregates of mesenchymal cells located on each side of the main body axis at the level of the third somite. As the fin buds grow, the first skeletal elements start to condensate by 37hpf in the center of the fin bud, and will give rise few hours later to the cartilaginous endochondral disk.[6] This chondrogenic condensation divides the mesenchymal cell population into a ventral and a dorsal half which will give rise to the muscles of the fin. The proximal part of the chondrogenic condensation will differentiate into the larval endoskeletal girdle and the distal part will develop as the endochondral disc which will give rise to the fin endoskeleton. At 28hpf, the bud is covered by a two-layered epidermis composed of one basal stratum and one flat peridermal cell layer. At about 31hpf, the apical epidermal cells lining the anterior-posterior axis of the bud thicken to form a transient ridge which is similar to the AER of the tetrapod limbs. By 34hpf, the apical epidermal cells undergo a morphological change, detach from the underlying mesenchyme and progressively form an epidermal fold separated by a subepidermal space.[1,6] At 48hpf, this epidermal fold starts to elongate and mesenchymal cells start to invade the structure. The actinotrichia, collagenous fibers, are the first supportive elements to form within the larval fin fold, as no fin rays have yet appeared at that stage.

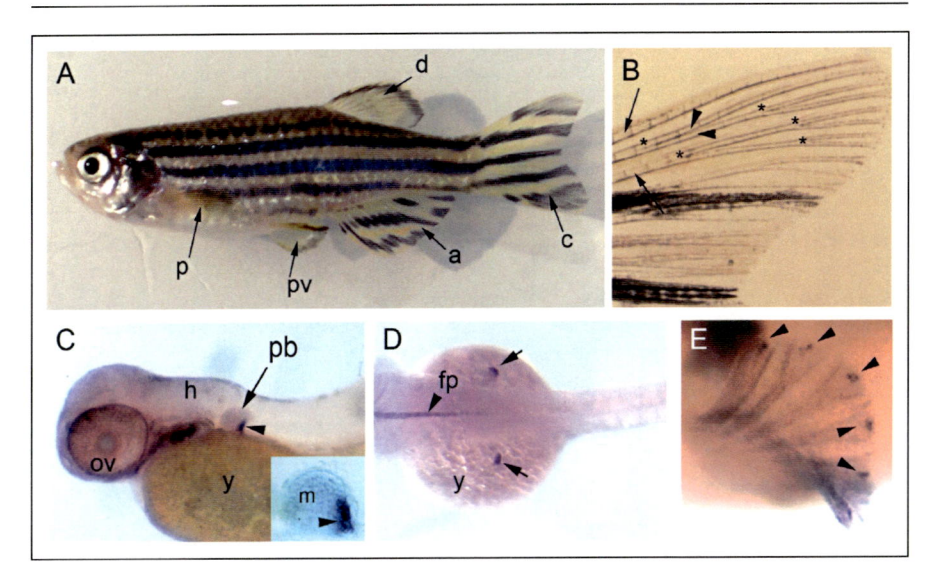

Figure 2. Zebrafish fins and *shh* expression during embryonic and larval development. A) lateral view of an adult zebrafish showing the five sets of fins (p: pectoral, pv: pelvic, d: dorsal, a: anal, c: caudal). B) dorsal part of the caudal fin showing the rays separated by interray tissue. Each ray or lepidotrichia (arrows) is made of two hemirays, each composed of a series of concave segments joined to each other by ligaments. These rays regularly bifurcate, except for the outermost. *: bifurcation point; arrowheads: sister ray branches. C-D) Expression of *shh* detected by in situ hybridization using an antisense RNA probe. (C, lateral view and D, dorsal view.) At 72hpf, *shh* is strongly expressed in the posterior mesenchyme of the fin bud (arrowheads in C,D). Note: *shh* expression in the floor plate is visible in (D). E) pectoral fin of a 4 week-old larva showing *shh* expression at the distal tip of each developing fin ray. fp: floor plate, h: hindbrain, m: fin mesenchyme, pb: pectoral fin bud, y: yolk sac, ov: optic vesicle.

Adult Fin Formation

During the third week of development, the larval fin undergoes massive rearrangement of the endoskeleton and musculature, as it switches from larval to adult shape. Following the rotation of the larval fin, drastic remodeling of the endoskeleton occurs. Then, specialized cells from the distal mesenchyme, the scleroblasts, intercalate between the actinotrichia and the basement membrane and start to secrete the bone matrix in a proximal to distal fashion, forming the fin rays or lepidotrichia. This type of bone (called dermal or intramembranous) is directly mineralized in the subepidermal space, unlike the endochondral bone which is formed through a cartilaginous precursor. Fin ray structure and morphogenesis are further described below (Figs. 2B, 5A).

Shh and Twhh Expression during Fin Bud Development

The first sign of *shh* expression appears by 26-28hpf, in very few cells of the posterior region of the early pectoral fin bud.[1,7] By 30hpf, *shh* expression intensifies and is now clearly restricted to the posterior margin of the fin bud, in a position which is considered to be analogous to the zone of polarizing activity (ZPA) of the tetrapod limb bud (see next paragraph). By 48hpf, *shh* is expressed in a wider domain but limited to the posterior part of the fin bud. This expression is maintained until approximately 3 day-post-fertilization (dpf) (Fig. 2C,D), and is then progressively downregulated until *shh* transcripts become undetectable by in situ hybridization by 4dpf.[13-15] *twhh* and *shh* have very similar expression patterns in structures of the embryonic axis such as the notochord, the floor plate and the branchial arches.[2,16] However, *twhh* expression is not detected as early as *shh* in the fin bud but starts at

48hpf in a group of cells corresponding to *shh*-expressing cells.[6,17] During morphogenesis of the rays of all fin types which initiates around the fourth week post-fertilization, *shh* is reexpressed in the basal epidermal layer (BEL) at the tip of each forming lepidotrichia (Fig. 2E), in a pattern reminiscent to that observed during fin regeneration (see below).[15]

Shh, Retinoic Acid Regulation and the ZPA

The ZPA is a signalling center located in the posterior mesenchyme of the tetrapod limb and is responsible for the patterning of the anteroposterior (A/P) axis of the limb.[18] Shh and retinoic acid are two signalling molecules involved in this process.[1,7,8,19-22] In chick embryos, grafting experiments of cells of the ZPA as well as ectopic expression of shh or local application of retinoic acid (RA) at the anterior margin of the limb bud result in the formation of a mirror-image duplication of the digits,[8,22,23] thus connecting the polarizing activity of this region to the role of shh and RA. In zebrafish, retinoic acid treatment of embryos at 24-30hpf causes the formation of an ectopic *shh* expression domain in the anterior region of the pectoral fin bud between 2-3dpf,[7,14] and later (by 4dpf), signs of a duplication of the fin bud have been observed, reminiscent of the digit duplication observed in chick.[13] It is interesting to note that although RA was delivered to whole zebrafish embryos in these experiments, while locally applied to the chick limb buds,[22] both treatments lead to a similar ectopic expression of *shh* and limb/fin structure duplication. This suggests that very few cells have the potential to form a polarizing zone. Hoffman et al observed that a two hour RA treatment of 30hpf zebrafish embryos causes a transient downregulation of *shh*, followed by the reappearance, a few hours later, of the posterior *shh* domain.[14] This domain progressively extends towards the anterior region of the bud. Once *shh* is activated in cells at the anterior margin, its expression is down-regulated in cells of the center of the bud, therefore leaving, 24 hours after the end of the treatment, two discrete anterior and posterior *shh* domains in the bud. Similar RA treatments of mutants of the *shh* gene (*syu*, see next paragraph) lead to the same result indicating that this anterior ectopic expression of *shh* is independent of shh signalling but could depend on factors secreted by the AER. *hoxd-11* and *hoxd-12* which are normally expressed in the posterior mesenchyme of the fin bud, present an anterior extension of their expression domain after early RA treatment (at 5hpf) of wild-type embryos, but not of *syu* mutant embryos.[24] Thus, the anterior expansion of these genes by RA is independent of shh signalling suggesting that *hox* genes could represent of the intermediate factors between *shh* and RA.[14,24] Surprisingly, a DR5-type retinoic acid receptor binding sequence (or retinoic acid response element, RARE) has been identified in the promoter region of the *shh* gene in zebrafish only.[25] This element is functional in vitro, and could directly link retinoic acid to shh. However, no mutation experiment has brought clear evidence about the activity of this element in vivo, and further investigation will clarify its potential function during zebrafish fin bud development.

Mutants of the Hh Pathway and Fin Bud Development

Shh function in zebrafish has been studied using a group of mutants presenting somite formation defects (the *you*-type mutants, see Table 1), including the *syu* mutant in which the *shh* gene is disrupted (Table 2).[26] All the mutated genes of the *you*-type family identified so far are involved in the Hh signalling pathway. Interestingly, mutation of individual genes of the Hh pathway in zebrafish leads to relatively mild and variable phenotypes that are not lethal before several days of development. Two mutants, *syu* and *smu* present defects of fin bud development[26-28] providing a valuable tool to study the function of the Hh pathway in the development and morphogenesis of the zebrafish fins.

The most obvious phenotypes of the *syu* (and *smu*) mutants are defects of the embryonic axis: the embryos show a strong body curvature, U-shaped somites, underdeveloped eyes and jaw-related structures. In addition to these defects, the *syu* embryos present a wide range of allele-dependent alterations of pectoral fin development.[26,29] The two weak alleles (*syu^{tq252}*, *syu$^{tq b70}$*) provoke moderate and variable reduction of the fin fold and the fin endoskeleton,

Table 1. Zebrafish you-type genes and mutants

Mutant Name	Early Fin Defects*	Mutated Gene	Reference
You-too (yot)	N	gli2	57
Sonic-you (syu)	Y	shh	26
Chameleon (con)	N	dispatched-1	58
u-boot (ubo)	N**	prdm1	59
Slow-muscle-omitted (smu)	Y	smoothened	60
Iguana (igu)	N	dzip1	61
detour (dtr)	N	gli1	62
you	?	?	27

*N: no published data.; **ubo mutant presents later fin degeneration defect.[27]

Table 2. Alleles of the zebrafish syu mutants

Allele Name	Mutation Type	Mutation Location	Relative Strength	Shh Expression
tq252	Substitution (ENU)	promoter	weak	reduced
Tqb70	Substitution (ENU)	unknown	weak	reduced
tbx392	Substitution (ENU)	splice donor junction	strong	almost no expression
T4	spontaneous	7.5 kb deletion encompassing shh coding sequence	strong	no expression

whereas the strong allele syu^{tbx392} causes a drastic reduction of both the fin bud and fin fold. Embryos homozygous for the deletion allele syu^{t4} initiate fin bud development, but fin growth is not sustained and the mutant completely lack pectoral fins (Table 2 and Fig. 3). At the molecular and cellular levels the disruption of shh activity in the early fin bud has three major consequences: disorganization of the A/P patterning, failure to develop and maintain a proper distal epidermis and a decrease in cell proliferation. The expression of the posterior *hox* genes, which are involved in the A/P patterning, is perturbed in *syu* embryos. For example, *hoxd-13* expression seems to be completely dependent on shh activity as this gene fails to be expressed in *syu* mutants. In contrast, *hoxd-11* and *hoxd-12* transcription initiates in a *shh*-independent fashion but requires the shh signal to be maintained in the posterior region of the fin bud and *hoxd-10* expression seems totally independent of shh signalling.[24] This raises the question of the factors on which these genes rely to initiate their expression.[b] Twhh is unlikely to be one of these factors, as its expression is activated later than shh.[17] RA has been shown to induce *shh* expression, even in the absence of the shh signal, showing that it is an important factor for the early specification of the fin bud and could be one of these molecules.

Another phenotype of the *syu* mutant fin is the shortening or the absence of the pectoral fin fold (Fig. 3, left panel), and in the early syu^{t4} the total absence of a normal apical ectoderm (AE). Phenotypically, the early AE forms normally in *syu* mutants carrying the moderate or the

---

[b] This point is further described in another chapter of this book.

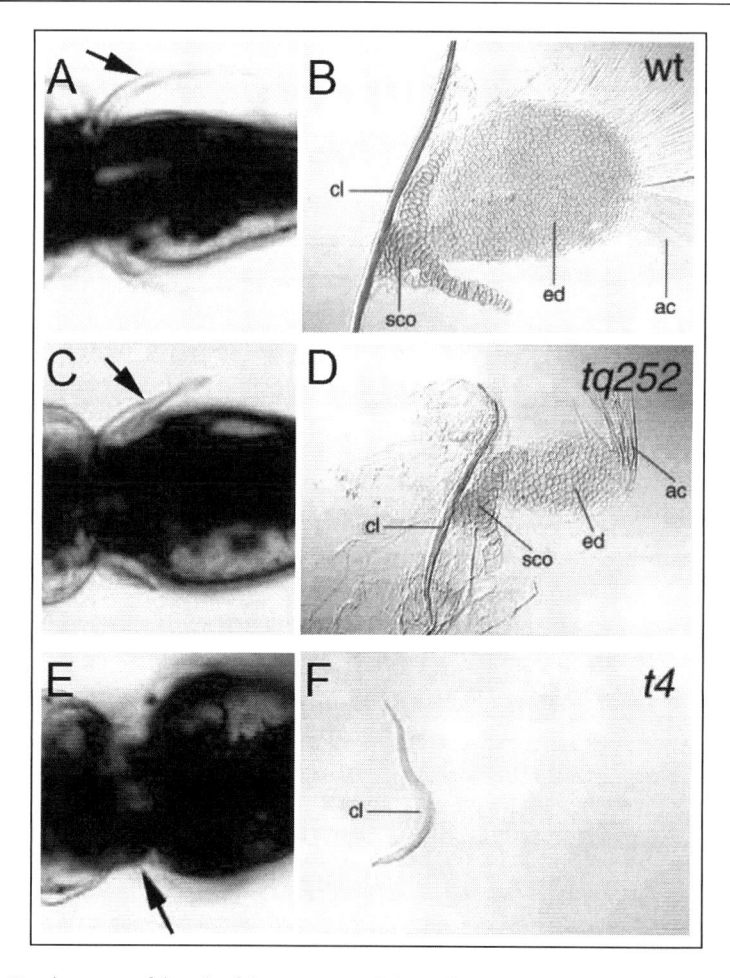

Figure 3. Fin phenotype of the zebrafish *syu* mutant alleles. Left panel: fin phenotype of the *syu* alleles. A) At 48hpf, pectoral fins of wild type larvae are developing and present an elongated fin fold clearly visible from a dorsal view (arrow). C) The hypomorphic allele *syu*^{tq252} causes the reduction of the fin and (E) the deletion allele *syu*^{t4} a complete lack of the pectoral fin buds. Right panels: B) fin skeleton of a wild type larva at 6dpf after alcian blue staining of cartilage elements. D) the hypomorphic allele *syu*^{tq252} causes an overall reduction of all skeletal elements while (F) the *syu*^{t4} mutants are lacking most of fin skeleton except part of the embryonic pectoral girdle (cl). cl: cleithrum, sco: scapulocoracoid, ed: endochondral disk, ac: actinotrichia. Reprinted with permission from Schauerte HE, van Eeden FJ, Fricke C et al. Sonic hedgehog is not required for the induction of medial floor plate cells in the zebrafish. Development 1998; 125(15):2983-2993. © The Company of Biologists Limited 1998, and Neumann CJ, Grandel H, Gaffield W et al. Transient establishment of anteroposterior polarity in the zebrafish pectoral fin bud in the absence of sonic hedgehog activity. Development 1999; 126(21):4817-4826 © The Company of Biologists Limited 1999.

weak allele, but degenerates a few hours later. Expression of the transcription factor dlx2 or the secreted factor fgf2 normally initiates during the early formation of the AE in *syu* mutants, but it is quickly downregulated at the stage corresponding to fin fold elongation in wild type embryos.[24] These observations show that shh activity is not necessary to initiate the expression of distal markers or to specify the distal epidermis, but it is required to maintain the AE integrity and the expression of the distal markers.[24] For instance, fgf8, a late marker of the AE, whose expression depends on fgf2 expression, completely fails to be expressed in the *syu* mutants.[24,30] In addition to the problems caused to the A/P patterning and the formation of the distal epidermis, *shh* disruption causes a decrease in cell proliferation throughout the whole fin bud that leads to defects in cartilage condensation and bone formation (Fig. 3, right panel). The decrease in cell proliferation is detected before the formation of the AE, suggesting that the proliferation problems in *syu* mutants are initially not due to the absence of factors secreted by the distal epidermis. However, the subsequent downregulation of the factors expressed in the AE is likely to aggravate the proliferation defect.

Fin Ray Morphogenesis and Regeneration

The fin rays or lepidotrichia are the skeletal elements of the external part of the adult fins (Fig. 2B) which develop relatively late, during the fourth week of life.[15,31] The base of the lepidotrichia is attached to the fin endoskeleton via muscles and ligaments. The lepidotrichia are composed of two hemirays, shaped like parenthesis and facing each other (Fig. 4A). They are segmented and periodically bifurcate along the proximodistal axis. Each segment is attached to the next one by a collagenous ligament, forming a joint that gives flexibility to the fin ray. Blood vessels, nerves, pigment cells and connective tissue are located between the two hemirays and also in the inter-ray region.

The lepidotrichia are formed in two steps: first, the bone matrix is secreted in the subepidermal space by specialized cells, the scleroblasts, adjacent to the basal epidermal cell layer. Then, this matrix is mineralized and forms a bone devoid of cells. This type of bone is called dermal, or intramembranous, and contrarily to endochondral bone, no cartilage precursor precedes its formation. The matrix secretion and mineralization follow the proximal to distal progression of the fin growth.

Although zebrafish fin constitutes a good model for fin ray morphogenesis analysis during larval development, it has been more studied during another process, regeneration. Teleost fish, like zebrafish, possess the ability to regenerate their fins, and the ablation of any part of the fin distal to the first segment will trigger a regeneration program that will give rise to a new structure identical to the amputated one.

Regeneration in zebrafish is epimorphic, which means it involves cell proliferation and creation of a regeneration-specific structure, the blastema. In many aspects, the regeneration process is reminiscent of the development, and most of the genes expressed during embryonic or larval development are reexpressed during regeneration.

The regeneration process can be divided into three main steps (Fig. 4B-D):[32-35]

1. Wound healing (0-24hpa[c]): Within the 6 hours post amputation (hpa), an epithelial layer completely covers the wound, followed in the next hours by several additional layers of epidermal tissue. This forms the apical epidermal cap (AEC, Fig. 4B) in a process that does not involve cell proliferation, but migration of epithelial cells from the unamputated region.[36,37] The innermost cell layer (the basal epidermal layer) located against the mesenchyme, recognizable by the cuboidal shape of cells, differentiates quickly after the formation of the AEC and is the source of factors regulating epithelial-mesenchymal interactions which will control the regenerate outgrowth.[15,34,38,39]
2. Blastema formation (24-72hpa). Following the formation of the AEC, fibroblast-like cells located up to two segments proximally to the amputation plane start to disorganize and migrate to the distal region, at the site where the blastema will form by cell proliferation.[34,36,37] It is still unclear whether these cells originate from the dedifferentiation of preexisting mesenchymal cells or from a population of progenitor cells. The blastema becomes clearly

---

[c] hpa: hour-post-amputation and dpa: day-post-amputation at 28.5°C. Regeneration can be conducted at 33°C and results in speeding up the process up to two times compared to the standard temperature used to raise zebrafish.

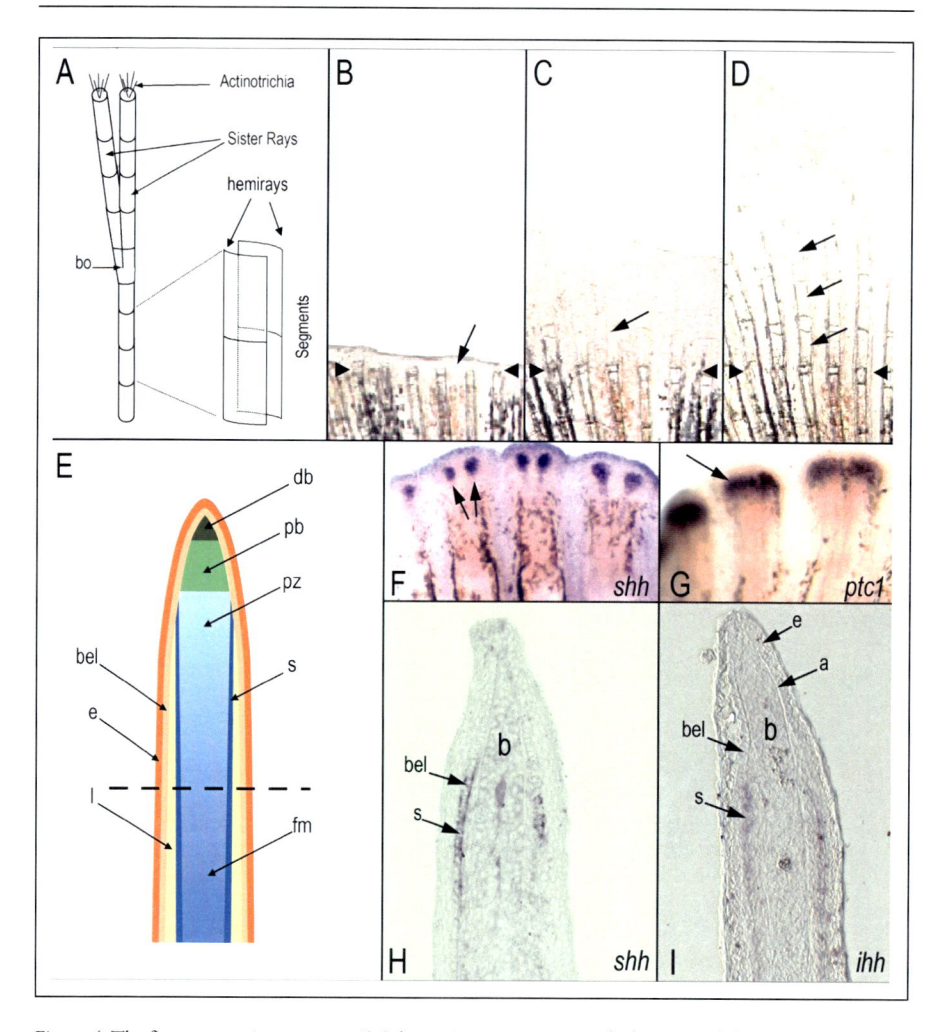

Figure 4. The fin regeneration process. A) Schematic representation of a fin ray. Each hemiray is composed of a succession of segments attached to each other by ligament-like joints. Periodically, lepidotrichia birfucate creating two sister rays. Both segmentation and bifurcation occur at the same level on each hemiray. B-D) Morphology of the regenerating caudal fin at 1dpa (B), 4dpa (C) and 6dpa (D). Few hours after amputation an epidermal layer covers the wound, followed by additional layers of epidermis migrating from the unamputated region. This forms the apical epidermal cap (AEC, arrow on B) by 1dpa. By 4dpa, as regenerative outgrowth occurs, the regenerated lepidotrichia is visible and a few segment limits have already formed (arrow on C). Later, at 6dpa (D) the morphology of the proximal region of the regenerate is very similar to the unamputated part and new ray segments are added distally (arrows on D). E) Schematic representation of a fin regenerate during outgrowth phase: db: distal blastema, pb: proximal blastema, pz: patterning zone, s: scleroblast layer, bel: basal epidermal layer, e: epidermis, l: lepidotrichia. F-I: in situ hybridization on whole mount (F-G) and sectioned (H-I) 4 dpa regenerates. *shh* (F-G) is expressed in the basal epidermal layer (bel) in two domains on each hemiray (arrows on F), preceding the morphological bifurcation of the lepidotrichia whereas *ptc1* (H) expression domain spans the entire hemiray width (arrow on G). Histological section allows us to localize *shh* expressing cells in the bel (H) whereas *ihh* is expressed in the scleroblasts (I) at the level of the patterning zone.

visible by 2dpa. At that time, mesenchymal cell division mostly occurs in the blastema region, whereas proliferating epithelial cells are restricted to more proximal regions of the fin.[37]

3. Blastema maturation and regenerative outgrowth (72hpa and later). Immediately after the blastema formation, mesenchymal cells segregate into three populations (Fig. 4E): First, a small population of slow-cycling cells is located in the distal blastema (DB).[37,40] It has been proposed that this population would constitute a pool of undifferentiated progenitor cells for the second population in the proximal blastema region (PB) which shows an intense and rapid cell cycling, twice as fast as during blastema formation.[37] Finally, the most proximal part of the regenerate, the patterning zone (PZ), is mostly composed of differentiating mesenchymal cells in the core of the regenerate and scleroblasts in the periphery, adjacent to the basal epidermal layer. Cells of the PZ show little or no cell division. As regeneration continues, the blastema constantly remains distally located, driven by cell proliferation occurring in the PB, while cells of the PZ progressively differentiate into new structures which replace the amputated part of the fin (Fig. 4C,D). Complete regeneration is achieved within 3 weeks depending on the amputation level.

The Hedgehog Pathway and Fin Ray Patterning: Role of the Epithelial-Mesenchymal Interactions

shh-expressing cells are detected in a broad domain covering the distal tips of each ray at about 30hpa. At 2dpa, the expression is localized to cells of the basal epidermal layer (BEL) that covers the whole surface of the amputated ray. *ptc1* is also detected by 40hpa in the distal BEL, consistent with its role as a mediator of the Hh signal. During the outgrowth phase, starting by 4dpa, *shh* expression becomes restricted to a subset of cells of the BEL adjacent to the newly formed lepidotrichia, at the level of the PB and the PZ of the regenerate in each hemiray (Fig. 4H).[15,39] As regeneration proceeds, *shh* domain of expression regularly splits into two discrete cell populations in each hemiray (Fig. 4F). This event always precedes the morphological bifurcation of the fin ray, suggesting a possible role for shh in the specification of the bifurcation. However, *ptc1*, which is expressed at that stage in the BEL at the level of the shh-expressing cells and also in the adjacent scleroblasts, always shows a single domain of expression spanning the entire width of the hemiray (Fig. 4G). This raises the possibility that factors expressed in between the two shh domains would inhibit shh signal in the central region of the hemiray in the early steps of branching formation.

A second *Hh* gene, coding for an orthologue of the mammalian Indian hedgehog (Ihh), was recently isolated in our laboratory.[4] *ihh*-expressing cells are observed at 4dpa in the scleroblasts expressing *ptc1* and adjacent to the *shh*-expressing cells of the BEL (Fig. 4I). The fact that *ihh* is transcribed in the differentiating scleroblasts may suggest a more direct role for this factor in bone formation than shh. Furthermore, this expression of *ihh* in scleroblasts is unexpected, as this gene has previously been shown to be expressed in cartilage cells during endochondral bone formation only.[41,42]

Due to its easy access and relative simplicity, the zebrafish fin ray is a good model to perform functional and genetic studies of the regeneration process. However, as only few mutants survive long enough to be studied during fin regeneration, it has been necessary to develop and adapt new methods to manipulate gene activity in this system. Chemical treatments,[15,34,43] cell transfection by microinjection[43] and temperature-inducible mutants[32,37,44] have revealed the requirement of Hh signalling for proper patterning of the bony rays during fin regeneration.

RA treatments of zebrafish undergoing fin regeneration cause an inhibition of the regenerate growth followed by ray patterning defects.[45] Treatments as short as 12h transiently inhibit regenerate outgrowth and downregulate *shh* expression. Thus, as in the embryonic fin buds,[14] RA treatments of regenerating fins lead to a rapid downregulation of *shh* expression, supporting the idea of a direct role of RA in *shh* transcription via the RARE located in 5' region of the *shh* locus. After the end of RA treatment, the regenerate outgrowth resumes almost immediately whereas it takes 3 days for *shh* transcripts to be detected again. When reinitiated, the distal limit of the *shh* expression domain corresponds to the distal limit of the bone matrix deposition suggesting that *shh* expression may determine some aspects of scleroblast differentiation and patterning.[15]

Further evidence of the role of the Hh pathway in bone patterning has been demonstrated by Quint et al who developed a method of gene transfection based on microinjection of plasmid DNA into the blastema (Fig. 5).[43] Ectopic expression of shh following injection of plasmid constructs coding for the active peptide of shh between ray branches induced an ectopic expression of *ptc1* in this region and the fusions of the two branches. These fusions are caused by deposition of ectopic bone material between the basal epidermal layer and the mesenchyme of the interray region (Fig. 5C-D). However, no bone forms in the deeper mesenchyme of the blastema where the *shh* transgene is also expressed, as indicated by the induction of *ptc1*. This suggests that only cells at the epithelial-mesenchyme interface have the potential to differentiate into scleroblasts. Bone morphogenetic proteins (BMP), members of the transforming growth factor β (TGF-β) family, are able to promote bone formation in both in vivo and in vitro systems.[46,47] During fin regeneration, *bmp2b* is expressed in the distal BEL in a pattern similar to that of *shh*, as well as in the adjacent scleroblasts.[15] A second member of the BMP family, *bmp4*, is restricted to the distal mesenchyme. *bmp2b* ectopic expression analysis using the approach described above leads to bone fusions similar to those obtained following ectopic *shh* expression (Fig. 5E-F). However, cotransfection of *shh* and *chordin*, an inhibitor of the BMP signal,[48] fails to produce any fusion. Altogether, these results indicate that the effect of *shh* ectopic expression is mediated by BMP signalling, which would act downstream of shh.[43] As *bmp2b* ectopic expression does not induce *shh*-dependent *ptc1* expression, no feedback loop mechanism between BMP and Hh seems to exist in the regenerate.

The effects of a loss of Hh signalling in the fin regenerate were analyzed using the steroidal alkaloid, cyclopamine, an inhibitor of Hh signalling.[49] Treatments of regenerating fins with cyclopamine initially cause a proximal extension of the expression domain of *shh*, with a slight reduction of the regenerate outgrowth. In a second step, the outgrowth is completely inhibited and *shh* is no longer expressed. After 5 days of treatment, the regenerate is much shorter compared to an untreated fin, shows an accumulation of pigment cells in the distal region and no ray bifurcation. The initial upregulation of *shh* is suggestive of a feedback mechanism that would normally restrict *shh* expression at the level of the PB and PZ. The progressive arrest of fin regeneration, correlates with an inhibition of blastema cell proliferation in the regenerate epidermis and mesenchyme suggesting that the Hh pathway is necessary for blastema maintenance and outgrowth. Interestingly, bone deposition is still taking place but with abnormal patterns suggesting that inhibition of Hh signalling does not affect already-differentiated scleroblasts, but may rather affect the proliferation of undifferentiated blastema cells, their survival, and/or differentiation into specialized cell types, including scleroblasts. Another possible role for the Hh signals could be the regulation of the distribution of scleroblasts, i.e., their alignment against the basal epidermal layer. This role is further suggested by the phenotype of the temperature-sensitive regeneration mutant, *emmental* (*emm*). This mutant, in which *sly1* (a gene coding for a protein involved in protein trafficking) is disrupted when fish are subjected to a heat shock at 33°C, presents blastema formation defects and a downregulation of *shh* expression.[50] Interestingly, scleroblasts of regenerating *emm* fins are no longer ordered against the BEL but randomly dispersed throughout the blastema, a phenotype which connects scleroblast alignment and shh signalling.

It is likely that the fibroblast growth factors (Fgf) play an important role in fin regeneration, possibly through interaction with the Hh pathway. Wfgf/fgf24[51] a Fgf ligand of the fgf8/fgf17/fgf18 subclass, and the receptor fgfr1[34] are expressed during fin regeneration. *fgfr1* is expressed at 18hpa in the forming blastema, then during the outgrowth phase in the distal BEL (including *shh*-expressing cells) and the distal blastema.[34] *wfgf* expression is restricted to the epithelium at the distal part of the regenerate and appears relatively late, at 48hpa, suggesting

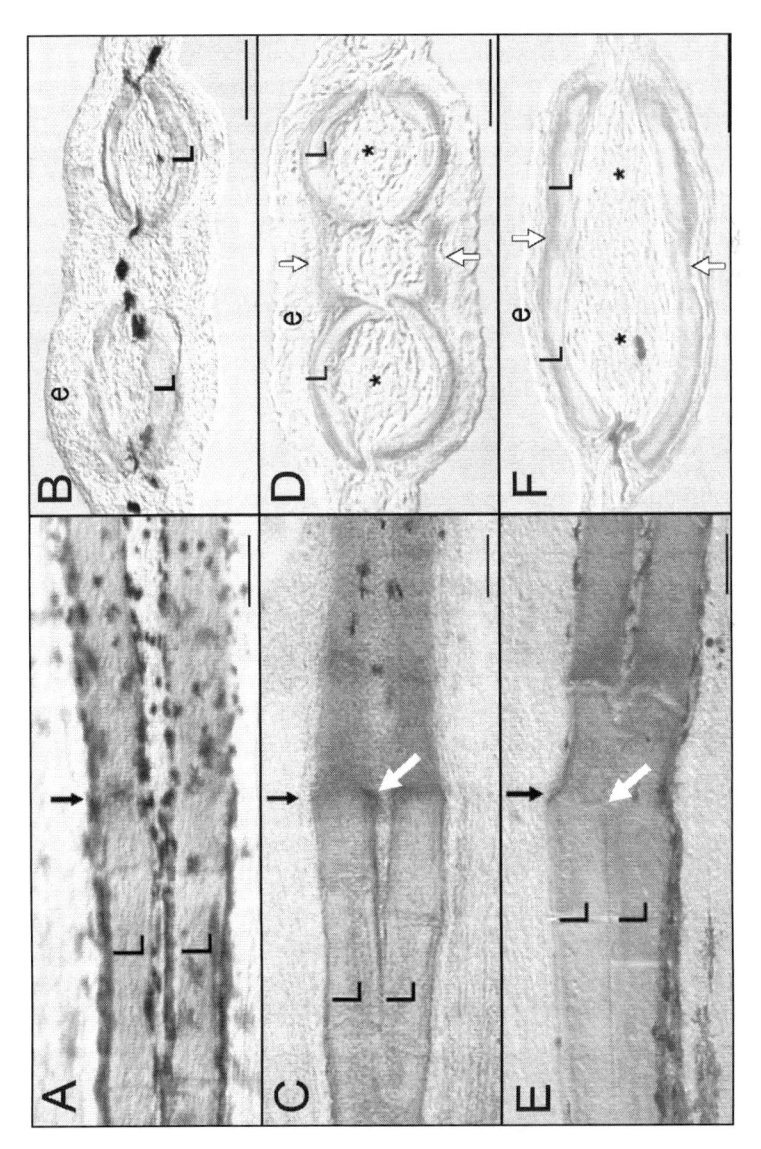

Sonic Hedgehog Signalling in the Developing and Regenerating Fins of Zebrafish   103

Figure 5. Effect of shh and bmp2b ectopic expression on bone patterning. A,B) Wild type fin rays. Transfection of blastema cells with shh (C,D) or bmp2b (E,F) causes ray fusions due to ectopic bone matrix deposition (white arrows on C-F) in between the sister rays. Cryosection of the lepidotrichia at the level of the fusion (right panel) shows that ectopic bone deposition is restricted to the epithelial-mesechymal interface, and no bone forms in the inner ray mesenchyme (*). Left panel: whole mount view of individual rays stain with alcian blue. Right panel, cryosection of the ray at the level indicated with a black arrow on the left panel. L: lepidotrichia, e: epidermis, *: intra-ray mesenchyme. Reprinted with permission from: Quint E, Smith A, Avaron F et al. Bone patterning is altered in the regenerating zebrafish caudal fin after ectopic expression of sonic hedgehog and bmp2b or exposure to cyclopamine. Proc Natl Acad Sci USA 2002; 99(13):8713-8718. ©2002 National Academy of Sciences, U.S.A.

that an additional, yet unidentified, Fgf ligand could be expressed at earlier stage concurrently with fgfr1. Interactions between the Fgf-Hh signalling pathways are probable, as treating regenerating fins with the Fgf signal inhibitor SU5402 causes effects reminiscent of cyclopamine treatments: blastema outgrowth inhibition and down-regulation of *shh* without affecting the wound epidermis or bone deposition.

The Wnt pathway is also involved in regeneration and is likely interacting with the Hh pathway. Wnt factors are secreted molecules, unrelated to the Hh proteins but sharing a lot of similarities with them at the structural level and their mode of action.[52,53] Complex regulations exist between the Wnt and Hh pathways depending on the tissue in which they are expressed. For example, ectopic activation of β-*catenin*, which transduces the wnt signal to the nucleus, induces *shh* expression in mouse epidermis,[54] whereas wnt3 is able to counteract the effect of shh overexpression in chick neural tube explants.[55] The Wnt pathway is also involved in bone formation as β-catenin is required for osteoblast (bone forming cells of the endoskeleton) differentiation in the mouse embryo, possibly by acting downstream of the Hh pathway.[56] Several members of the Wnt signalling pathway, β-catenin, wnt3a, wnt5 and the transcription factor lef1, are expressed during fin regeneration.[39] In the early stages, β-catenin is expressed in the distal blastema whereas *wnt5* and *lef1* transcripts are located in proximal cells of the wound epidermis. During fin outgrowth, *wnt5* is expressed in the distal BEL, and *lef1* is found in most of the BEL including *shh*-expressing cells. Both RA treatment and inhibition of Fgf signalling using SU5402 during fin outgrowth downregulate *lef1* expression, in the same way as *shh* expression.[39] This coregulation suggests that the Hh and Wnt pathways may participate in similar processes in the basal epidermal layer during the outgrowth phase of fin regeneration.

Concluding Remarks and Future Prospects

This chapter presented the main data related to the role of the Hedgehog pathway during fin bud development and fin regeneration. The early steps of zebrafish pectoral fin development are highly reminiscent of tetrapod limb development, and shh function appears to be conserved in this process. An in-depth observation of the fin phenotype of *you*-type mutants is likely to provide new insights into the role and the regulation of the Hh pathway, as it will allow us to dissect the Hh pathway, and analyze the effects of its disruption at various levels. In the regenerating fin, shh plays a role in blastema maintenance and the patterning of the regenerating bony ray, probably through the correct differentiation and alignment of the scleroblasts. Future studies will further investigate the role of the Hh pathway during regeneration and its interaction with the major signalling pathways that have been described during embryonic development.

References

1. Krauss S, Concordet JP, Ingham PW. A functionally conserved homolog of the Drosophila segment polarity gene hh is expressed in tissues with polarizing activity in zebrafish embryos. Cell 1993; 75(7):1431-1444.
2. Ekker SC, Ungar AR, Greenstein P et al. Patterning activities of vertebrate hedgehog proteins in the developing eye and brain. Curr Biol 1995; 5(8):944-955.
3. Currie PD, Ingham PW. Induction of a specific muscle cell type by a hedgehog-like protein in zebrafish. Nature 1996; 382(6590):452-455.
4. Avaron F, Hoffman L, Guay D et al. Characterization of two new zebrafish members of the hedgehog family: atypical expression of the zebrafish indian hedgehog gene in skeletal elements of both endochondral and dermal origins. Dev Dyn 2006; 235(2):478-89.
5. Dane PJ, Tucker JB. Modulation of epidermal cell shaping and extracellular matrix during caudal fin morphogenesis in the zebra fish Brachydanio rerio. J Embryol Exp Morphol 1985; 87:145-161.
6. Grandel H, Schulte-Merker S. The development of the paired fins in the zebrafish (Danio rerio). Mech Dev 1998; 79(1-2):99-120.
7. Akimenko MA, Ekker M. Anterior duplication of the Sonic hedgehog expression pattern in the pectoral fin buds of zebrafish treated with retinoic acid. Dev Biol 1995; 170(1):243-247.
8. Riddle RD, Johnson RL, Laufer E et al. Sonic hedgehog mediates the polarizing activity of the ZPA. Cell 1993; 75(7):1401-1416.

9. Fallon JF, Lopez A, Ros MA et al. FGF-2: Apical ectodermal ridge growth signal for chick limb development. Science 1994; 264:104-107.
10. Vogel A, Rodriguez C, Izpisúa-Belmonte J-C. Involvement of FGF-8 in initiation, outgrowth and patterning of the vertebrate limb. Development 1996; 122:1737-1750.
11. Loomis CA, Harris E, Michaud J et al. The mouse Engrailed-1 gene and ventral limb patterning. Nature 1996; 382:360-363.
12. Riddle RD, Ensini M, Nelson C et al. Induction of the LIM homeobox gene Lmx1 by WNT7a establishes dorsoventral pattern in the vertebrate limb. Cell 1995; 83:631-640.
13. Bruneau S, Rosa FM. Dynamo, a new zebrafish DVR member of the TGF-beta superfamily is expressed in the posterior neural tube and is up-regulated by Sonic hedgehog. Mech Dev 1997; 61(1-2):199-212.
14. Hoffman L, Miles J, Avaron F et al. Exogenous retinoic acid induces a stage-specific, transient and progressive extension of Sonic hedgehog expression across the pectoral fin bud of zebrafish. Int J Dev Biol 2002; 46(7):949-956.
15. Laforest L, Brown CW, Poleo G et al. Involvement of the sonic hedgehog, patched 1 and bmp2 genes in patterning of the zebrafish dermal fin rays. Development 1998; 125(21):4175-4184.
16. Chandrasekhar A, Warren Jr JT, Takahashi K et al. Role of sonic hedgehog in branchiomotor neuron induction in zebrafish. Mech Dev 1998; 76(1-2):101-115.
17. Du SJ, Dienhart M. Zebrafish tiggy-winkle hedgehog promoter directs notochord and floor plate green fluorescence protein expression in transgenic zebrafish embryos. Dev Dyn 2001; 222(4):655-666.
18. Ng JK, Tamura K, Buscher D et al. Molecular and cellular basis of pattern formation during vertebrate limb development. Curr Top Dev Biol 1999; 41:37-66.
19. Capdevila J, Izpisua Belmonte JC. Patterning mechanisms controlling vertebrate limb development. Annu Rev Cell Dev Biol 2001; 17:87-132.
20. Lu H-C, Revelli J-P, Goering L et al. Retinoid signaling is required for the establishment of a ZPA and for the expression of Hoxb-8, a mediator of ZPA formation. Development 1997; 124:1643-1651.
21. Helms J, Thaller C, Eichele G. Relationship between retinoic acid and sonic hedgehog, two polarizing signals in the chick wing bud. Development 1994; 120:3267-3274.
22. Tickle C, Alberts BM, Wolpert L et al. Local application of retinoic acid to the limb bud mimics the action of the polarizing region. Nature 1982; 296:564-565.
23. Chang DT, Lopez A, von Kessler DP et al. Products, genetic linkage, and limb patterning activity of a murine hedgehog gene. Development 1994; 120:3339-3353.
24. Neumann CJ, Grandel H, Gaffield W et al. Transient establishment of anteroposterior polarity in the zebrafish pectoral fin bud in the absence of sonic hedgehog activity. Development 1999; 126(21):4817-4826.
25. Chang B-E, Blader P, Fischer N et al. Axial (HNF3β) and retinoic acid receptors are regulators of the zebrafish sonic hedgehog promoter. EMBO J 1997; 16:3955-3964.
26. Schauerte HE, van Eeden FJ, Fricke C et al. Sonic hedgehog is not required for the induction of medial floor plate cells in the zebrafish. Development 1998; 125(15):2983-2993.
27. van Eeden FJ, Granato M, Schach U et al. Mutations affecting somite formation and patterning in the zebrafish, Danio rerio. Development 1996; 123:153-164.
28. Barresi MJ, Stickney HL, Devoto SH. The zebrafish slow-muscle-omitted gene product is required for Hedgehog signal transduction and the development of slow muscle identity. Development 2000; 127(10):2189-2199.
29. van Eeden FJM, Granato M, Schach U et al. Genetic analysis of fin formation in the zebrafish, Danio rerio. Development 1996; 123:255-262.
30. Grandel H, Draper BW, Schulte-Merker S. Dackel acts in the ectoderm of the zebrafish pectoral fin bud to maintain AER signaling. Development 2000; 127(19):4169-4178.
31. Borday V, Thaeron C, Avaron F et al. Evx1 transcription in bony fin rays segment boundaries leads to a reiterated pattern during zebrafish fin development and regeneration. Dev Dynamics 2001; 220:91-98.
32. Johnson SL, Weston JA. Temperature-sensitive mutations that cause stage-specific defects in zebrafish fin regeneration. Genetics 1995; 141:1583-1595.
33. Poss KD, Keating MT, Nechiporuk A. Tales of regeneration in zebrafish. Dev Dynamics 2003; 226(2):202-210.
34. Poss KD, Shen J, Nechiporuk A et al. Roles for Fgf signaling during zebrafish fin regeneration. Dev Biol 2000; 222(2):347-358.
35. Akimenko MA, Mari-Beffa M, Becerra J et al. Old questions, new tools, and some answers to the mystery of fin regeneration. Dev Dynamics 2003; 226(2):190-201.

36. Poleo G, Brown CW, Laforest L et al. Cell proliferation and movement during earlin fin regeneration in zebrafish. Dev Dynamics 2001; 221:380-390.
37. Nechiporuk A, Keating MT. A proliferation gradient between proximal and msxb-expressing distal blastema directs zebrafish fin regeneration. Development 2002; 129(11):2607-2617.
38. Akimenko M-A, Johnson SL, Westerfield M et al. Differential induction of four msx homeobox genes during fin development and regeneration in zebrafish. Development 1995; 121:347-357.
39. Poss KD, Shen J, Keating MT. Induction of lef1 during zebrafish fin regeneration. Dev Dyn 2000; 219(2):282-286.
40. Santamaría JA, Marí-Beffa M, Santos-Ruiz L et al. Incorporation of bromodeoxyuridine in regenerating fin tissue of the goldfish Carassius auratus. J Exp Zool 1996; 275:300-307.
41. St-Jacques B, Hammerschmidt M, McMahon AP. Indian hedgehog signaling regulates proliferation and differentiation of chondrocytes and is essential for bone formation. Genes Dev 1999; 13(16):2072-2086.
42. Vortkamp A, Lee K, Lanske B et al. Regulation of rate of cartilage differentiation by Indian hedgehog and PTH-related protein. Science 1996; 273(5275):613-622.
43. Quint E, Smith A, Avaron F et al. Bone patterning is altered in the regenerating zebrafish caudal fin after ectopic expression of sonic hedgehog and bmp2b or exposure to cyclopamine. Proc Natl Acad Sci USA 2002; 99(13):8713-8718.
44. Poss KD, Nechiporuk A, Hillam AM et al. Mps1 defines a proximal blastemal proliferative compartment essential for zebrafish fin regeneration. Development 2002; 129(22):5141-5149.
45. Géraudie J, Brulfert A, Monnot M-J et al. Teratogenic and morphogenetic effects of retinoic acid on the regenerating pectoral fin in zebrafish. J Exp Zool 1994; 269:12-22.
46. Canalis E, Economides AN, Gazzerro E. Bone morphogenetic proteins, their antagonists, and the skeleton. Endocr Rev 2003; 24(2):218-235.
47. Yoon BS, Lyons KM. Multiple functions of BMPs in chondrogenesis. J Cell Biochem 2004; 93(1):93-103.
48. Piccolo S, Sasai Y, Lu B et al. Dorsoventral patterning in Xenopus: Inhibition of ventral signals by direct binding of chordin to BMP-4. Cell 1996; 86(4):589-598.
49. Incardona JP, Gaffield W, Kapur RP et al. The teratogenic Veratrum alkaloid cyclopamine inhibits sonic hedgehog signal transduction. Development 1998; 125(18):3553-3562.
50. Nechiporuk A, Poss KD, Johnson SL et al. Positional cloning of a temperature-sensitive mutant emmental reveals a role for sly1 during cell proliferation in zebrafish fin regeneration. Dev Biol 2003; 258(2):291-306.
51. Fischer S, Draper BW, Neumann CJ. The zebrafish fgf24 mutant identifies an additional level of Fgf signaling involved in vertebrate forelimb initiation. Development 2003; 130(15):3515-3524.
52. Nusse R. Wnts and Hedgehogs: Lipid-modified proteins and similarities in signaling mechanisms at the cell surface. Development 2003; 130(22):5297-5305.
53. Watt FM. Unexpected Hedgehog-Wnt interactions in epithelial differentiation. Trends Mol Med 2004; 10(12):577-580.
54. Callahan CA, Oro AE. Monstrous attempts at adnexogenesis: Regulating hair follicle progenitors through Sonic hedgehog signaling. Curr Opin Genet Dev 2001; 11(5):541-546.
55. Robertson CP, Braun MM, Roelink H. Sonic hedgehog patterning in chick neural plate is antagonized by a Wnt3-like signal. Dev Dyn 2004; 229(3):510-519.
56. Hu H, Hilton MJ, Tu X et al. Sequential roles of Hedgehog and Wnt signaling in osteoblast development. Development 2005; 132(1):49-60.
57. Karlstrom RO, Talbot WS, Schier AF. Comparative synteny cloning of zebrafish you-too: Mutations in the Hedgehog target gli2 affect ventral forebrain patterning. Genes Dev 1999; 13(4):388-393.
58. Nakano Y, Kim HR, Kawakami A et al. Inactivation of dispatched 1 by the chameleon mutation disrupts Hedgehog signalling in the zebrafish embryo. Dev Biol 2004; 269(2):381-392.
59. Baxendale S, Davison C, Muxworthy C et al. The B-cell maturation factor Blimp-1 specifies vertebrate slow-twitch muscle fiber identity in response to Hedgehog signaling. Nat Genet 2004; 36(1):88-93.
60. Varga ZM, Amores A, Lewis KE et al. Zebrafish smoothened functions in ventral neural tube specification and axon tract formation. Development 2001; 128(18):3497-3509.
61. Sekimizu K, Nishioka N, Sasaki H et al. The zebrafish iguana locus encodes Dzip1, a novel zinc-finger protein required for proper regulation of Hedgehog signaling. Development 2004; 131(11):2521-2532.
62. Karlstrom RO, Tyurina OV, Kawakami A et al. Genetic analysis of zebrafish gli1 and gli2 reveals divergent requirements for gli genes in vertebrate development. Development 2003; 130(8):1549-1564.

CHAPTER 10

Hedgehog Signalling in T Lymphocyte Development

Susan Outram,* Ariadne L. Hager-Theodorides and Tessa Crompton

Abstract

T cell development occurs in the thymus, which is seeded by multipotential lymphocyte progenitor cells. These cells then move through a sequence of clearly defined developmental stages at the end of which they become a fully functional mature T cell. For correct organogenesis and T cell development to occur the thymic stroma and the developing thymocytes must interact with one another. Thymocyte development is regulated by factors produced by the thymic stroma. Sonic hedgehog (Shh) is secreted by the thymic stroma and Patched (Ptc), Smoothened (Smo) and the Gli transcription factors are expressed by thymocytes. In the mouse, Shh is involved in the proliferation and efficient progression through the differentiation process, as well as maintaining normal thymic cellularity. In the human, Shh signals to progenitor cells in a paracrine fashion to instruct these cells to maintain the precursor cell pool by increasing their cell viability and inhibiting their expansion and concomitant progression to the next stage in development. Thus, Shh plays an important role in T cell development in both human and mouse.

Introduction to T Cell Development

Central to the development of the T cell is the thymus. The thymus provides the optimal environment required for maturation of functional T cells. The adult thymus consists of several lobes of tissue made up of a central medullary region surrounded by an outer cortex. The thymus is formed during foetal development by the seeding of the thymic primordium by T cell progenitors and requires stage specific interactions between the epithelial cells and the developing thymocytes (Fig. 1).[1] In the adult thymus the T cell progenitors arrive via the corticomedullary blood vessels. These progenitor cells first seed the subcapsula, the most external thymic compartment, where they start out on a complex but carefully regulated developmental pathway. This pathway requires the interaction of thymocytes with the thymic stroma made up of thymic epithelial cells and mesenchyme derived cells. These interactions are bidirectional between the thymic stroma and developing lymphocyte (Fig. 2).[2]

During T cell development thymocytes pass through a series of stages which can be defined by the cell surface expression of CD4 and CD8. CD4$^-$8$^-$ (DN) thymocytes progress to the CD4$^+$8$^+$ (DP) stage in development and then to mature CD4$^+$8$^-$ or CD4$^-$8$^+$ single positive (SP) T cells. Here we will address the effects of hedgehog (Hh) signalling on thymocyte development in mouse and human separately.

*Corresponding Author: Susan Outram—Department of Biological Sciences, Imperial College London, Sir Alexander Fleming Building, South Kensington Campus, London SW7 2AZ, U.K. Email: s.outram@ic.ac.uk

Shh and Gli Signalling and Development, edited by Carolyn E. Fisher and Sarah E.M. Howie. ©2006 Landes Bioscience and Springer Science+Business Media.

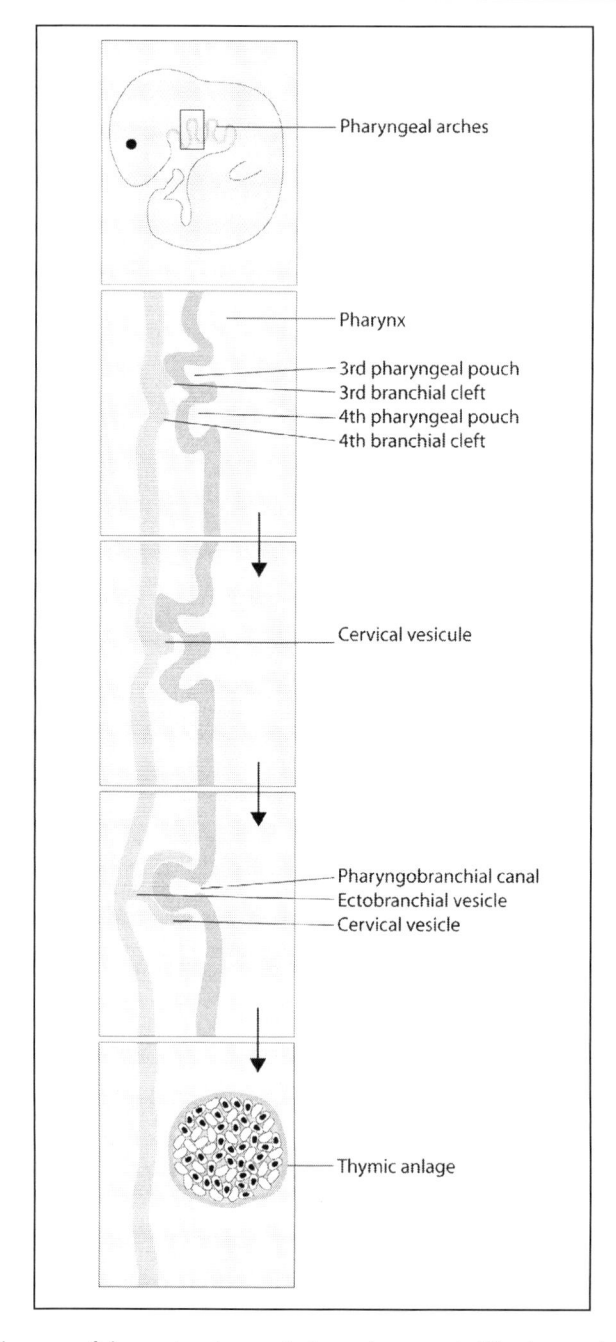

Figure 1. Development of the murine thymus during embryogenesis. The thymus develops from the endoderm of the third pharyngeal pouch and the ectoderm of the third branchial cleft (first to fourth panels). The thymic rudiment buds from the endoderm at around embryonic day E11 and starts being seeded with common lymphoid progenitors, originating from the fetal liver, at day E11-11.5 (fifth panel), at which stage the development and patterning of the thymus begins depending on interactions between epithelial cells and developing thymocytes.

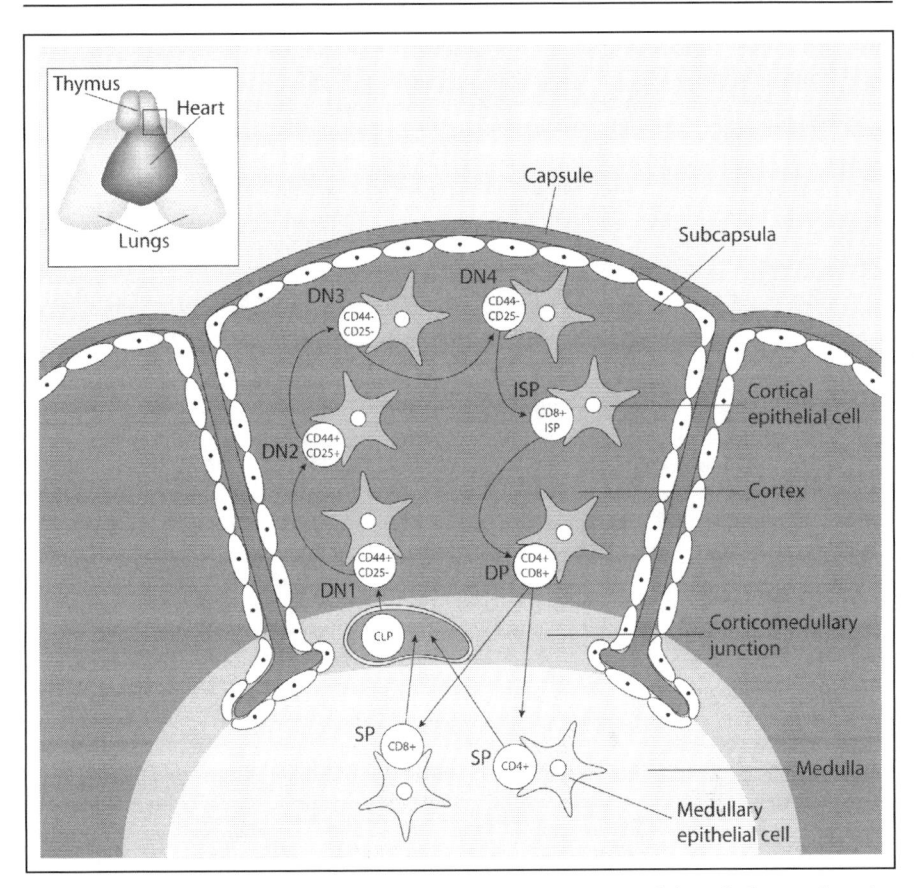

Figure 2. Thymocyte development in murine thymus. The thymus is situated above the heart and consists of several lobules each containing cortical (outer) and medullary (inner) regions, separated by the corticomedullary junction region. Common lymphoid progenitors (CLP) enter the thymus via the corticomedullary junction blood vessels and migrate through the cortex to the subcapsula and then back to the medulla. During this migration they gradually develop from the CLP stage to the mature CD4⁺ or CD8⁺ single positive (SP) stage. CD4⁺ and CD8⁺ SP cells migrate to the periphery via the blood vessels of the corticomedullary junction.

Effect of Hh Signalling on Thymocyte Development in the Mouse

In the mouse the DN population of thymocytes may be further subdivided into four developmental stages based on the expression of the cell surface markers CD44 and CD25. The earliest thymic subset (DN1) is positive for expression of CD44 and negative for expression of CD25. This cell then acquires CD25 expression and is known as DN2. CD44 expression is then downregulated and the cell becomes DN3. Finally CD25 expression is lost and the cell is negative for both CD44 and CD25 (DN4). This may be summarised as follows $CD44^+CD25^-$ (DN1) ▶ $CD44^+CD25^+$ (DN2) ▶ $CD44^-CD25^+$ (DN3) ▶ $CD44^-CD25^-$ (DN4). In order for the cell to make the transition from DN3 to DN4 the TCR β chain must be rearranged and expressed at the cell surface in a complex with the invariant preTα chain in the form of the preTCR. Signalling through the preTCR complex allows for allelic exclusion at the TCRβ locus, thus preventing the T cell from expressing more then one TCR β chain at the cell surface, as well as proliferation and differentiation. This checkpoint in the developmental process is known as β selection.[3] Following a signal received through the preTCR the thymocyte progresses to the DP stage in development usually via an intermediate population most commonly expressing CD8. This cell is known as the intermediate single positive (ISP). Differentiation from the DP cell to the mature SP cell is dependent on expression and positive selection of the αβTCR which consists of the TCR β chain in a complex with a rearranged TCR α chain.

The DN1 population contains cells that are still multipotential, and may give rise to T cells, B cells, NK or dendritic cells,[4,5] but as the cell acquires the expression of CD25 and moves through the developmental program it becomes progressively more committed to the T cell lineage. At the DN3 stage in development the cell has become irreversibly committed to the T cell lineage. This developmental process is illustrated in Figure 2.

The first study to show an involvement of Hh signalling in murine thymic development was from our laboratory.[6] Analysis of expression of the molecules involved in Hh signalling revealed that RNAs encoding Sonic hedgehog (Shh) and Indian Hedgehog (Ihh) are both present in the thymus. Shh transcripts were found to be expressed by the thymic stroma and both Shh and Ihh proteins were detected by immunofluoresense staining on frozen sections of adult murine thymus. Shh was detected in epithelial cells and Ihh was detected associated with blood vessels located in the thymic medulla. Analysis of expression of Desert Hedgehog in the mouse thymus has not yet revealed this molecule to be present. Analysis of expression of the receptors for Hh, Patched (Ptc) and smoothened (Smo), showed that both these receptors are detectable in the adult murine thymus. Similar findings were reported by Li et al.[7,8] Transcripts for the Ptc molecule were detected in DN, DP and CD8 single positive thymocytes whereas Smo transcripts were detected in DN adult murine thymocytes only. More detailed analysis of Smo expression by cell surface staining of the DN subsets revealed that Smo was most highly expressed on the surface of CD44⁺CD25⁺ DN2 subset and that cell surface expression gradually decreased in each subsequent DN population. The downstream effector molecules for the Hh signalling pathway are the zinc finger transcription factors Gli 1-3.[9] Transcripts for Gli1, Gli2 and Gli3 are all detectable in the adult thymus. Analysis of these same molecules at day E14.5 in embryogenesis revealed that transcripts for Ihh, Shh, Ptc, Smo and Glis 1-3 are all present.

In this study, the function of Hh signalling in murine thymic development was studied by treating fetal thymic organ cultures (FTOC) with the human recombinant Shh protein and the anti-Hh neutralising antibody, 5E1. FTOC provides an ideal in vitro culture system in which to study the effects of addition of exogenous molecules to the process of thymic development. We found that treatment of FTOC with anti-Shh neutralising antibody accelerated differentiation from DN to DP thymocyte and treatment of FTOC with a high concentration of recombinant Shh protein inhibited this differentiation. This arrest of differentiation occurred at the CD25⁺ stage of thymocyte development after initiation of TCR β gene rearrangement. However, treatment of FTOC with the neutralising Hh antibody did not replace the requirement for a preTCR signal. We used a system in which thymocyte development in the genetically modified RAG1-/- mouse is arrested at the DN CD25⁺ stage in development.[10] Thymocytes in these mice are unable to rearrange their TCR β locus with the result that they cannot express a preTCR at the cell surface. However it is possible to mimic the preTCR signal in these mice by administering anti-CD3 antibody to Rag -/- FTOC.[11] Treatment of FTOC with anti Hh antibody instead of anti-CD3 antibody did not induce thymocyte differentiation, one of the downstream consequences of preTCR signalling. However it did accelerate anti-CD3 induced differentiation. Conversely, addition of Shh protein after anti-CD3 treatment partially inhibited differentiation. These data suggested that Shh might function to maintain CD25⁺ DN thymocytes as nonproliferating cells while they arrange their TCR β genes. We also showed that an immediate consequence of preTCR signalling was a downregulation of Smo expression. The subsequent inability of the cell to signal through Hh might allow the cell to reenter cell cycle. However, although these studies provided evidence that Hh signalling regulates T cell development, they did not define the physiological role of each Hh species in the thymus. Also, their interpetation was complicated by the fact that the neutralising antibody will bind both Shh and Ihh, both of which are expressed in the thymus.

To address this, a second study from our laboratory analysed genetically modified mice in which the Shh gene was disrupted.[12] This study revealed that Shh does indeed regulate foetal thymus cellularity and thymocyte differentiation.[13] Thymi were isolated from Shh-/- mice at a number of different stages in embryogenesis and cell number, differentiation status, cell survival and proliferation status of the thymocytes were analysed. It was found that Shh was involved at three distinct stages in thymocyte development.

In Shh-/- mice the proportion of lymphocyte-lineage cells early in development was decreased relative to littermates, suggesting that Shh might be involved in the maintenance/expansion of prethymic progenitor cells in the foetal liver, that it might function as a chemoattractant in the seeding process, or might be involved in the maintenance/expansion of the earliest DN1 thymocytes. Interestingly, Shh has been shown to regulate the expansion of primitive human haematopoietic progenitor cells in an autocrine manner,[14] and has recently been shown to function as a chemoattractant in neural development.[15]

The transition from DN1 to DN2 cell was severely impeded in Shh knockout thymi, suggesting that Shh is necessary for differentiation to DN2 or for T cell lineage commitment. It is at this stage in development that T cell fate becomes specified. It has previously been reported that expression of the Notch 1 molecule is involved in T cell fate specification, by regulating specification to the T versus B cell lineage,[16-18] $\alpha\beta$ T versus $\gamma\delta$ T[19] and CD4 versus CD8 lineage commitment.[20,21] However, it appears that Hh signalling is not involved in these lineage commitment decisions as the percentage of B220+ B cells, NK1.1+ NK cells and TCR $\gamma\delta$+ cells was unchanged in thymi isolated from Shh-/- mice.

The DN3 population in Shh-/- embryos then seems to partially recover but a second arrest in development occurs at the transition from DN to DP thymocyte, with an increase in cell death at the DN4 stage. Also, the overall cellularity of thymi isolated from Shh-/- mice was greatly reduced at all developmental stages.

The finding that thymocyte differentiation to the DP stage was reduced in Shh-/- thymi was surprising given our previous finding that treatment of FTOC with 5E1 accelerated thymocyte differentiation from DN to DP cell. There are a number of different possibilities to explain this finding. Firstly it is possible that, as Shh is absent throughout thymic development in the Shh-/- embryo, the effect of removal of Shh is acting on an earlier stage of development than that in the earlier in vitro study. This could result in different target cells being affected allowing for a different outcome in development. Secondly, we may be observing a dose effect with different concentrations of Shh inducing a different outcome. When analysing the Shh-/- thymi, no Shh is present whereas removal of Shh from the FTOC system using a neutralising antibody may leave low levels of Shh still present.

In summary these data suggest that Shh produced by the thymic stroma has a role in the control of thymocyte development in vivo in the mouse. Shh is involved in the proliferation and efficient progression through the developmental process as well as maintaining normal thymic cellularity (summarised in Fig. 3).

Effect of Hh Signalling in Human Thymic Development

In humans, thymocyte development is also characterised by a DN-DP-SP set of transitions. The progenitor cell that seeds the thymus is CD4⁻8⁻CD34⁺CD1a-. As in the mouse, this early progenitor cell is multipotential and may become a T cell, NK cell, Dendritic cell or monocyte. As this DN cell progresses through thymocyte development it acquires CD1a at the cell surface and becomes committed to the T cell lineage. The cell then gradually loses CD34 expression and gains CD4 expression followed by CD8α and then CD8 β expression. In humans, TCR β chain rearrangement occurs mainly at this developmental stage. After β-selection DP thymocytes begin to rearrange their TCR α locus allowing for expression of an $\alpha\beta$ TCR at the cell surface. This cell is now a target for positive selection.[22] Positively selected cells then upregulate CD3, CD69 and CD27 and down regulate either CD4 or CD8 becoming a mature SP thymocyte.

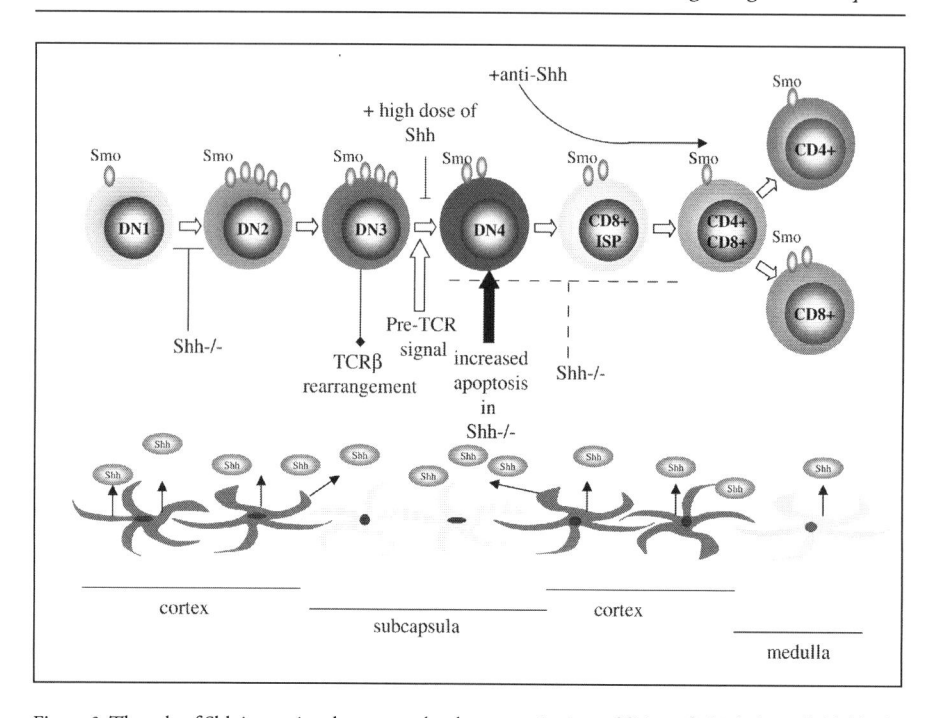

Figure 3. The role of Shh in murine thymocyte development. In vitro addition of a high dose of Shh blocks transition of DN3 to DN4 cell. In contrast, addition of anti-Shh antibody accelerates thymocyte development to the DP stage. Shh deficiency results in severely reduced transition of DN1 cells to the DN2 stage. Furthermore it increases apoptosis in the DN4 subset and causes a partial block in the transition of DN to DP stage. Shh is normally expressed by thymic epithelial cells in the subcapsulla and the medulla and is present in the cortical area as well. Smo is expressed in all thymocyte subsets at varying levels, its highest expression being in the DN2 subset.

Analysis of expression of the component parts of the Hh signalling pathway revealed that these molecules are all present in the human thymus. Thymic samples from children aged 1 month to three years undergoing corrective cardiovascular surgery were analysed.[23] RNA transcripts for Shh, Ihh and Dhh were all detected in thymic epithelium but not in thymocytes. Immunostaining studies revealed that the localisation of Shh expressing cells was restricted to the subcapsular and medullary areas whereas Ihh and Dhh producing epithelial cells were randomly distributed throughout the thymic parenchyma. Analysis of expression of the Hh receptors, Ptc 1 and Smo, revealed their presence in $CD34^+$ progenitor cells, immature $CD4^+8^+$ cells and mature $CD4^+8^-$ and $CD4^-8^+$ cells as well as thymic epithelium. Ptc 2 was expressed only in $CD34^+$ cells and thymic epithelium. Cell surface staining studies showed that on average, 35% of total thymocytes expressed the Smo receptor at their cell surface. In the human thymus, DP and to a lesser extent, DN thymocytes contained the highest proportion of Smo+ cells. Immunostaining revealed that Smo expression was associated with cell clusters composed of epithelial cells and thymocytes. These clusters were located in the subcapsular, cortical and medullary areas suggesting the existence of niches in which Hh signalling is taking place. RNA transcripts for Gli1, Gli2 and Gli 3 were found to be present in the $CD34^+$ early progenitor cell and the thymic epithelium only. Gli1 and Gli3 were also detected in $CD4^-8^+$ SP thymocytes but all Glis were absent or below levels of detection in DP or $CD4^+8^-$ thymocytes.

A role for Hh signalling in the human thymus was also reported.[24] In this study it was shown that Shh significantly increased the viability of $CD34^+$ precursor cells. $CD34^+$ thymocytes were cultured for 48 hours with different doses of Shh and cell viablilty was assessed.

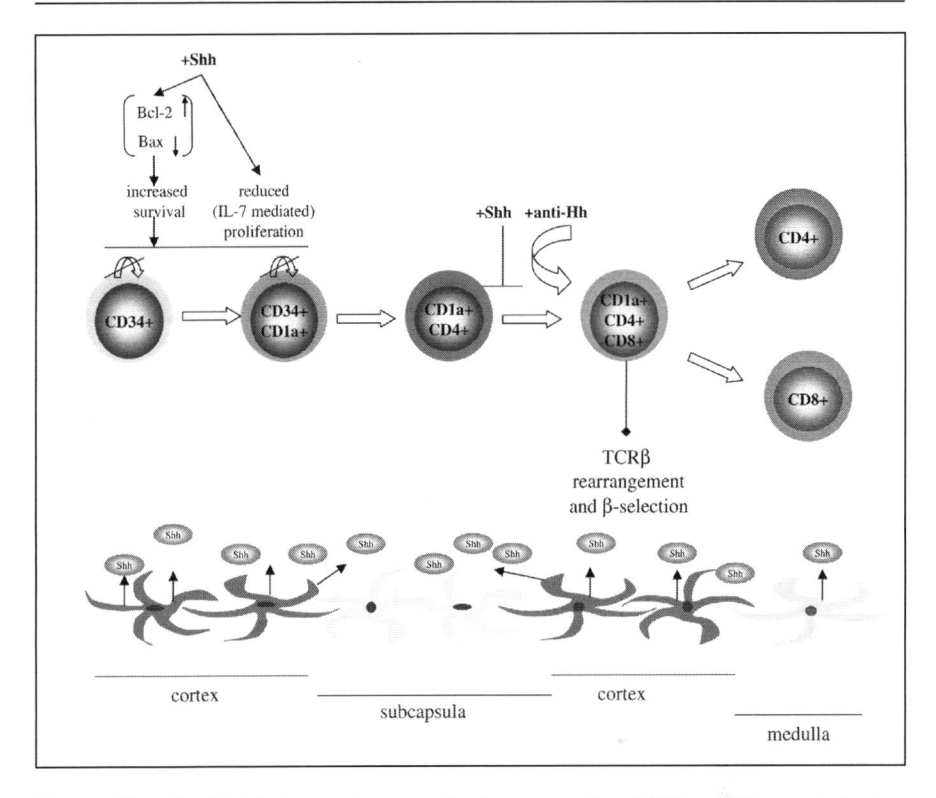

Figure 4. The role of Shh in human thymocyte development in vitro. Addition of Shh protein in vitro increased viability of CD34+ human thymocyte progenitors, possibly via the upregulation of Bcl-2 and simultaneous downregulation of Bax. It also reduced proliferation of IL-7 treated CD34+ thymocytes. Furthermore, addition of Shh in vitro severely impaired differentiation to the DP stage whereas addition of anti-Hh antibody promoted differentiation to the DP stage.

Doses of Shh ranging from 0.05ng/ml to 500ng/ml all resulted in an increase in cell viability. A possible mechanism for this could be by modulating Bcl-2 and Bax expression. Shh induced an increase in the Bcl-2/Bax ratio due to upregulation of Bcl-2 expression and down regulation of Bax expression in CD34+ precursor cells. Such a change in ratio would result in an increase in cell viabilty. This effect was completely abrogated by application of the Hh neutralising monoclonal antibody 5E1.

Addition of Shh to these cultures also resulted in a dose-dependent inhibition of proliferation. By using thymic reaggregation assays it was shown that Shh treated CD34+ precursor cells could not properly reconstitute thymocyte development. The production of DP thymocytes was totally blocked after five days in culture, whereas treatment with the anti-Hh neutralising antibody had the opposite effect. Expansion and survival of these CD34+ progenitor cells is dependent on factors such as Interleukin 7 (IL-7).[25] In the same study, it was shown that addition of Shh to IL-7 treated CD34+ cultures or human mouse chimeric FTOC could completely inhibit proliferation and differentiation of the CD34+ cells.

To summarise this section, Hh signalling also plays an important role in early human T cell development. Shh may be provided to the target CD34+ progenitor cells in a paracrine fashion by the epithelial cells from the subcapsulary area. The subsequent Hh signalling in these cells may maintain the CD34+ precursor cell pool by increasing their cell viability and inhibiting their expansion and concomitant progression to the ISP CD4+ stage in development. This is summarised in Figure 4.

Conclusions

Overall these studies show that Hh signalling is an important regulator of early T cell development in human and mouse. Hh is likely to act in concert with other morphogens such as the Bone morphogenetic (Bmp) and Wnt families of proteins. Bmp 4 for example is known to be a Hh target gene.[26] Both BMP and Wnt families of proteins have already been shown to be involved in the regulation of T cell development.[27-29] So, the role of morphogens such as Hh in T cell development will be as part of a complex web of signalling events which still remains to be fully characterised.

References

1. Manley NR. Thymus organogenesis and molecular mechanisms of thymic epithelial cell differentiation. Semin Immunol 2000; 12(5):421-428.
2. Anderson G, Jenkinson EJ. Lymphostromal interactions in thymic development and function. Nat Rev Immunol 2001; 1(1):31-40.
3. von Boehmer H, Aifantis I, Feinberg J et al. Pleiotropic changes controlled by the preT-cell receptor. Curr Opin Immunol 1999; 11(2):135-142.
4. Akashi K, Reya T, Dalma-Weiszhausz D et al. Lymphoid precursors. Curr Opin Immunol 2000; 12(2):144-150.
5. Shortman K, Vremec D, Corcoran LM et al. The linkage between T-cell and dendritic cell development in the mouse thymus. Immunol Rev 1998; 165:39-46.
6. Outram SV, Varas A, Pepicelli CV et al. Hedgehog signaling regulates differentiation from double-negative to double-positive thymocyte. Immunity 2000; 13(2):187-197.
7. Li CL, Toda K, Saibara T et al. Estrogen deficiency results in enhanced expression of Smoothened of the Hedgehog signaling in the thymus and affects thymocyte development. Int Immunopharmacol 2002; 2(6):823-833.
8. Li CL, Zhang T, Saibara T et al. Thymosin alpha1 accelerates restoration of T cell-mediated neutralizing antibody response in immunocompromised hosts. Int Immunopharmacol 2002; 2(1):39-46.
9. Ruiz i Altaba A, Sanchez P, Dahmane N. Gli and hedgehog in cancer: Tumours, embryos and stem cells. Nat Rev Cancer 2002; 2(5):361-372.
10. Mombaerts P, Iacomini J, Johnson RS et al. RAG-1-deficient mice have no mature B and T lymphocytes. Cell 1992; 68(5):869-877.
11. Levelt CN, Mombaerts P, Iglesias A et al. Restoration of early thymocyte differentiation in T-cell receptor beta-chain-deficient mutant mice by transmembrane signaling through CD3 epsilon. Proc Natl Acad Sci USA 1993; 90(23):11401-11405.
12. Chiang C, Litingtung Y, Lee E et al. Cyclopia and defective axial patterning in mice lacking Sonic hedgehog gene function. Nature 1996; 383(6599):407-413.
13. Shah DK, Hager-Theodorides AL, Outram SV et al. Reduced thymocyte development in sonic hedgehog knockout embryos. J Immunol 2004; 172(4):2296-2306.
14. Bhardwaj G, Murdoch B, Wu D et al. Sonic hedgehog induces the proliferation of primitive human hematopoietic cells via BMP regulation. Nat Immunol 2001; 2(2):172-180.
15. Charron F, Stein E, Jeong J et al. The morphogen sonic hedgehog is an axonal chemoattractant that collaborates with netrin-1 in midline axon guidance. Cell 2003; 113(1):11-23.
16. Pui JC, Allman D, Xu L et al. Notch1 expression in early lymphopoiesis influences B versus T lineage determination. Immunity 1999; 11(3):299-308.
17. Radtke F, Wilson A, Stark G et al. Deficient T cell fate specification in mice with an induced inactivation of Notch1. Immunity 1999; 10(5):547-558.
18. Izon DJ, Punt JA, Pear WS. Deciphering the role of Notch signaling in lymphopoiesis. Curr Opin Immunol 2002; 14(2):192-199.
19. Washburn T, Schweighoffer E, Gridley T et al. Notch activity influences the alphabeta versus gammadelta T cell lineage decision. Cell 1997; 88(6):833-843.
20. Robey E, Chang D, Itano A et al. An activated form of Notch influences the choice between CD4 and CD8 T cell lineages. Cell 1996; 87(3):483-492.
21. Yasutomo K, Doyle C, Miele L et al. The duration of antigen receptor signalling determines CD4+ versus CD8+ T-cell lineage fate. Nature 2000; 404(6777):506-510.
22. Spits H. Development of alphabeta T cells in the human thymus. Nat Rev Immunol 2002; 2(10):760-772.
23. Sacedon R, Varas A, Hernandez-Lopez C et al. Expression of hedgehog proteins in the human thymus. J Histochem Cytochem 2003; 51(11):1557-1566.

24. Gutierrez-Frias C, Sacedon R, Hernandez-Lopez C et al. Sonic hedgehog regulates early human thymocyte differentiation by counteracting the IL-7-induced development of CD34+ precursor cells. J Immunol 2004; 173(8):5046-5053.
25. Okamoto Y, Douek DC, McFarland RD et al. IL-7, the thymus, and naive T cells. Adv Exp Med Biol 2002; 512:81-90.
26. Hammerschmidt M, Brook A, McMahon AP. The world according to hedgehog. Trends Genet 1997; 13(1):14-21.
27. Hager-Theodorides AL, Outram SV, Shah DK et al. Bone morphogenetic protein 2/4 signaling regulates early thymocyte differentiation. J Immunol 2002; 169(10):5496-5504.
28. Tsai PT, Lee RA, Wu H. BMP4 acts upstream of FGF in modulating thymic stroma and regulating thymopoiesis. Blood 2003; 102(12):3947-3953.
29. Mulroy T, McMahon JA, Burakoff SJ et al. Wnt-1 and Wnt-4 regulate thymic cellularity. Eur J Immunol 2002; 32(4):967-971.

Chapter 11

Hedgehog Signalling in Prostate Morphogenesis

Marilyn L.G. Lamm* and Wade Bushman

Abstract

The prostate gland has not traditionally been a popular model system in developmental biology, and mechanistic studies of prostate morphogenesis have generally lagged behind work in other well-characterised systems. The mesenchymal-epithelial interactions in prostate development and the role of testosterone as an inducer of prostate morphogenesis have certainly been a subject of enduring interest, but the lack of molecular markers for prostate differentiation and of transgenic models with prostate-specific mutations have hindered molecular studies. This is changing, and recent findings have catalysed rapid advances in our understanding of prostate development. Studies have shown striking parallels between morphogenetic signals that regulate prostate morphogenesis and paradigms developed from work done in classic developmental model systems. Several growth factors such as fibroblast growth factor 10, bone morphogenetic protein 4 and transforming growth factor β1 apparently play similar roles in the foetal prostate as in other embryonic structures. A major signalling molecule in diverse developmental systems, Sonic hedgehog (Shh) has emerged as a subject of paramount interest in prostate biology. This is in part because of its key role in prostate ductal morphogenesis and differentiation but, largely, because Shh has recently been identified as a factor that promotes human prostate cancer growth. Therefore, the hedgehog signalling pathway is a promising target for therapies to slow or arrest prostate tumour growth.

Prostate Morphogenesis

The prostate is a male accessory sex gland that develops from the urogenital sinus (UGS), a simple tubular endodermal derivative of the embryonic hindgut. The UGS consists of epithelial cells that line its lumen and mesenchymal cells that envelope the epithelium. The outgrowth of the UGS epithelium into the surrounding mesenchyme to form bud-like structures is the earliest discernible morphological evidence of prostate development, and this occurs during embryonic development: at around 10 to 12 weeks of gestation in humans, 17.5 and 18.5 embryonic days in mice and rats, respectively (embryonic day 0 or E0 = day of vaginal plug).[1-3] Prostate development continues as UGS epithelial buds grow into elongated solid tube-like structures that eventually differentiate into a network of canalised branched ducts with secretory functions. The temporal pattern of prostate ductal morphogenesis differs among mammalian species: ductal branching occurs during foetal development in humans but largely during postnatal life in rodents.[1-3] Additionally, the overall architectural organisation of prostatic

*Corresponding Author: Marilyn L.G. Lamm—Department of Pediatrics, Northwestern University Feinberg School of Medicine, Children's Memorial Research Center, Chicago, Illinois 60614, U.S.A. Email: mlamm@northwestern.edu

Shh and Gli Signalling and Development, edited by Carolyn E. Fisher and Sarah E.M. Howie. ©2006 Landes Bioscience and Springer Science+Business Media.

ducts is different: distinct paired ductal lobes (anterior, dorsolateral, and ventral lobes) with characteristic branching networks in rodents and a tubuloalveolar gland in humans.[1,4-7] However, the key morphogenetic events of prostatic epithelial budding, ductal branching, and ductal differentiation are strikingly conserved, suggesting common paradigms of regulation in prostate development among mammalian species.

Mesenchymal-Epithelial Signalling in Prostate Morphogenesis: Role of Androgens

Normal organ development is predicated upon appropriate, often at times reciprocal, interactions between mesenchyme and epithelium, and deregulation of such signalling pathways has been associated with significant birth defects and disease. The initial trigger for prostate morphogenesis is androgen-dependent and originates from the UGS mesenchyme. Testosterone is secreted by the foetal testes shortly before the onset of prostate morphogenesis, i.e., at around 9 weeks gestation in humans and about E13 in rodents, then declines postnatally.[8-10] Testosterone is converted to 5-α dihydrotestosterone (DHT) in the UGS by 5-α reductase and DHT is considered to be the major active androgen that promotes prostate morphogenesis.[3,11] In the presence of exogenous DHT, embryonic male and female rodent UGS form prostatic buds in vitro.[12] Conversely, loss of androgens during foetal development, either through surgical or chemical castration, or loss of androgen sensitivity such as in testicular feminization (Tfm mice), inhibits prostate development.[3,13-15] The UGS mesenchyme (UGM) expresses androgen receptors during gestation and it is the direct tissue target of androgen signalling during foetal prostate development.[16-17] Several experimental approaches, most notably tissue recombination studies, have established the absolute requirement for an androgen-dependent inductive signal from the UGM to the UGS epithelium (UGE) to initiate formation of epithelial prostatic buds.[18] This (these) inductive factor(s) must (1) be a downstream target of androgen signalling in the UGM, (2) be a secreted ligand that can travel from the UGM to the UGE, (3) have functional receptors in the UGE, and (4) directly participate in the process of epithelial bud formation, the morphological event heralding prostate development. The identity of the UGM-derived inductive factor(s) for prostate morphogenesis remains unknown, although several growth factors have been proposed as likely candidates.

Epithelial-Mesenchymal Signalling in Prostate Morphogenesis: Sonic Hedgehog-Gli Pathway

Hedgehog (Hh) proteins are secreted ligands that play critical roles in vertebrate embryonic development. Hh signalling promotes cell proliferation, cell survival, and cell differentiation in several developing organs (see other chapters in this book). There are three known vertebrate *Hh* genes: *Desert hedgehog* (*Dhh*), *Indian hedgehog* (*Ihh*) and *Sonic hedgehog* (*Shh*). *Dhh* is most closely related to the homolog gene *hedgehog* in *Drosophila*; *Ihh* and *Shh* are more related to one another.[19]

In rodents, the *Shh* gene is expressed in the UGS epithelium at E11.5 (the earliest day examined) which is at least 6 days prior to prostatic bud formation.[20] A time course analysis shows that *Shh* gene expression increases during the prebudding period (i.e., prior to E17.5 or E18.5) and remains relatively high throughout the period of prostatic epithelial budding, during late gestation through to birth.[21-23] *Shh* gene and protein expression gradually diminish through the first 10 days after birth, a period characterised by continued bud formation and outgrowth, and additionally, by intense ductal branching in all three distinct lobes of the rodent prostate.[21,22,24-25] Between postnatal days 20 and 30, when the ductal branching process is nearly complete and the initially prominent sheath of prostatic mesenchyme surrounding the prostatic epithelium has considerably thinned out to form the stromal layer around the distinctly and highly branched prostatic ducts, *Shh* gene expression declines to very low levels characteristic of the adult. *Dhh* expression in the UGS is not observed and *Ihh* expression is

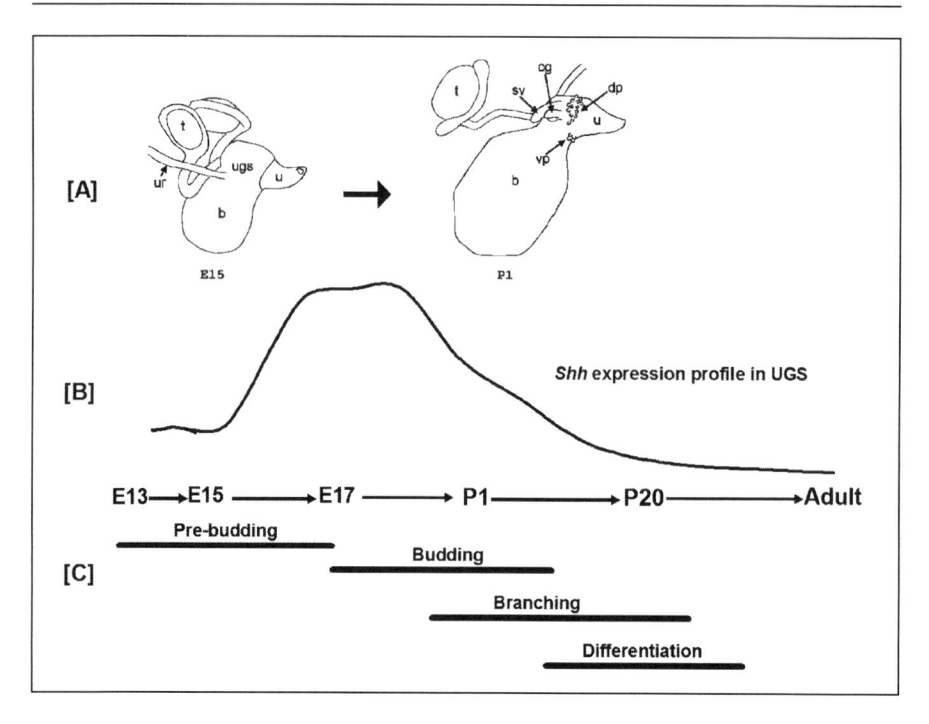

Figure 1. A) Drawings illustrating the gross morphology of the mouse lower urogenital tract at embryonic day 15 (E15) prior to formation of buds from the prostatic anlage in the urogenital sinus (ugs) to its appearance at birth (P1) with nascent buds in the anterior lobe or coagulating gland (cg), dorsal prostate (dp), and ventral prostate (vp). Reprinted with permission from: Lamm MLG, Catbagan WS, Laciak RJ et al. Dev Biol 2002; 249(2):349-366, ©2002 Elsevier. B) Schematic illustration of the expression profile of the *Shh* gene in the mouse UGS during prostate morphogenesis. *Shh* expression increases during the prebudding phase (i.e., prior to E17.5) and remains relatively high during the period of epithelial budding at late gestation through to birth, gradually diminishing through the first 10 days after birth to very low levels in the adult. C) Diagram identifying key events of epithelial budding, ductal branching, and ductal differentiation during a timeline of prostate morphogenesis. t: testis; ur: ureter; b: bladder; u: urethra; sv: seminal vesicle.

very low.[21] A schematic illustration of how *Shh* expression fits in the timeline of key morphogenetic events in prostate development is presented in Figure 1.

As in the developing prostate in rodents, Shh expression (demonstrated at the protein level) in the human foetal prostatic epithelium also increases coincident with the onset of ductal budding and outgrowth: from 9.5 weeks (earliest time point examined) through to 13 weeks of gestation, with expression particularly robust in newly formed prostatic buds.[24] Unlike the time course of Shh expression in rodent prostate development, however, Shh is down-regulated prior to birth i.e., expression gradually decreases from week 16 through to week 20, and is absent at 34 weeks of gestation.[24] This period of diminishing Shh expression in foetal human prostate coincides with extensive prostatic ductal branching,[1] a curious similarity with events during the early postnatal period in rodents when Shh levels are also declining. In contrast to the very low level of expression in mouse prostate, *Shh* message in the adult human prostate is surprisingly high and this might be attributable to a wide range of histopathologic conditions to which the human prostate is exposed throughout its adult lifespan.[26]

Shh gene and protein expression is localised strictly in the epithelium in both rodent and human developing prostate.[21-25] In situ hybridisation analysis, as shown in Figure 2, reveals a pattern of distribution that begins with uniform *Shh* expression throughout the UGS

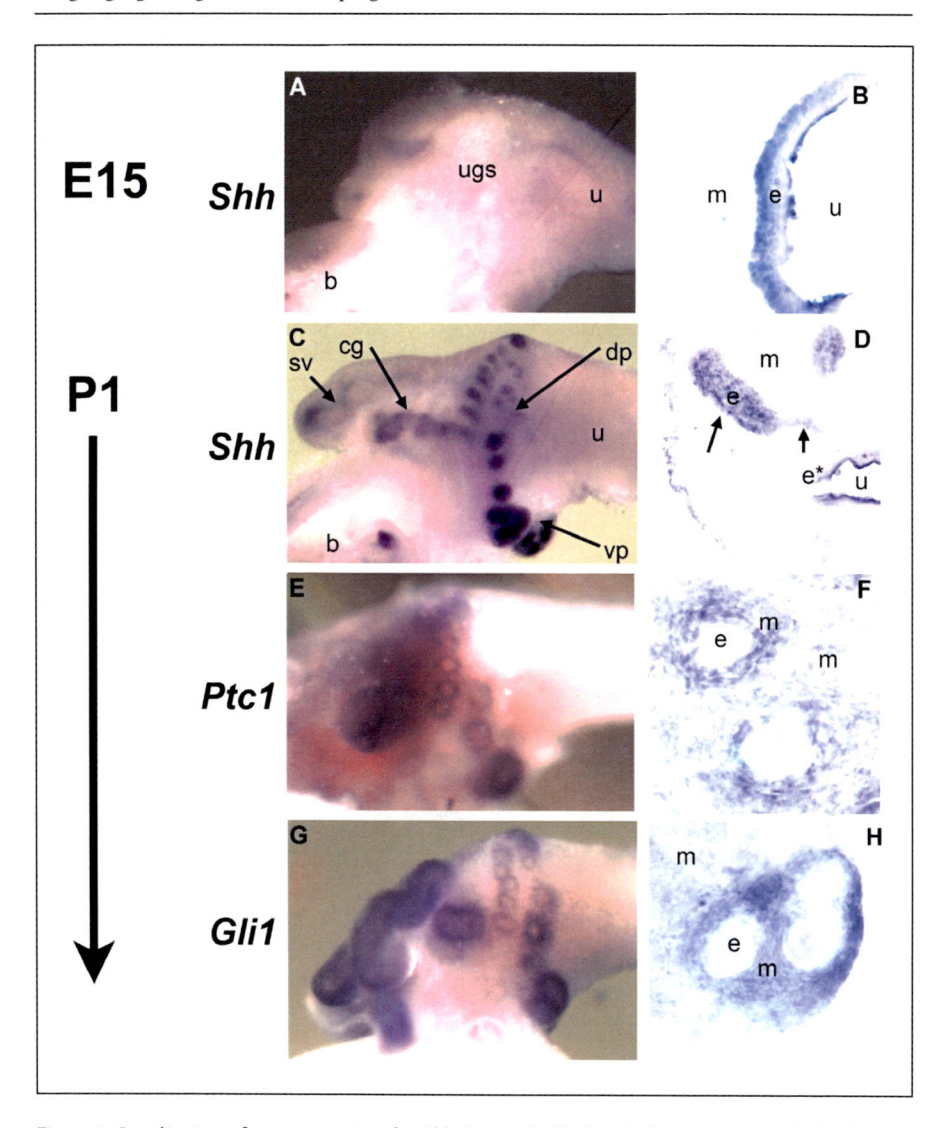

Figure 2. Localisation of gene expression for *Shh*, *Ptc1* and *Gli1* by whole mount in situ hybridisation. Although staining for *Shh* expression is not visible in whole mounts of E15 mouse UGS (A), uniform expression is evident in epithelium (e) lining the lumen of the urethra (u) in whole mount sections (B). C) At P1, *Shh* expression is focused to the nascent buds of the dorsal prostate (dp), coagulating gland (cg), and ventral prostate (vp). D) Apparent concentration of *Shh* expression is exhibited in the epithelium (e) of the distal duct (long arrow) relative to the proximal duct (short arrow), with diminished expression in the epithelium (e*) of the urethra, u. No *Shh* expression is detected in the mesenchyme, m, at any stage of prostate development. Expression of *Ptc1* (E) and *Gli1* (G) surround the prostatic buds, and expression of both genes is more concentrated in the mesenchyme immediately surrounding the epithelium source of the Shh ligand (F,H). Low level expression of both genes is also observed in the prostatic epithelium suggesting the possibility of autocrine signalling. Reprinted with permission from: Lamm MLG, Catbagan WS, Laciak RJ et al. Dev Biol 2002; 249(2):349-366, ©2002 Elsevier.

epithelium during the prebudding phase, transitions to greater localisation in epithelial clusters or buds that evaginate into the mesenchyme accompanied by diminished expression in the UGS luminal epithelium, and becomes more restricted in the advancing apical or distal regions of prostatic ducts.[22,25]

Consistent with paracrine signalling, the genes for the Shh receptor *Patched* (*Ptc1*) and the *Gli* family of known transcriptional activators of hedgehog signalling (*Gli1* and *Gli2*) are highly localised in the mesenchyme of the UGS immediately surrounding the epithelial source of the *Shh* ligand.[22] The spatial relationship in expression patterns of *Shh*, *Ptc1*, and *Gli1* in the UGS is shown in Figure 2. The expression of *Gli3*, another transcriptional regulator of hedgehog signalling, is diffuse throughout the UGM.[22] Expression of *Ptc1*, *Gli1* and *Gli3*, albeit low, was also detected in the UGE, suggesting some degree of autocrine signalling interaction.[22,25] As with *Shh*, levels of expression for *Ptc1* and the *Gli* transcription factors increase coincident with onset of prostatic budding in mice and gradually decrease postnatally.[22,25] As *Shh* expression becomes localised to the apical regions of elongating ducts, the expression for *Ptc1* appears to be strongest in the mesenchyme surrounding the distal ducts relative to the proximal ducts.[22,25] Likewise, the expression of the three *Gli* genes in the ductal mesenchyme exhibits a proximodistal gradient.[25] This asymmetric distribution of elements of the Shh-Gli pathway during embryonic ductal morphogenesis may signal the early establishment of a proximodistal heterogeneity in the morphology and function of the adult prostatic ducts.[27]

Shh Signalling during the Budding Phase of Prostate Morphogenesis

There is evidence that epithelial-mesenchymal interaction via the Shh-Gli pathway occurs as early as the prebud stage in foetal prostate development. Exogenous Shh peptide exerts an inductive effect on both *Ptc* and *Gli1* gene expression (known downstream targets of the pathway) in isolated mouse male E14 UGS; this effect is direct and inhibited by cyclopamine, a specific and potent chemical inhibitor of hedgehog action.[22] Indeed, cyclopamine inhibition of hedgehog signalling in E14 UGS inhibits epithelial and mesenchymal cell proliferation.[22]

Concurrent with onset of prostate morphogenesis marked by bud formation, the rodent UGS at late gestation and during the early postnatal period is characterised by relatively high levels of expression of *Shh*, *Ptc* and the *Gli* transcription factors. However, functional studies of Hh signalling using antibody blockade, chemical inhibition and genetic loss of function models, have yielded somewhat conflicting data on the requirement for Shh signalling in normal prostate morphogenesis. Antibody blockade using a polyclonal antibody to Shh appeared to block prostate development in a subcapsular renal graft model.[21] Studies of the Shh null transgenic mouse, however, showed that the UGS from this mutant could undergo budding morphogenesis in organ culture and, when transplanted under the renal capsule of an adult male host mouse, could undergo glandular morphogenesis with apparently normal prostatic morphology. Explants of UGS from Shh null mutant male mice can be induced to form prostatic buds when grown in the presence of androgenic support.[23,28] Since a quantitative comparison of prostatic ducts between androgen-treated explants from Shh null mutant mice and their wild-type counterparts was not available, a possible role for hedgehog signalling in the formation of a full compliment of ductal buds cannot be unequivocally excluded. Indeed, the mean total number of prostate buds formed in E18.5 UGS of Shh null mutant male and female mice exposed to DHT in utero appeared to be less than those in wild-type controls.[28] A key concern with these genetic studies is that the Shh null only abrogates Shh function as opposed to globally blocking Hh signalling, as cyclopamine does. Since Shh is not the only hedgehog ligand expressed in the urogenital sinus, the potential for functional redundancy in Hh ligands exists.

Chemical inhibition of Hh signalling with cyclopamine has produced a variety of observations, including inhibition of ductal budding, altered ductal bud morphology, increased ductal branching, and changes in epithelial cell proliferation and differentiation. The seemingly contradicting results from cyclopamine inhibition studies may be a function of the stage in prostate development when signalling is disrupted. When cyclopamine inhibition of hedgehog signalling is initiated in the prebud E14 mouse UGS, epithelial cell proliferation is decreased and total number of prostatic buds is apparently reduced.[22] These data, together with results showing abrogation of growth and glandular morphogenesis following Shh antibody blockade in E15 UGS,[21] suggest an early requirement for hedgehog signalling in prostate growth and morphogenesis. However, when initiated later in development i.e., in E16.5 mouse UGS or the neonate rat ventral prostate (VP), cyclopamine treatment produces opposite effects: epithelial cell proliferation is increased, prostate growth is enhanced, and number of ducts is either increased or not significantly affected.[28,29] Conversely, exogenous Shh inhibits cell proliferation and decreases the number of prostatic ducts.[28,29] In addition, exogenous Shh promotes terminal differentiation of luminal epithelial cells and appears to pattern slender elongated prostatic ducts.[23,28,29] Collectively, these data may indicate a possible shift in the role for Shh signalling in prostate morphogenesis: from promoting bud formation and outgrowth via increased epithelial cell proliferation during early development to, later, a role in branching morphogenesis and differentiation which entails inhibition of epithelial cell proliferation.

Shh Signalling in Prostatic Ductal Branching

There is some evidence that Shh signalling regulates postnatal branching morphogenesis. Exogenous Shh treatment of neonate rat VP explants leads to a reduction in ductal branching revealing a more expansive mesenchymal area, whereas cyclopamine inhibition of Shh signalling increases the formation of ductal branches into the thinning mesenchyme.[23,25,29] Given this inhibitory action of Shh, the postnatal decline in expression levels of *Shh*, *Ptc* and the *Gli* genes can be viewed as permissive for intensive branching activities.

The inhibitory action of Shh on ductal branching has been linked to factors that appear to be downstream of the pathway: see Figure 3. Shh upregulates the expression of *Transforming Growth Factor-β1 (TGF-β1)* and *activin A* which are both expressed in prostatic mesenchyme in spatial association with distal epithelial ducts, and which are both known to inhibit prostate branching morphogenesis.[18,29,30] *Bone morphogenetic protein 4 (Bmp4)* is another member of the TGF family that is expressed in the prostatic mesenchyme, particularly strongly in areas separating nascent buds, and *Bmp4* restricts prostatic ductal outgrowth and branching.[31] However, whether *Bmp4* is a direct target of Shh signalling in the prostate remains to be resolved.[25,29] Shh has also been shown to down-regulate the expression of mesenchymal *Fibroblast growth factor 10 (Fgf10)*, and exogenous Fgf10 can reverse Shh-mediated inhibition of prostate growth and branching in rat ventral prostate.[25] A model for ductal branching that involves the interaction of Shh, FGF10, and Bmp4 was recently proposed.[25] Whether this model, which is based largely on the dichotomous branching pattern of the VP, will stand up to rigorous experimental challenge remains to be determined; however, it has introduced an important discussion of possible signalling interactions regulating ductal morphogenesis and will serve as a testable hypothesis for future mechanistic studies.

The prostate branching architecture in rodents is lobe-specific and hints at unique pathways of regulation. However, the postnatal rodent ventral prostate has been used almost exclusively to study branching morphogenesis. A recent study indicates prostate lobe-specific responses in both Shh signalling and branching morphogenesis to high-dose oestrogen exposure.[25] Thus, a clear understanding of prostatic ductal morphogenesis and the role that Shh plays in this process requires studies targeted at all prostatic lobes.

Shh Signalling during Ductal Outgrowth and Differentiation

Shh has been implicated in the patterning of prostatic ducts as they continue to grow and extend into the mesenchyme. UGS exhibited slender ducts when treated with exogenous Shh and enlarged blunt-ended ducts when exposed to cyclopamine.[28,29] These effects maybe explained, in part, by the anti-proliferative action of Shh on epithelial cells during this phase in prostate morphogenesis resulting in thinner ducts. Shh may also regulate the surrounding mesenchyme, and cyclopamine inhibition of signalling could disrupt mesenchymal/stromal organisation and contribute to altered ductal morphology. That the morphology of ductal tips

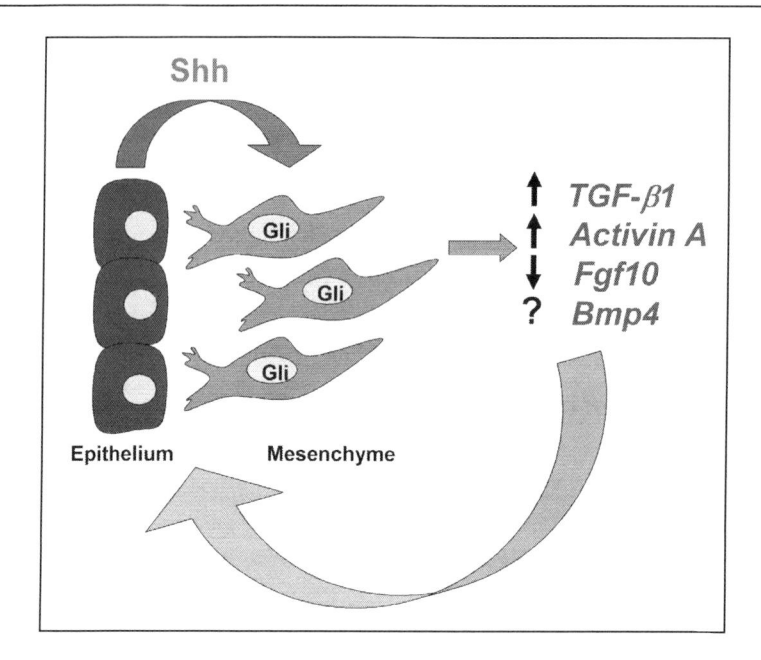

Figure 3. Schematic illustrating Shh-driven epithelial-mesenchymal interactions in the developing prostate. Recent studies have identified some downstream target genes of the Shh-Gli signalling pathway in the mesenchyme of the developing prostate. *TGF-β1* and *activin A* are postulated to inhibit epithelial cell proliferation and facilitate prostate branching morphogenesis.[29] A decrease in *Fgf10* expression is postulated as the proximate cause for Shh-mediated growth inhibition in the prostate.[25] *Bmp4* has been shown to inhibit prostate ductal budding and morphogenesis,[31] but conflicting data exist as to whether it is a downstream target of the Shh pathway in the prostate.

is altered in the absence of Hh signalling is significant in light of observations that Shh expression is more focused in the apical distal areas of elongating ducts in association with high level expression of *Ptc*, *Gli1* and *Gli2*.[22,25]

Prostatic buds grow out into the UGM initially as solid cords of epithelial cells. Concurrent with ductal canalisation, epithelial cells differentiate into basal and luminal cells which exhibit distinctive expression patterns of cytokeratins (CKs) and p63.[32] The link between Shh and the terminal differentiation of ductal epithelial cells has been investigated recently. Exogenous Shh increased the proportion of epithelial cells that did not express CK14 and p63, indicative of increased luminal cell differentiation; conversely, cyclopamine inhibited differentiation.[29] In another study, however, cyclopamine accelerated both ductal canalisation and epithelial cell differentiation, suggesting that Shh has an inhibitory effect on these processes.[23] Since lesions in ductal cell differentiation manifest themselves in prostatic diseases including cancer, a clear understanding of the role of Shh signalling in this morphogenetic event needs to be established.

Concluding Remarks

Studies to date have established an important role for hedgehog signalling in prostate development. However, the picture is far from complete. At least five important questions remain to be answered.

1. How is the expression of *Shh* scripted in a process that is fundamentally androgen dependent? Prostate development is absolutely dependent on testosterone. However, the exact mechanism of action(s) of testosterone remains almost a complete mystery. Despite considerable effort, no factor that plays an important growth-inducing role in prostate development has been shown to be strictly androgen dependent. *Shh* expression may be somewhat

increased by testosterone, but the effect is not robust enough to be a trigger for prostate development.[21] What is intriguing is that the spatial pattern of *Shh* expression in the prebud phase seems to be influenced by testosterone[22] suggesting that one action of testosterone may be to specify or pattern the expression of factors at sites of future epithelial ductal outgrowth.

2. Is there functional redundancy in Hh peptides that mitigates the effect of genetic loss of Shh function? There is the potential that functional redundancy in Hh ligand may complicate the interpretation of experiments performed with the Shh null mutant. Further work is necessary to determine whether Ihh could provide some degree of functional compensation in the absence of Shh.

3. What are the signalling interactions that regulate Shh expression and action during ductal budding and ductal morphogenesis? Studies to date suggest that Shh exerts dichotomous actions in ductal budding and ductal morphogenesis that may be explained by differential responses of target cells at specific stages in prostate development. Shh, Fgf10, Bmp4, TGF-β1 and activin, all appear to have important roles to play during ductal branching morphogenesis. It remains to be determined whether these signalling interactions also regulate ductal budding.

4. What are the targets of Shh activation during prostate development? While some apparently conserved Hh target genes such as *Insulin-like Growth Factor Binding Protein-6* are expressed in the developing prostate in a Hh dependent fashion,[33] the full complement of Shh activated target genes is unknown. In particular, it remains to be determined whether Shh induces the expression of any genes that are unique to the prostate.

5. What are the mechanisms that integrate the actions of Shh in ductal morphogenesis and the process of terminal differentiation? The expression of Shh during the continuum of activities from bud formation to ductal growth and branching is characterised by a dynamic evolution that correlates with morphologic changes and coordinate differentiation. Several studies suggest that inhibition of Shh action during postnatal development affects both ductal growth and cell differentiation. Understanding how growth and differentiation are linked to the actions of Shh—whether they are both regulated directly by Shh, linked in an epistatic hierarchy, or are both down-stream of a single Shh-controlled regulator—is an important and answerable question.

Studies on the role of Shh in prostate development have assumed added significance due to recent findings showing an important role for Shh signalling in prostate cancer growth and progression. A better understanding of the actions of Shh in normal prostate morphogenesis may clarify its role in the genesis of prostate cancer and its role in tumour progression and provide insights into the potential therapeutic uses of pharmacological Hh antagonists.

References

1. Lowsley OS. The development of the human prostate gland with reference to the development of other structures at the neck of the urinary bladder. Am J Anat 1912; 13:299-349.
2. Kellokumpu-Lehtinen P. Development of sexual dimorphism in human urogemital sinus complex. Biol Neonate 1985; 48:157-167.
3. Cunha GR, Donjacour AA, Cooke PS et al. The endocrinology and developmental biology of the prostate. Endocr Rev 1987; 8:338-362.
4. McNeal JE. Regional morphology and pathology of the prostate. Am J Clin Pathol 1968; 49:347-357.
5. Sugimura Y, Cunha GR, Donjacour AA. Morphogenesis of ductal networks in the mouse prostate. Biol Reprod 1986; 34:961-971.
6. Hayashi N, Sugimura Y, Kawamura J et al. Morphological and functional heterogeneity in the rat prostatic gland. Biol Reprod 1991; 45:308-321.
7. Timms BG, Mohs TJ, Didio LJA. Ductal budding and branching patterns in the developing prostate. J Urol 1994; 151:1427-1432.
8. Siiteri PK, Wilson JD. Testosterone formation and metabolism during male sexual differentiation in the human embryo. J Clin Endocrinol Metab 1974; 38:113-125.

9. Pointis G, Latreille MT, Mignot TM et al. Regulation of testosterone synthesis in the fetal mouse testis. J Steroid Biochem 1979; 11:1609-1612.
10. Corpechot C, Baulieu EE, Robel P. Testosterone, dihydrotestosterone and androstanediols in plasma, testes and prostates of rats during development. Acta Endocrinol (Copenh) 1981; 96:127-135.
11. Tsuji M, Shima H, Terada N et al. 5α-reductase activity in developing urogenital tracts of fetal and neonatal male mice. Endocrinology 1994; 134:2198-2205.
12. Lasnitzki I, Mizuno T. Induction of the rat prostate gland by androgens in organ culture. J Endocrinol 1977; 74:47-55.
13. Jost A. Problems of fetal endocrinology: The gonadal and hypophyseal hormones. Recent Prog Horm Res 1953; 8:379-418.
14. Wells LJ, Cavanaugh MW, Maxwell EL. Genital abnormalities in castrated fetal rats and their prevention by means of testosterone propionate. Anat Rec 1954; 118:109-133.
15. Neumann F, Elger W, Kramer M. Development of vagina in male rats by inhibiting androgen receptors with an antiandrogen during the critical phase of organogenesis. Endocrinology 1966; 78:628-632.
16. Takeda H, Mizuno T, Lasnitzki I. Autoradiographic studies of androgen-binding sites in the rat urogenital sinus and postnatal prostate. J Endocrinol 1985; 104:87-92.
17. Cooke PS, Young P, Cunha GR. Androgen receptor expression in developing male reproductive organs. Endocrinology 1991; 128:2867-2873.
18. Cunha GR, Alarid ET, Turner T et al. Normal and abnormal development of the male urogenital tract: Role of androgens, mesenchymal-epithelial interactions, and growth factors. J Androl 1992; 13:465-475.
19. Ingham PW, McMahon AP. Hedgehog signaling in animal development: Paradigms and principles. Genes Dev 2001; 15:3059-3087.
20. Bitgood MJ, McMahon AP. Hedgehog and Bmp genes are coexpressed at many diverse sites of cell-cell interaction in the mouse embryo. Dev Biol 1995; 172:126-138.
21. Podlasek CA, Barnett DH, Clemens JQ et al. Prostate development requires sonic hedgehog expressed by the urogenital sinus epithelium. Dev Biol 1999; 209:28-39.
22. Lamm MLG, Catbagan WS, Laciak RJ et al. Sonic hedgehog activates mesenchymal Gli1 expression during prostate ductal bud formation. Dev Biol 2002; 249:349-366.
23. Freestone SH, Marker P, Grace OC et al. Sonic hedgehog regulates prostatic growth and epithelial differentiation. Dev Biol 2003; 264:352-362.
24. Barnett DH, Huang H-Y, Wu X-R et al. The human prostate expresses sonic hedgehog during fetal development. J Urol 2002; 168:2206-2210.
25. Pu Y, Huang L, Prins GS. Sonic hedgehog-patched Gli signaling in the developing rat prostate gland: Lobe-specific suppression by neonatal estrogens reduces ductal growth and branching. Dev Biol 2004; 273:257-275.
26. Fan L, Pepicelli CV, Dibble CC et al. Hedgehog signaling promotes prostate xenograft tumor growth. Endocrinology 2004; 145:3961-3970.
27. Lee C. Biology of the prostatic ductal system. In Naz RK, ed. Prostate: Basic and Clinical Aspects. Boca Raton: CRC Press, 1997:53-71.
28. Berman DM, Desai N, Wang X et al. Roles for hedgehog signaling in androgen production and prostate ductal morphogenesis. Dev Biol 2004; 267:387-398.
29. Wang B, Shou J, Ross S et al. Inhibition of epithelial ductal branching in the prostate by sonic hedgehog is indirectly mediated by stromal cells. J Biol Chem 2003; 278:18506-18513.
30. Cancilla B, Jarred RA, Wang H et al. Regulation of prostate branching morphogenesis by activin A and follistatin. Dev Biol 2001; 237:145-158.
31. Lamm MLG, Podlasek CA, Barnett DH et al. Mesenchymal factor bone morphogenetic protein 4 restricts ductal budding and branching morphogenesis in the developing prostate. Dev Biol 2001; 232:301-314.
32. Hayward SW, Baskin LS, Haughney PC et al. Epithelial development in the rat ventral prostate, anterior prostate and seminal vesicle. Acta Anat (Basel) 1996; 155:81-93.
33. Lipinski RJ, Cook CH, Barnett DH et al. Sonic hedgehog signaling regulates the expression of Insulin-like growth factor binding protein-6 during fetal prostate development. Dev Dyn 2005; 233(3):829.

CHAPTER 12

Sonic Hedgehog Signalling in Visceral Organ Development

Huimin Zhang, Ying Litingtung and Chin Chiang*

Abstract

The secreted signalling molecule encoded by *Sonic hedgehog* (*Shh*) has been shown to play an indispensable role in mammalian organogenesis. During embryonic development, one of the prominent sites of *Shh* expression is in the tubular gut endoderm and its derivatives such as the esophagus, lung, stomach and intestine. Loss of *Shh* function results in profound growth and patterning defects of the gastrointestinal tract and associated organs. Furthermore, misregulation of Shh signalling in human patients has been implicated in a variety of gastrointestinal tumors. In this chapter, we will discuss studies that reveal the critical roles of Shh signalling in mammalian visceral organ development and homeostasis.

Introduction

Gut morphogenesis in mouse begins around embryonic day 8 (E8.0) when the lateral edges of the flat endodermal sheet begin to converge medio-ventrally by a complex process of differential growth and embryonic folding beginning at the cephalic and lateral regions and progressing caudally. As gut tube closure is completed by E9.0, complex patterning events involving inductive interactions between gut endoderm and surrounding mesoderm begin along the anterior-posterior (AP) axis, regionalizing the gut into defined organ segments such as esophagus, lung, stomach, spleen, liver, duodenum, pancreas, intestines and rectum (Fig. 1). *Shh* is expressed broadly in the developing gut endoderm but its expression becomes gradually regionalized during gut differentiation. *Shh* expression is mostly excluded from mature organs, however, focal expression in specific compartments of the stomach and intestine can be detected.[17,32] Shh provides the instructive signal essential for the proliferation and differentiation of the gut mesoderm. This is achieved through binding to its receptor Patched (Ptch), permitting activation of downstream target genes mediated by the Gli family of zinc-finger transcription factors.[1] Genetic and biochemical studies have revealed that Gli1 and Gli2 function as activators,[2,3] while Gli3 possesses both activator and repressor functions.[3-6] Several factors expressed in the gut mesoderm are known to directly influence Shh activity. In particular, the secreted Hedgehog-interacting protein, Hip, functions to inhibit Shh from binding to its receptor.[7] In this chapter, we will review the critical roles of Shh signalling in the development and pathogenesis of the gastrointestinal tract and associated organs.

*Corresponding Author: Chin Chiang—Department of Cell and Developmental Biology, Vanderbilt University Medical Center, 465 21st Ave. South, Nashville, Tennessee 37232, U.S.A. Email: chin.chiang@vanderbilt.edu

Shh and Gli Signalling and Development, edited by Carolyn E. Fisher and Sarah E.M. Howie. ©2006 Landes Bioscience and Springer Science+Business Media.

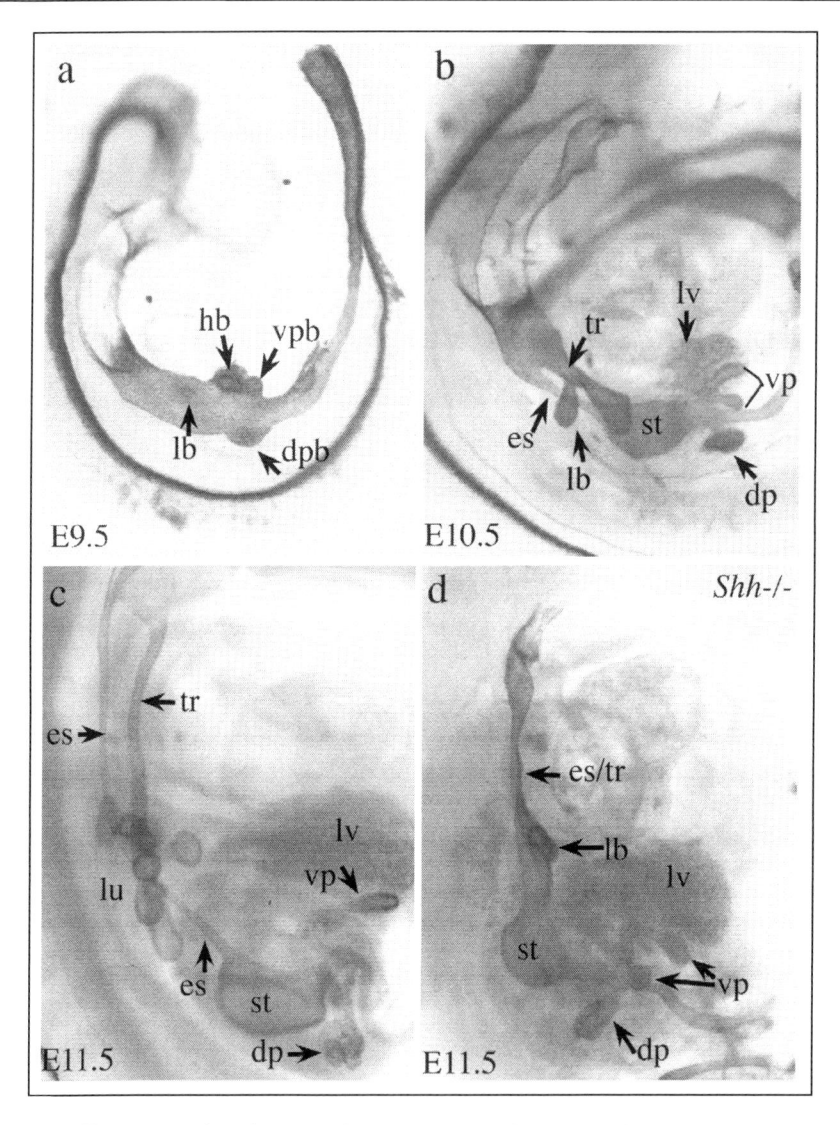

Figure 1. Differentiation of gut derivatives during early stages of embryonic development. Whole-mount immunohistochemistry on E9.5 (a), E10.5 (b) and E11.5 (c,d) wild-type (a-c) and *Shh-/-* (d) embryos using an antibody specific for Hnf3β (adapted, with permission, from Litingtung et al. Nature Genetics 1998; 20:58-61, ©1998 Nature Genetics). Emerging buds of lung (lb), liver (hb), ventral and dorsal pancreas (vpb and dpb) from gut endoderm are clearly evident at E9.5 (arrows). By E11.5, the wild-type esophagus (es) and trachea (tr) are completely separated, whereas *Shh-/-* trachea and esophagus remain attached to each other (d). Note that the growth of *Shh-/-* lung (lb) is severely affected. Pancreas (vp and dp) and liver (lv) development appear to be normal at E11.5.

Esophagus

The esophagus differentiates from the dorsal foregut endoderm as the trachea, with a pair of lung buds, emerges ventrally (Fig. 1). *Shh* expression in the foregut endoderm can be detected as early as E8.5 when closure of the tubular gut begins rostrocaudally. *Shh* expression is excluded from the dorsal foregut endoderm prior to the separation of the esophagus and trachea.[8] Remarkably, loss of *Shh* function results in shortening and severe narrowing of the esophagus which also fails to separate from the trachea (Fig. 1).[8,9] These characteristics are reminiscent of a spectrum of human foregut congenital malformations known as esophageal atresia (EA) and tracheoesophageal fistula (TEF).[10] These EA/TEF phenotypes can be recapitulated when *Gli2* and *Gli3* functions are both eliminated,[11] consistent with the critical role of Shh signalling in patterning foregut derivatives. The mechanism by which absence of *Shh* disrupts esophageal development is not well understood. Given that *Shh* is not expressed in the early dorsal foregut endoderm, it is possible that Shh signalling may have an earlier role in the maintenance of endodermal progenitor cells that contribute to the foregut endoderm. Disruption of retinoid acid (RA) signalling, by ablating the functions of several members of the retinoid acid receptor family, leads to similar EA/TEF phenotypes,[12] raising the possibility that Shh signalling may interact with RA signalling during foregut morphogenesis.

Notably, *Shh* expression is excluded from the adult esophagus. Recent studies have revealed that the expressions of *Shh* and its pathway components are activated in several primary esophageal tumor cell lines.[13] The ability of Hh pathway inhibitor, cyclopamine, to block growth of these epithelial tumor cell lines suggests that Shh functions as a mitogen and/or survival factor via autocrine signalling. Whether Shh pathway activation is required for the initiation and/or maintenance of esophageal tumor phenotype remains to be determined.

Lung

Lung morphogenesis in the mouse starts around E9.0 when a lung primordium can be distinguished on the ventral side of the upper foregut (Fig. 1). This newly formed lung primordium divides laterally into two buds as they invade the surrounding splanchnic mesenchyme.[14] Starting around E10.5, a sequential and highly ordered patterning event, termed branching morphogenesis, occurs in the epithelium to generate the bronchial tree and the proximal-distal axis of the lung.[15] Concomitant with bronchial tree morphogenesis, the surrounding splanchnic mesenchyme also undergoes a series of differentiation events resulting in the generation of airway smooth muscle which is juxtaposed to the proximal bronchial tubules, blood vessels and neural networks.[16]

During murine lung branching morphogenesis, *Shh* transcripts are localized throughout the developing respiratory epithelium with high levels at the distal tips. This expression pattern is maintained until E16.5, when Shh protein becomes more localized to nonciliated cells in the bronchiolar and bronchial epithelium.[8,17,18] Low level *Shh* expression remains detectable in the alveolar and bronchial epithelia up to postnatal day 24, while adult lung epithelial cells are devoid of *Shh* expression.[19] Detailed *Shh* expression patterns have also been reported in rat and human lungs indicating great similarities across species.[20] Targeted deletion of *Shh* or its signalling components leads to severe retardation in lung growth and branching morphogenesis (Fig. 2). Remarkably, mutant mice lacking both *Gli2* and *Gli3* show absence of lung,[11] a phenotype that is much more severe than that of *Shh* mutant.[8,9] This observation suggests that either *Ihh* partially compensates for the loss of *Shh* or that Gli family proteins may have other functions independent of *Shh* signalling. Further studies are necessary to distinguish among these possibilities. Disruption of *Shh* signalling in the lung also causes defects in mesenchymal cell proliferation and differentiation, leading to reduced mesenchymal cell numbers as well as disrupted vasculogenesis and bronchial myogenesis.[9,21] By contrast, excessive proliferation of lung mesenchymal cells is observed in transgenic lungs in which *Shh* signalling is upregulated by either *Shh* overexpression in the endoderm[18] or removal of the *Shh* inhibitor, Hip1, in the mesenchyme.[22] Taken together, these observations indicate that *Shh* signalling is crucial for normal epithelial branching as well as mesenchymal cell proliferation and differentiation during lung development. The mechanism by which *Shh* regulates branching is not well understood. This is in part complicated by the fact that *Shh* is also required for lung mesenchymal cell proliferation and differentiation. Fibroblast growth factor 10 (Fgf10) has been shown to be a key

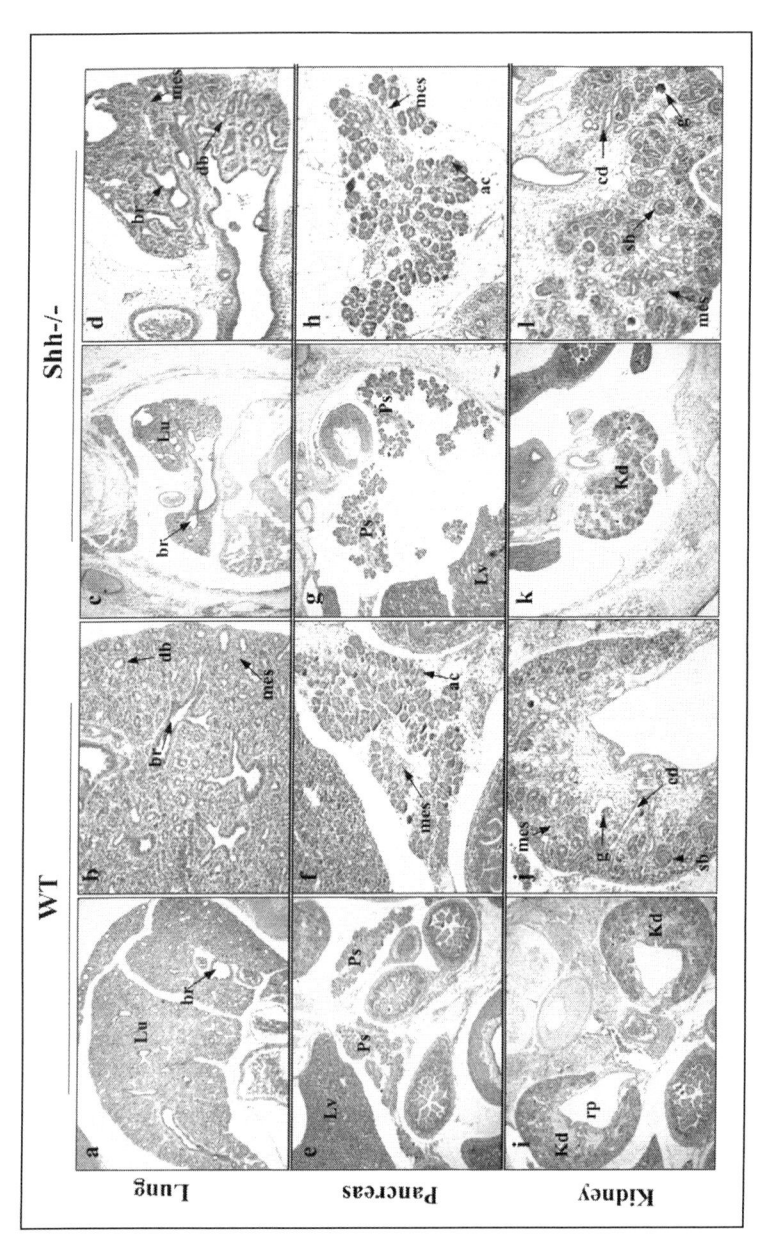

Figure 2. Hematoxylin and eosin staining of E16.5 WT (a,b,e,f,i,j) and Shh-/- (c,d,g,h,k,l) cross-sections of lung (a-d), pancreas (e-h) and kidney (i-l). The Shh-/- lung (Lu) displays absence of bilateral asymmetry and defective branching morphogenesis (c, d). The Shh-/- kidney (Kd) is severely hypoplastic and exhibit fusion of left and right kidney as well as loss of renal pelvis (rp) structure (k,l). The morphology of Shh-/- pancreas (Ps: g,h) appears to be normal when compared with WT (e,f). Abbreviations: ac: acinar cells; br: bronchus; cd: collecting ducts; db: distal bronchioles; g: glomerular; Kd: kidney; Lu: lung; Lv: liver; mes: mesenchyme; Ps: pancreas; rp: renal pelvis; sb: S-shape body. Magnification: 40x for panel a, c, e, g, i and k; 100x for panel b, d, f, h, j and l.

secreted factor expressed in the distal lung mesenchyme that regulates branching morphogenesis.[23,24] It has been reported that exogenous Shh protein can repress the expressions of several Fgfs in culture.[25] This finding combined with the observation that *Fgf10* expression domain is expanded in *Shh* mutant lungs,[8,9] suggest that *Shh* signalling may regulate focal budding process by restricting *Fgf10* expression in the lung mesenchyme.[26]

Recently, some progress has been made in unraveling the molecular mechanism of *Shh*-mediated regulation of cell proliferation and differentiation in the lung. As mentioned earlier, Gli3 is a bipartite transcription factor capable of functioning as an activator (Gli3A) or as a repressor (Gli3R) upon cleavage of full-length Gli3.[4] Recent studies have revealed that *Shh* controls the balance of Gli3R and Gli3A species in the developing lung. Abrogation of *Shh* function as in *Shh-/-* mutant lung or Shh signalling blockade in lung explants significantly shifts the balance in favor of Gli3R.[21] The accumulation of Gli3R species appears to contribute significantly to the *Shh-/-* lung phenotype, as removal of *Gli3* can partially restore growth potential and vascular differentiation in *Shh-/-* lung.[21] However, it is not clear to what extent Gli3A contributes to proper lung development. In *Gli3* mutants, defective lung lobulation has been reported.[27]

Although *Shh* expression is not detectable in normal adult lungs, the Shh pathway appears to be involved in maintaining lung homeostasis. Recently, considerable attention has been given to the role of Shh signalling in airway epithelium remodeling and lung disease progression. For instance, activation of Shh pathway has been documented during repair of acute airway injury.[28] Moreover, Shh pathway activation has been detected in several small-cell lung cancer (SCLC) cell lines.[28] The growth of these tumor cell lines in nude mice xenografts appears to be dependent on Hh pathway activation, as administration of cyclopamine, a Hh pathway-specific inhibitor, can completely block tumor formation.[28] The Shh signalling pathway has also been implicated in the pathogenesis of interstitial lung fibrosis, a disease caused by the presence of hyperproliferative interstitial fibroblast cells due to injury to the airway epithelium.[29]

Stomach

The stomach is a distinct and specialized compartment of the gastrointestinal tract formed by regionalization and differentiation of the most distal part of the foregut (Fig. 3). The developing stomach, like the rest of the gut, is lined by an endodermal epithelium which is surrounded by mesenchymal cells of splanchnic mesodermal origin. While these mesenchymal cells differentiate into tissues such as smooth muscle, by contrast, the stomach endodermal layer remains relatively undifferentiated until late gestation when cytodifferentiation occurs to generate the gastric epithelium with gastric unit primordia or buds. These gastric buds undergo complex morphogenesis postnatally to generate tubular invaginations, known as gastric units, into the lamina propria.[30] The gastric epithelium of the adult mouse stomach can be subdivided based on distinct morphological and functional characteristics; the proximal forestomach is composed of stratified squamous epithelium while the distal portion of the stomach is composed of glandular epithelium which can be further subdivided into three zones: the zymogenic, mucoparietal and pure mucous zones.[31] Gastric units in the zymogenic zone are highly organized vertical structures with compartmentalized regions consisting of the apical pit followed by the isthmus, a neck and a base. The neck and base are situated in the lower part and constitute the gland region of the gastric unit.[32] Within each gastric unit is a distinct arrangement of cells including mucus-producing pit cells, acid-producing parietal cells and pepsinogen-producing zymogenic cells.[31] The distinct epithelial cell types of the zymogenic zone have been shown to be continuously self-renewed and replenished by proliferating stem cells in the isthmus of the gastric unit.[30,31,33]

Shh is expressed in the developing mouse stomach epithelium with high expression in the forestomach and lower expression level in the hindstomach. Strong epithelial expression of *Shh* has been found to be associated with high expression of *Bmp4* in the adjacent mesenchyme, as

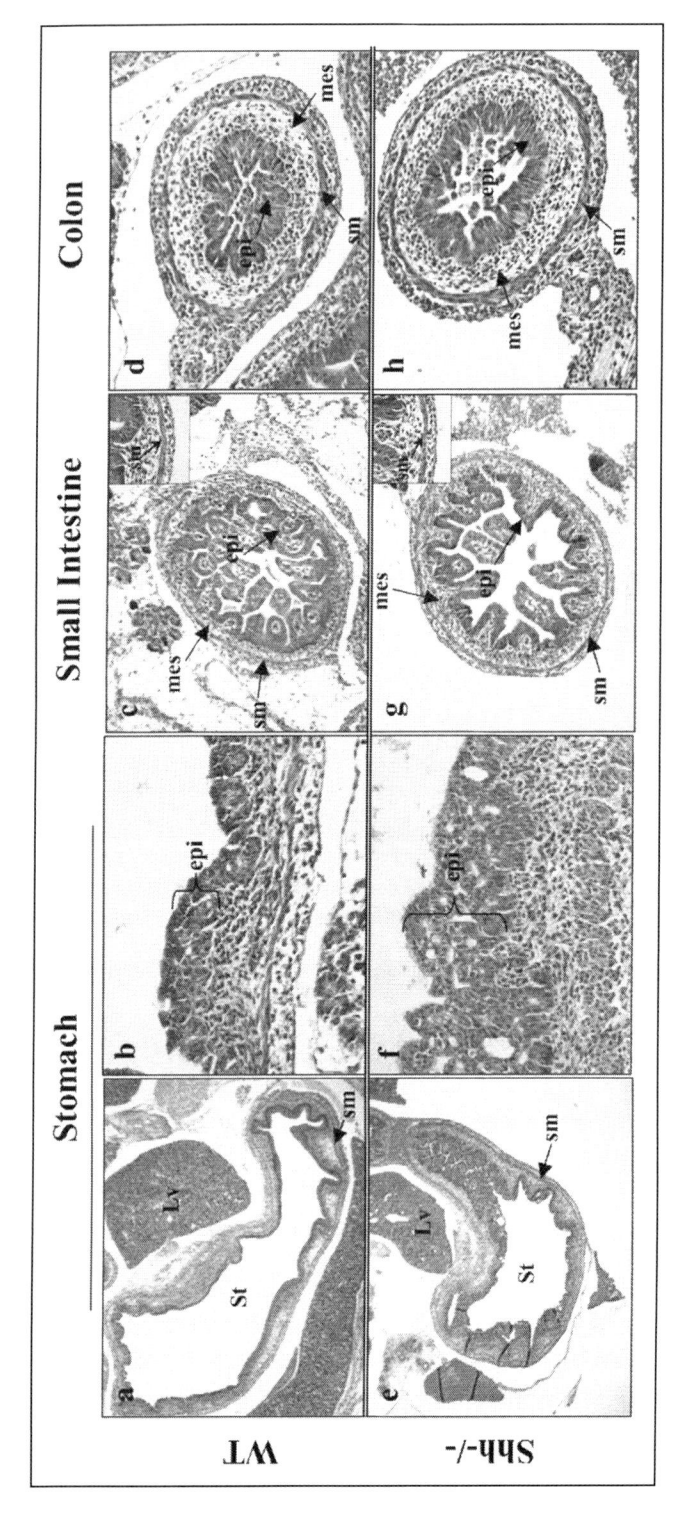

Figure 3. Hematoxylin and eosin staining of E16.5 WT (a-d) and *Shh-/-* (e-h) cross-sections of stomach (a-b,e-f), small intestine (c,g) and colon (d,h). Development of anterior part of the stomach appears to be normal in *Shh-/-*, whereas the glandular epithelium of the stomach shows hyperplasia in *Shh-/-* when compared to WT (a-b,e-f). Development of small intestine and colon in *Shh-/-* is relatively normal, although there is a slight reduction in circular smooth muscle (sm) layers in the small intestine as previously reported (see insets in c and g). Magnification: 40x for panels a and e; 100x for panels b and f; 200x for panel c, d, g and h.

has been observed at other sites of *Shh* expression and epithelial-mesenchymal interactions.[17] *Shh* expression is maintained in the zymogenic zone epithelium of the glandular stomach during embryogenesis and likely throughout life in both mouse and human.[32,34,35] In the human stomach, *Shh* appears to be expressed exclusively in parietal cells with particularly high expression in parietal cells closer to pit cells and lower expression in parietal cells closer to the base of the gastric tubular unit; however, in the mouse, *Shh* expression can be detected in both parietal and zymogenic cells.[32]

Patterning of the stomach epithelium into nonglandular and glandular zones appeared normal in *Shh*-/- mouse mutants, however, a substantial overgrowth of the stomach epithelium was observed (Fig. 3b,f). The *Shh*-/- glandular stomach epithelium displays partial intestinal metaplasia as demonstrated by the expression of intestinal markers within patches of the stomach.[34] In agreement, it was found that inhibition of Shh using cyclopamine, a potent inhibitor of hedgehog signalling, considerably enhanced gastric glandular epithelial proliferation in the murine stomach accompanied by a switch from gastric to intestinal cell fate.[32] However, whether or how downregulation of Shh target genes in the stomach is associated with depletion of parietal and zymogenic cells with concomitant replacement by overproliferating intestinal-type cells remains to be fully elucidated. It has been suggested that Shh may function to induce or maintain a stomach character, however, specific downstream target genes of Shh that are likely important in mediating proper mesenchymal-epithelial signalling remain to be elucidated.

By contrast, upregulation of Shh signalling activity has been reported in human stomach tumors growing in vivo and in stomach tumor cell lines suggesting that Shh hyperactivity contributes to uncontrolled gastric epithelial proliferation.[13] This finding is also consistent with the mitogenic role of Shh in many organ systems. While the distinct roles of Shh signalling in the stomach appears to be conflicting, these findings are likely revealing differences in the intricate molecular circuitry that directs normal stomach morphogenesis during embryogenesis as opposed to a response to epithelial injury in adulthood.[36] The findings that absence of Shh function in the glandular stomach can lead to intestinal metaplasia and Shh hyperactivity appears to be associated with gastric tumor growth, underscore the importance of controlling proper level of Shh signalling during embryogenesis and throughout life.

Pancreas

Morphogenesis of the pancreas in mouse is initiated as soon as the gut tube is formed around E9.0. Three primordial pancreatic buds protrude from the gut endoderm at the foremidgut boundary, with one bud located dorsally and two buds, ventrally (Fig. 1). As development progresses, one of the ventral pancreatic buds regresses while the other fuses with the dorsal bud to form the pancreas. During this period, epithelial-mesenchymal interactions result in extensive organ morphogenesis and cell differentiation within the pancreas area leading to the formation of endocrine and exocrine cell compartments. The endocrine cells organize into islets of Langerhans, a cluster of hormone-secreting cells that regulate glucose homeostasis while the exocrine cells secrete digestive enzymes into the duodenum.

While Shh promotes the development of several gut derivatives, it functions as a negative regulator during development of the pancreas.[37] Although *Shh* is expressed broadly in the gut endoderm, its expression is initially excluded from the dorsal endoderm.[8] However, at the foremidgut boundary, *Shh* expression is excluded from both the dorsal and ventral pancreas tissues during development.[38,39] Based on chick notochord extirpation studies, it was proposed that a factor such as fibroblast growth factor 2 (Fgf2) secreted from the notochord could inhibit *Shh* expression in the dorsal endoderm.[39,40] This inhibition appears to play a role in pancreas development. Ectopic expression of *Shh* in the pancreatic primordium under the regulation of pancreatic and duodenal homeobox gene 1 (*Pdx1*) promoter, in mice, leads to severe disruption of pancreatic architecture and significant reduction of both endocrine and exocrine cells.[38] Additionally, the pancreatic mesenchyme of *Pdx1-Shh* transgenic embryos was found to be partially transformed into contractile muscle with characteristics of duodenal mesoderm. Similarly, loss of Hh inhibitor, Hhip, also leads to impairment of pancreatic growth and endocrine cell differentiation.[41] These observations raise a critical question as to whether *Shh* functions to restrict and define the pancreatic primordium boundary. Initial observation in chick appears to support this model; inhibition of Hh signalling using cyclopamine in ovo leads to ectopic formation of epithelial buds and scattered insulin-positive cells in the distal stomach and duodenum.[42] However, analysis of *Shh* mutants did not reveal an expansion of pancreatic tissue (Fig. 2e,f), even in the *Shh-/-;Ihh+/-* background.[43]

Hh pathway activation has been implicated in the pathogenesis of pancreatic cancer. It has recently been reported that Hh pathway is activated in several human pancreatic tumor cell lines and pancreatic tumors.[13,44] The growth of these tumor cell lines in nude mice xenografts appears to be dependent on Hh pathway activation as administration of cyclopamine can significantly inhibit tumor growth. Furthermore, pancreata of *Pdx1-Shh* transgenic mice show abnormal ductal epithelial growth resembling precursor stages of human pancreatic cancer.

Intestine

During embryogenesis, regionalization of the midgut and hindgut gives rise to the duodenum, small and large intestines, rectum and anus. Unlike the invaginations in the glandular stomach, the small intestinal epithelium evaginates into the lumen to form villi which are finger-like projections that function to increase the gut surface area for nutrient absorption. Both the small and large intestines (colon) contain glands known as crypts where stem cells are thought to reside.[45] Differentiation of intestinal mesenchyme gives rise to the smooth muscle layer that surrounds the gut epithelium.

Shh is initially expressed throughout the endodermal epithelium of the developing midgut and hindgut. Later in embryogenesis, *Shh* expression is confined to the base of villi and crypts in the small intestine and base of crypts in the colon.[17] The smooth muscle layer in the intestine develops a few cells away from the *Shh*-expressing endoderm, separated by a *Bmp4*-expressing domain in the submucosal mesenchyme. Inhibition or ectopic activation of *Bmp4* expression in chick embryonic gut explants had no apparent patterning effects on smooth muscle development, suggesting that the lack of smooth muscle in the submucosal mesenchyme is not due to *Bmp4* expression.[46] By contrast, ectopic activation of *Shh* expression in chick gut explants inhibits smooth muscle differentiation. Additionally, reduction of Shh signalling in chick gut explants using cyclopamine leads to ectopic expression of smooth muscle marker including the subepithelial domain.[46] These observations suggest that Shh directly inhibits smooth muscle differentiation. However, we need to be circumspect about this interpretation since cyclopamine-treated gut explants also show reduction in the subepithelial mesenchymal cell population. In fact, it appears that the level of smooth muscle marker expression in these treated gut explants is reduced to various extents depending on the level of Hh pathway inhibition.[46] Furthermore, mice lacking either *Shh* or *Ihh* have reduced number of smooth muscle cells. At E18.5, *Shh*$^{-/-}$ small intestine shows about 20% reduction in thickness of the circular smooth muscle layer (Fig. 3c,g).[34] Similar reduction in smooth muscle differentiation is observed when Hh signalling is knocked down by Villin-driven ectopic expression of hedgehog inhibitor, Hip, in the intestinal epithelium.[47] Taken together, it appears that Shh may not directly inhibit smooth muscle differentiation, but could be required for the proliferation of smooth muscle progenitor cells, a reduction in which could affect the level or timing of smooth muscle differentiation. However, it remains possible that low level Shh signalling may have a direct role in the differentiation of intestinal smooth muscle.

In addition to smooth muscle defects, absence of Shh signalling also leads to hyperproliferation of the small intestinal epithelium, leading to extensive villi formation.[34] Interestingly, this effect appears to be due to ectopic activation of Wnt pathway in the epithelium, as the expression of several Wnt target genes are enhanced in the small intestinal epithelium of transgenic embryos with reduced Hh signalling.[47] The repressive effect of Hh signalling on the Wnt pathway has also been proposed in the adult colon where expression of Ihh, in mature colonic enterocytes (absorptive cells) at the tip of crypts, is thought to counteract Wnt signalling at the base of crypts.[48] As Wnt pathway activation is intimately associated with colon carcinoma,[49] loss of Ihh may have a significant impact on colon homeostasis. Interestingly, Ihh expression is lost from the colonic epithelium of many patients with familial adenomatous polyps.[48] However, mutations in human IHH cause brachydactyly type A1 syndrome which is often associated with short statures;[50] these patients do not appear to display a higher occurrence of colon cancer.

The lack or reduction of Shh signalling also has profound consequences in the morphogenesis of the distal hindgut. In *Shh-/-* or *Gli2* and *Gli3* compound mutants, the distal hindgut and lower urinary tract share a common outlet resulting in a severe form of anorectal malformation known as persistent cloaca. In *Gli2* or *Gli3* mutants, a milder spectrum of hindgut defects such as narrowing of the anus (anal stenosis) and abnormal connection between rectum, anus and urethra (rectal-urethral fistula) is observed.[51] In fact, these phenotypic characteristics are highly reminiscent of a spectrum of human anorectal malformations (ARM). Recently, embryos with ARM have been generated by exposing pregnant mice to all-trans retinoic acid (ATRA).[52] In these embryos, the expressions of *Shh* and its putative target, *Bmp4*, are significantly downregulated in the hindgut, suggesting the involvement of Shh signalling in ARM.[52] The question remains as to how a teratogenic dose of ATRA leads to downregulation of Shh signalling in the hindgut epithelium. Further studies are required to establish whether ARM in these embryos are indeed caused by disruption of Shh signalling or by elevated RA signalling that is independent of Shh signalling.

Kidney

Kidney organogenesis in the mouse begins at E11.0 with the outgrowth of a ureteric bud epithelium from the mesonephric (Wolffian) duct into the surrounding metanephric mesenchyme, a distinct population of mesoderm-derived cells. The ureteric bud subsequently divides and undergoes branching morphogenesis which is dependent on reciprocal signalling interactions between epithelium and mesenchyme.[53] A host of signalling molecules and transcription factors have been genetically shown to be involved in crucial ureteric bud epithelial and metanephric mesenchymal interactions during kidney morphogenesis.[54] For example, Glial-derived neurotrophic factor (GDNF) secreted by the surrounding mesenchyme binds to c-ret receptor tyrosine kinase expressed in the ureteric bud epithelium to promote migration and invasion of the ureteric bud into the surrounding mesenchyme.[55] The existing ureteric bud-derived epithelium gives rise to collecting ducts while signals emanating from the ureteric bud induce a subset of metanephric mesenchymal cells to aggregate, forming renal vescicles. These vesicles undergo extensive sequential morphogenesis (tubulogenesis) to form S-shaped bodies that eventually convert into polarized tubular epithelia of nephrons which are the basic functional units of the mature kidney involved in filtration. These nephron epithelia fuse with the collecting tubules to form a complete kidney ductal system.[53]

Shh is expressed in the branching ureteric bud epithelium of the developing mouse embryo at E11.5. As development advances, *Shh* expression becomes restricted to the ureteric epithelium of the distal, nonbranching medullary collecting ducts and ureter, a urinary tract connecting the kidney with the bladder. Accordingly high levels of *Ptch* expression, a readout for Shh signalling, and *Bmp4* were detected in mesenchymal cells adjacent to the *Shh*-expressing epithelium of the distal collecting ducts and ureter, indicating paracrine signalling.[17,56] Several molecules have been suggested as candidate targets of Shh signalling in the metanephric mesenchyme, however, their definite roles in mediating Shh function remain to be elucidated. Strong *Shh* expression has also been detected in the newborn mouse kidney in the inner medullary collecting ducts, the renal pelvic and ureter epithelia.[56] *Shh-/-* kidneys exhibit hypoplasia and fusion as a result of midline defects (Fig. 2i-l). It has also been reported that *Gli2-/-Gli3+/-* mouse mutants, which are deficient in the Shh signalling pathway, display renal anomalies.[57] In order to understand the role of Shh specifically in the kidney, a *HoxB7*-driven Cre transgenic mouse line, which is activated in the mesonephric duct and its derivatives, was used to conditionally ablate *Shh* function specifically in the kidney primordium.[56] These *Shh* conditional mutant mice displayed abnormal kidney development postnatally including severe renal hypoplasia and dilated proximal ureter likely due to reduction in ureteral smooth muscle leading to a condition known as hydroureter in which urine abnormally accumulates in the ureter due to its inefficient transport to the bladder. Detailed analysis of the *Shh* conditional mutants revealed that Shh function is required for proliferation of the kidney mesenchyme as well as the normal timing of smooth muscle differentiation in the ureter.[56] Although *Bmp4* expression in the ureteral mesenchyme is dependent on Shh signalling, it does not appear to mediate the mitogenic function of Shh. Bmp4 has also been shown to promote smooth muscle differentiation in the kidney and ureter, however, it does not appear to be absolutely essential for the process.[56,58] Therefore, the precise role of *Bmp4* in kidney development and function remains to be elucidated. It has been suggested that, in *HoxB7-Cre* conditional *Shh* mutants, the severe reduction in medullary kidney and ureteral mesenchymal cell proliferation is likely the underlying cellular defect leading to kidney hypoplasia and shortening of the ureter. Although Shh has been implicated in visceral smooth muscle differentiation, the precise mechanism remains unclear. While bronchial[9,21] and intestinal smooth muscle myogenesis[34] appear to be dependent, at least in part, on Shh signalling, smooth muscle differentiation in the ureter appears to be inhibited by Shh, consistent with a negative role of Shh in the generation of smooth muscle in the chick gut.[46] It has been suggested that Shh may be required to promote proliferation of smooth muscle progenitor populations.[56] However, increase in the mesenchymal cell population in *Shh* and *Gli3* double mutant lung did not restore bronchial myogenesis which is absent in *Shh* mutant lung with severe defect in mesenchymal proliferation,[21] suggesting that, as in ureteral myogenesis, a direct role of Shh in smooth muscle differentiation remains possible.

Conclusion

Over the past decade, extensive knowledge has been gained in understanding the critical function of *Shh* during embryonic development. It is apparent that basic mechanisms governing cell survival, proliferation, differentiation and tissue patterning share great similarities among different tissues and organs. However, much less is known about how Shh signalling regulates these diverse cellular events during development. Therefore, the future challenge will be to identify and functionally characterize downstream effectors of Shh signalling during gut morphogenesis. During organ maturation, *Shh* expression becomes compartmentalized in specialized glands of the stomach and the intestine; however, the roles of *Shh* in these glands remain elusive. The availability of conditional mutants in Hh signalling pathway and the ever-increasing tissue-specific Cre mouse lines should facilitate our understanding of Shh function in these glands.

The observation that constitutive Shh pathway activation is associated with many forms of visceral organ malignancies has generated excitement and provided challenge for future investigations. It is thought that the majority of gastrointestinal tumors arise from repetitive injury to the epithelial lining of visceral organs, leading to unregulated proliferation of epithelial cells within a stem cell niche. Given that Shh is involved in the proliferation of adult stem cells in the brain,[59-61] it will not be surprising that Shh can act directly on stem cell niches during tissue repair and tumor growth.

References

1. Ingham PW, McMahon AP. Hedgehog signaling in animal development: Paradigms and principles. Genes Dev 2001; 15:3059-87.
2. Bai CB, Auerbach W, Lee JS et al. Gli2, but not Gli1, is required for initial Shh signaling and ectopic activation of the Shh pathway. Development 2002; 129:4753-61.
3. Motoyama J et al. Differential requirement for Gli2 and Gli3 in ventral neural cell fate specification. Dev Biol 2003; 259:150-61.

4. Wang B, Fallon JF, Beachy PA. Hedgehog-regulated processing of Gli3 produces an anterior/posterior repressor gradient in the developing vertebrate limb. Cell 2000; 100:423-34.
5. Litingtung Y, Dahn RD, Li Y et al. Shh and Gli3 are dispensable for limb skeleton formation but regulate digit number and identity. Nature 2002; 418:979-83.
6. Bai CB, Stephen D, Joyner AL. All mouse ventral spinal cord patterning by hedgehog is Gli dependent and involves an activator function of Gli3. Dev Cell 2004; 6:103-15.
7. Chuang PT, McMahon AP. Vertebrate Hedgehog signalling modulated by induction of a Hedgehog-binding protein. Nature 1999; 397:617-21.
8. Litingtung Y, Lei L, Westphal H et al. Sonic hedgehog is essential to foregut development. Nat Genet 1998; 20:58-61.
9. Pepicelli CV, Lewis PM, McMahon AP. Sonic hedgehog regulates branching morphogenesis in the mammalian lung. Curr Biol 1998; 8:1083-6.
10. Skandalakis JE, Gray SW, Ricketts R. Embryology for surgeons. Baltimore: Williams and Wilkins, 1994.
11. Motoyama J et al. Essential function of Gli2 and Gli3 in the formation of lung, trachea and oesophagus [see comments]. Nat Genet 1998; 20:54-7.
12. Mendelsohn C et al. Function of the retinoic acid receptors (RARs) during development (II). Multiple abnormalities at various stages of organogenesis in RAR double mutants. Development 1994; 120:2749-71.
13. Berman DM et al. Widespread requirement for Hedgehog ligand stimulation in growth of digestive tract tumours. Nature 2003; 425:846-51.
14. Wessells NK. Mammalian lung development: Interactions in formation and morphogenesis of tracheal buds. J Exp Zool 1970; 175:455-66.
15. Perl AK, Whitsett JA. Molecular mechanisms controlling lung morphogenesis. Clin Genet 1999; 56:14-27.
16. Tollet J, Everett AW, Sparrow MP. Spatial and temporal distribution of nerves, ganglia, and smooth muscle during the early pseudoglandular stage of fetal mouse lung development. Dev Dyn 2001; 221:48-60.
17. Bitgood MJ, McMahon AP. Hedgehog and Bmp genes are coexpressed at many diverse sites of cell-cell interaction in the mouse embryo. Dev Biol 1995; 172:126-38.
18. Bellusci S et al. Involvement of Sonic hedgehog (Shh) in mouse embryonic lung growth and morphogenesis. Development 1997; 124:53-63.
19. Miller LA, Wert SE, Whitsett JA. Immunolocalization of sonic hedgehog (Shh) in developing mouse lung. J Histochem Cytochem 2001; 49:1593-604.
20. Unger S, Copland I, Tibboel D et al. Down-regulation of sonic hedgehog expression in pulmonary hypoplasia is associated with congenital diaphragmatic hernia. Am J Pathol 2003; 162:547-55.
21. Li Y, Zhang H, Choi SC et al. Sonic hedgehog signaling regulates Gli3 processing, mesenchymal proliferation, and differentiation during mouse lung organogenesis. Dev Biol 2004; 270:214-31.
22. Chuang PT, Kawcak T, McMahon AP. Feedback control of mammalian Hedgehog signaling by the Hedgehog-binding protein, Hip1, modulates Fgf signaling during branching morphogenesis of the lung. Genes Dev 2003; 17:342-7.
23. Bellusci S, Grindley J, Emoto H et al. Fibroblast growth factor 10 (FGF10) and branching morphogenesis in the embryonic mouse lung. Development 1997; 124:4867-78.
24. Min HS, Danilenko DM, Scully SA. Fgf-10 is required for both limb and lung development and exhibits striking functional similarity to Drosophila branchless. Gene Dev 1998; 12:3156-3161.
25. Cardoso WV, Itoh A, Nogawa H et al. FGF-1 and FGF-7 induce distinct patterns of growth and differentiation in embryonic lung epithelium. Dev Dyn 1997; 208:398-405.
26. Hogan BLM et al. Branching morphogenesis of the lung: new models for a classical problem. Cold Spring Horbor Symp Quant Biol 1997; 12:249-256.
27. Grindley JC, Bellusci S, Perkins D et al. Evidence for the involvement of the Gli gene family in embryonic mouse lung development. Dev Biol 1997; 188:337-48.
28. Watkins DN et al. Hedgehog signalling within airway epithelial progenitors and in small-cell lung cancer. Nature 2003; 422:313-7.
29. Stewart GA et al. Expression of the developmental Sonic hedgehog (Shh) signalling pathway is up-regulated in chronic lung fibrosis and the Shh receptor patched 1 is present in circulating T lymphocytes. J Pathol 2003; 199:488-95.
30. Karam SM, Li Q, Gordon JI. Gastric epithelial morphogenesis in normal and transgenic mice. Am J Physiol 1997; 272:G1209-20.
31. Rubin DC, Swietlicki E, Gordon JI. Use of isografts to study proliferation and differentiation programs of mouse stomach epithelia. Am J Physiol 1994; 267:G27-39.

32. van den Brink GR et al. Sonic hedgehog regulates gastric gland morphogenesis in man and mouse. Gastroenterology 2001; 121:317-28.
33. Karam SM, Leblond CP. Dynamics of epithelial cells in the corpus of the mouse stomach. III. Inward migration of neck cells followed by progressive transformation into zymogenic cells. Anat Rec 1993; 236:297-313.
34. Ramalho-Santos M, Melton DA, McMahon AP. Hedgehog signals regulate multiple aspects of gastrointestinal development. Development 2000; 127:2763-72.
35. Motoyama J, Takabatake T, Takeshima K et al. Ptch2, a second mouse Patched gene is coexpressed with Sonic hedgehog [letter]. Nat Genet 1998; 18:104-6.
36. Beachy PA, Karhadkar SS, Berman DM. Mending and malignancy. Nature 2004; 431:402.
37. Hebrok M. Hedgehog signaling in pancreas development. Mech Dev 2003; 120:45-57.
38. Apelqvist A, Ahlgren U, Edlund H. Sonic hedgehog directs specialised mesoderm differentiation in the intestine and pancreas. Curr Biol 1997; 7:801-4.
39. Hebrok M, Kim SK, Melton DA. Notochord repression of endodermal Sonic hedgehog permits pancreas development. Genes Dev 1998; 12:1705-13.
40. Kim SK, Hebrok M, Melton DA. Notochord to endoderm signaling is required for pancreas development. Development 1997; 124:4243-52.
41. Kawahira H et al. Combined activities of hedgehog signaling inhibitors regulate pancreas development. Development 2003; 130:4871-9.
42. Kim SK, Melton DA. Pancreas development is promoted by cyclopamine, a hedgehog signaling inhibitor. Proc Natl Acad Sci USA 1998; 95:13036-41.
43. Hebrok M, Kim SK, St Jacques B et al. Regulation of pancreas development by hedgehog signaling. Development 2000; 127:4905-13.
44. Thayer SP et al. Hedgehog is an early and late mediator of pancreatic cancer tumorigenesis. Nature 2003; 425:851-6.
45. Potten CS. Stem cells in gastrointestinal epithelium: Numbers, characteristics and death. Philos Trans R Soc Lond B Biol Sci 1998; 353:821-30.
46. Sukegawa A et al. The concentric structure of the developing gut is regulated by Sonic hedgehog derived from endodermal epithelium. Development 2000; 127:1971-80.
47. Madison BB et al. Epithelial hedgehog signals pattern the intestinal crypt-villus axis. Development 2005; 132:279-89.
48. van den Brink GR et al. Indian Hedgehog is an antagonist of Wnt signaling in colonic epithelial cell differentiation. Nat Genet 2004; 36:277-82.
49. Clevers H. Wnt breakers in colon cancer. Cancer Cell 2004; 5:5-6.
50. Kirkpatrick TJ et al. Identification of a mutation in the Indian Hedgehog (IHH) gene causing brachydactyly type A1 and evidence for a third locus. J Med Genet 2003; 40:42-4.
51. Mo R et al. Anorectal malformations caused by defects in sonic hedgehog signaling. Am J Pathol 2001; 159:765-74.
52. Sasaki Y, Iwai N, Tsuda T et al. Sonic hedgehog and bone morphogenetic protein 4 expressions in the hindgut region of murine embryos with anorectal malformations. J Pediatr Surg 2004; 39:170-3, (discussion 170-3).
53. Dressler G. Tubulogenesis in the developing mammalian kidney. Trends Cell Biol 2002; 12:390-5.
54. Yu J, McMahon AP, Valerius MT. Recent genetic studies of mouse kidney development. Curr Opin Genet Dev 2004; 14:550-7.
55. Schedl A, Hastie ND. Cross-talk in kidney development. Curr Opin Genet Dev 2000; 10:543-9.
56. Yu J, Carroll TJ, McMahon AP. Sonic hedgehog regulates proliferation and differentiation of mesenchymal cells in the mouse metanephric kidney. Development 2002; 129:5301-12.
57. Kim PC, Mo R, Hui CcC. Murine models of VACTERL syndrome: Role of sonic hedgehog signaling pathway. J Pediatr Surg 2001; 36:381-4.
58. Raatikainen-Ahokas A, Hytonen M, Tenhunen A et al. BMP-4 affects the differentiation of metanephric mesenchyme and reveals an early anterior-posterior axis of the embryonic kidney. Dev Dyn 2000; 217:146-58.
59. Palma V et al. Sonic hedgehog controls stem cell behavior in the postnatal and adult brain. Development 2005; 132:335-44.
60. Machold R et al. Sonic hedgehog is required for progenitor cell maintenance in telencephalic stem cell niches. Neuron 2003; 39:937-50.
61. Lai K, Kaspar BK, Gage FH et al. Sonic hedgehog regulates adult neural progenitor proliferation in vitro and in vivo. Nat Neurosci 2003; 6:21-7.

CHAPTER 13

Shh/Gli Signalling during Murine Lung Development

Martin Rutter and Martin Post*

Abstract

Murine lung development is a complex process regulated by many factors guiding a carefully orchestrated series of events leading to mature lung formation. Many developmental pathways have been implicated in governing proper lung formation. Most notably, the Shh/Gli pathway shown to be crucial to the development of numerous other organ systems, is an absolute requirement for correct lung formation. Many interactions between the Shh pathway and other fundamental lung signalling molecules such as fibroblast growth factor 10 (Fgf10) have presented themselves. While the specifics of these interactions have yet to be elucidated, the consequence of their actions is paramount in guiding lung development.

Murine lung development begins with the out pocketing of two endodermal lung buds from the ventral region of the primitive foregut tube around 9.5 days post coitum (dpc). The two primary lung buds then start to extend in a posterior-ventral track into the splanchnic mesenchyme, each bud representing the future left and right sides of the mature lung. Concurrently, the single foregut tube at the primary branch point begins to pinch into two distinct tubes forming the dorsal esophagus and ventrally located trachea. The right lung bud (right primary bronchus) then undergoes a secondary branching event leading to the creation of four secondary bronchi, each denoting one of the four right lung lobes (lobar bronchi). From this point, both the primary left bronchus and the four secondary bronchi of the right lung bud will continue to undergo further generations of dichotomous branching until the mature network of airways is formed. However, this branching process is not a chaotic event, but rather a carefully controlled process. A highly structured series of interactions between the developing airway epithelium and mesenchyme guides proper lung development.[1,2] These interactions are directed by many tissue specific morphogenic signals.[3] Much like the development of other branching organs such as the kidney, the mammalian lung requires a carefully orchestrated symphony of genes to accomplish its end goal. Several gene families have been shown to be involved in lung development, including fibroblast growth factors (Fgf), bone morphogenic proteins (Bmp), as well as the primary focus of this chapter, the Hedgehog (Hh) family.

While many factors contribute to the formation of the mature lung, Sonic hedgehog (Shh) is absolutely required for functional lung formation. Evidence of Shh in lung development was first postulated with the observation of expression of *shh* transcripts throughout the epithelium of the developing mouse lung at 11.5 dpc, the highest levels occurring at the developing tips of

*Corresponding Author: Martin Post—Program in Lung Biology Research, The Hospital for Sick Children Research Institute, Institute of Medical Sciences, University of Toronto, 555 University Avenue, Toronto, Ontario, M5G 1X8, Canada. Email: martin.post@sickkids.ca

Shh and Gli Signalling and Development, edited by Carolyn E. Fisher and Sarah E.M. Howie. ©2006 Landes Bioscience and Springer Science+Business Media.

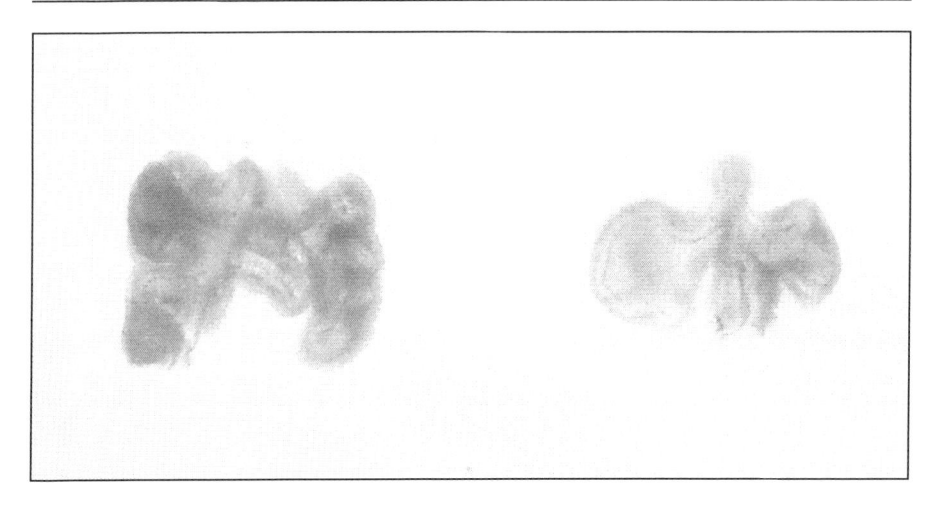

Figure 1. Photo showing a side by side ventral view comparison of a 12.5 dpc *shh* null lung (right) and its wild-type sibling counterpart (left). Note the *shh* null mutant only has a single right lobe, and does not show the same developmental complexity as the wild-type lung at the same stage.

the epithelial buds.[4] Expression was also detected in the tracheal diverticulum, the esophagus, and the developing trachea.[5] *Shh* expression was originally reported to be strongly expressed until about mid gestation, after which it decreased.[6] However, more recent evidence indicates *shh* expression increases towards birth peaking just prior to parturition.[7,8] Also interesting was the detection of high levels of the transcripts for *Patched* (*ptc*) and *Smoothened* (*smo*), the downstream Shh signal relaying proteins, in the mesenchyme adjacent to the *shh* expressing epithelial cells in the developing buds.[6] When the effect of *shh* over-expression in the lung epithelium using the surfactant protein (SP)-C promoter was examined, it was found that the ratio of interstitial mesenchyme to epithelial tubules had increased. More detailed analysis of these lungs revealed an abundance of mesenchyme and the absence of typical alveoli due to increased cellular proliferation in both the mesenchyme and the epithelium. Expression of *ptc* was also noticeably up-regulated in lungs of *shh* over-expressers, however no evidence of regulatory changes in other lung development related genes such as *bmp4* or *fgf7* was found.[6] Also interesting to note is the 2.5 fold increase in *gli1* expression in response to *shh* over-expression, while *gli2* and *gli3* expression levels remain unchanged.[9] If we now examine the other side of the coin, a similar picture presents itself that further supports the concept of *shh* as a regulator of lung proliferation. A knockout of the *shh* gene has been created in mice in which the second exon of *shh* has been replaced with a PGK-neo cassette resulting in a nonfunctional truncated protein upon translation.[10] While many developmental defects occur in this prenatal lethal model, we will focus on the pulmonary phenotype for the purposes of this chapter. *Shh* null lungs have a dramatically altered phenotype; most obvious is the complete lack of asymmetry as the secondary branching in the right lung is defective (see Fig. 1). The resulting single left and single right lung lobes are severely hypoplastic and fail to develop a vast network of mature air sacs. The trachea and esophagus do not divide into separate entities. Ultimately the lack of *shh* results in severely reduced mesenchymal proliferation and an extensive reduction in epithelial branching. When effects on gene regulation were examined, it was shown that *ptc*, *gli1* and *gli3* were all down-regulated in the lung mesenchyme.[11] Like the *shh* over-expression model, proximal-distal differentiation of the lung was unaffected while prominent proliferative defects were evident. However, *shh* over-expression with the SP-C promoter in the lungs of *shh*$^{-/-}$ mice showed a significant improvement in growth, branching morphogenesis and vascularization.[12] But, the peripheral over-expression failed to correct lobulation as well as cartilage defects in the trachea and bronchi seen in the *shh*$^{-/-}$ mice, signifying the importance for *shh* expression in areas to secondary branching and cartilage formation. More recently, a lung specific knockout of *shh* has been achieved in mice, which can also be temporally controlled through administration of doxycycline.[12,13] Shh expression appears to primarily be required for lung development prior to 13.5 dpc, after this point in development only mild defects in peripheral lung structure are observed when *shh* function is removed. However, removal of *shh* prior to 13.5 dpc resulted in severe malformations similar to the $shh^{-/-}$ null mouse, with defects in the trachea, bronchi and peripheral lungs, as well as many changes in gene expression levels as evident from microarray data. This study also demonstrated the localized spatial requirement for *shh* in proper cartilage formation in the conducting airways of the developing lung.[12]

Ptc is a twelve pass transmembrane receptor protein that is a fundamental component of the Shh signalling pathway. While Ptc itself is not a transcriptional regulator, nor a diffusible morphogen, its presence and function is absolutely critical for normal lung development. Ptc is expressed in high concentrations in the mesenchyme near the epithelial border of the developing tips neighboring to *shh* expression, as well as at lower concentrations in the distal epithelium.[6] A *ptc* null mutant has been created, however this embryonic lethal genetic defect offered no clues to the role of *ptc* in lung development as these mice die around 9.0 dpc to 10.5 dpc, at the start of lung formation.[14] While a mouse with lung specific over-expression of *ptc* has not been created, other experiments show that increased expression of *ptc* results in a reduction of Shh signalling, consequentially down-regulating expression of Shh responsive genes such as *gli1* and *ptc* itself.[14,15] This would suggest that over expression of *ptc* in the lung near the mesenchymal border would attenuate the Shh epithelial to mesenchymal signalling, resulting in somewhat of a less severe *shh* null phenotype.[16] Most likely proliferative and branching defects would present themselves, however depending on the onset of over-expression, early lung development may proceed to further stages than the *shh* null phenotype.

Smoothened (Smo), another trans-membrane protein essential to the Hh signalling pathway, has been targeted for gene deletion in mice. The resulting phenotype is very severe, and offers little insight into effects on lung develop with embryos dying prior to 9.5 dpc.[17] To speculate on possible effects of a *smo* null mutation on lung development, it must be taken into consideration that there is only one mammalian homologue of *smo*, and that it has been suggested that all three Hh signalling pathways would use *smo*.[17] Therefore one would expect a lung specific *smo* null mutant to have a phenotype at least as severe to that of the *shh* null lung. Furthermore, since the transcripts for Indian Hedgehog (Ihh) have more recently been found to be expressed in the lung as well, the resulting phenotype could be even more detrimental to lung formation as the effects of Ihh on lung development are not known.[18]

Another protein important in the regulation of Shh through a negative feedback loop is Hedgehog-interacting protein 1 (Hip1). *Hip1* is induced in Shh responsive cells upon Shh signalling and encodes a membrane-bound protein capable of directly binding to Shh, Ihh and Desert Hedgehog (Dhh).[19] *Hip1* expression has been found in the lung epithelium, as well as the underlying mesenchyme. Closer inspection shows *hip1* is transcribed in cells near sources of Hh signalling, in a domain that overlaps with *ptc* expression.[20] An increase in *hip1* expression is observed in *shh* over-expression models, and conversely *hip1* is decreased in *shh* null mutants. Experiments in which *hip1* was ectopically expressed in the developing endochondral skeleton where Ihh is accountable for Hh signalling, it was found to attenuate Hh signalling showing a similar phenotype to the *Ihh* null mutant.[20] More recently in continuing their investigation into Hip1 function, Chuang and coworkers (2003) have demonstrated that the *hip1* loss-of-function mutant mouse has increased Hh signalling, disrupting morphogenesis in the lung and skeleton. They indicate that increased Shh function due to lack of *hip1* function causes a misregulation of *fgf10* expression resulting in failure of secondary branching.[21] The sum of these observations implicates Hip1 as a negative feedback regulator of Hh signalling, crucial to normal development of the lung, as well as other organ systems.

Turning attention to the downstream transcription factors of the Shh pathway, we can see further evidence of the importance of Shh signalling in lung development. The Gli family of transcription factors are a group of three genes which encode proteins containing DNA binding zinc finger motifs. Early analysis suggested these genes would function as transcription factors with a relationship to cellular proliferation. The *glis* were found to be expressed in the splanchnic component of the lateral mesoderm in the developing gut amongst other places.[22] More detailed analysis of expression patterns revealed that all three *gli* genes are strongly expressed in different but over-lapping domains in the lung mesenchyme during the pseudoglandular stage with expression declining towards birth.[9] *Gli1* is expressed in the distal mesenchyme, mostly concentrated around the developing endodermal lung buds. *Gli2* on the other hand has a more dispersed mesenchymal expression pattern which is still more spatially restricted towards the distal regions of the lung, but has strong expression near the trachea as well. *Gli3* is not particularly concentrated in either proximal or distal mesoderm, however is not as widely dispersed as *Gli2*, lying in between the expression domains of *Gli1* and *Gli2*.[9] Expression of all three *gli* genes is dynamic and seems to correspond with branching morphogenesis in the developing lung lobes. The temporal down-regulation of the *gli* transcripts towards birth appears to occur in three separate phases. While expression of each *gli* is elevated early in lung development, *gli2* and *gli3* show a decrease in expression from 12.5 dpc to 16.5 dpc, at this point *gli2* mRNA expression stabilizes. *Gli1* and *gli3* continue to decrease (along with *shh* expression) until just prior to birth when *gli2* will also further diminish, resulting in down-regulation of all *gli* genes just before birth.[9]

Removal of the zinc finger coding region of the *gli1* gene from mice results in a loss-of-function mutation in the Gli1 protein which can no longer signal to other Shh targets. These mice are viable, show no physical abnormalities and display no observable behavioral traits.[23] *Gli2* null mice on the other hand have a very severe lung phenotype. While a heterozygous *gli2* deletion has no detectable effect on lung development, complete removal of *gli2* results in a lethal phenotype, with mice dying in-utero during late gestation.[24] The lungs of the *gli2* null mice are very hypoplastic in appearance and most notably show defective branching in the right lung with only one lobe forming. The left lung still forms one lobe but has a severe reduction in wet weight of approximately 60% at 13.5 dpc. This developmental trend continued to 18.5 dpc, when the left lung weight was 50% lighter than its wild-type counter part. The lungs show little sign of apoptosis but bromodeoxyuridine (BrdU) incorporation experiments demonstrated a 40% and 25% reduction in cellular proliferation in the mesenchyme and epithelium, respectively. Histological analysis revealed that the trachea and esophagus are both hypoplastic, but still separate. Smaller air sacs were also evident and they were surrounded by thicker than normal mesenchyme. When lung development associated growth factors such as *fgf1, fgf7, fgf10, bmp2, bmp4,* and *bmp6* were examined for changes in gene expression, no deviations were found. However, in-situ hybridization was able to detect decreases in both *ptc* and *gli1* expression, further demonstrating a reduced response to Shh signalling.[25] *Gli3* null mice have been around for many years. The null allele designated $Gli3^{Xtj}$, was discovered in the "Extra toes" mouse mutant to be a viable homozygous deletion.[26] Unlike the viable *gli1* null mouse, the *gli3* null mouse does show an altered pulmonary phenotype. While the homozygous $gli3^{Xtj}$ mouse embryo is actually heavier than its wild-type littermates at all stages of gestation, the lungs are typically smaller with an altered shape, most noticeably a reduction in lung width.[9] The wet lung weight when measured at 18.5 dpc was 35% lower than wild-type littermates. The gli3 heterozygous mice did not show any altered lung phenotype. When gene expression for other Shh pathway members was tested in the *gli3* null lung, no changes in expression levels or localization was detected for *shh, gli1, gli2,* or *ptc*, as well as *bmp4*, and *wnt2*.[9]

The evolution of the Glis from their common ancestor cubitus interruptus (Ci) in *Drosophila melanogaster* to the mammalian three part signalling system, suggests the evolution of separate roles for each Gli. Double mutant combinations of the *gli* genes have been created to help elucidate the possible functional roles for each Gli transcription factor during pulmonary development. Different combinations of *gli1* and *gli2* null alleles show that there is some level of redundancy between the two genes. While neither the $gli1^{-/-}$, $gli2^{+/-}$, nor the double gli1/gli2 heterozygous mouse, show an altered lung phenotype, the combined $gli1^{-/-};gli2^{+/-}$; genetic condition results in a variable lung phenotype with minor alterations in size and shape relative to its wild-type sibling.[23] While not as severe as the *gli2* null lung phenotype, this signifies some functional redundancy between Gli1 and Gli2 activities. Further supporting this notion is the fact that when one copy of *gli1* is removed from a *gli2* null lung which already has a severe phenotype on its own, the resulting lung is even smaller. Finally, if all functional Gli1 and Gli2 protein is removed from the developing lung, the result is a lung with two very small single lobes, smaller than the *gli2* null lung.[23] To date, the only *gli1/gli3* double mutant mouse analyzed has been the *gli1$^{-/-}$;gli3$^{+/-}$* mouse. These mice show an identical polydactyly phenotype to the *gli3* null mouse, and analysis of pulmonary phenotype effects were not performed.[23] So it appears that there is functional redundancy between Gli1 and Gli2, and not Gli1 and Gli3. However, the most severe lung phenotype indicating further functional redundancy between Glis reveals itself in the *gli2/gli3* double null mutant. While the *gli2/gli3* heterozygous double mutant shows no observable foregut malformations, the *gli2/gli3* double null mouse suffers an early embryonic lethal phenotype typically not surviving past 10.5 dpc. A few of the double null embryos will survive to 14.5 dpc and these mice fail to show any lung formation. In fact, no evidence of any trachea or lung primordia is found past 9.5 dpc.[25] While the *gli2$^{+/-}$;gli3$^{-/-}$* mouse has been generated, no comment on the pulmonary phenotype has been reported thus far. However, the *gli2$^{-/-}$;gli3$^{+/-}$* mutant mouse has a severe lung phenotype, resulting in an extremely hypoplastic, single lung lobe. This is suggested to result from an ectopic lung bud developing between the left and right lung buds fusing them together after the primary branching at the posterior end of the lung. These mice also develop a single tracheo-oesophageal tube connecting the lung directly to the stomach.[25] While each of the Gli proteins has evolved to function independently, as evident from unique expression patterns and null phenotypes, there is also a level of redundancy between them. This is evident from the observation of increased severity of developmental defects in combination null mutants. One interesting double knockout recently published was the combined *shh$^{-/-}$* and *gli3$^{-/-}$* double knockout mouse.[27] These mice actually show a pulmonary phenotype that is less severe than the *shh* null lung. The *shh/gli3* double null lungs showed enhanced vasculogenesis and growth potential. There was also an increase in Cyclin D1 expression in both the epithelium and mesenchyme compared with the *shh* null lung. *Wnt2, fox1, tbx2* and *tbx3* (via *fox1*), have also been shown to be de-repressed by the removal of *gli3* from the *shh* null lung. The de-repression of these genes could explain the less severe growth defects seen in these double transgenic lungs. Perhaps the most interesting finding was that the levels of the truncated repressor form of Gli3 (Gli3R), were much higher in *shh* null lungs than wild-type. This could explain the reduction in phenotypic severity of the *shh$^{-/-}$;gli3$^{-/-}$* lung as the Gli3R level would no longer increase but be absent. Therefore, since no more Gli3R is present, the effect of increase in Gli3 repressor function in the *shh* null lung over wild-type levels is abrogated, thus decreasing phenotypic severity.

It is worth taking a closer look at a few of the other signalling factors and how they relate to certain aspects of lung development and the potential for interaction with the Hedgehog signalling network. Several Bmps, members of the transforming growth factor-β (Tgf-β) super family, have been found in the lung and they include Bmp4, Bmp5, and Bmp7. *Bmp5* is expressed throughout the embryonic lung mesenchyme, however null mutants show no pulmonary aberrations, and over-expression of *Bmp5* in the lung has not yet been examined.[28] *Bmp7* was found to be ubiquitously expressed in the lung endoderm and the null phenotype is quite severe, however no lung defects were reported.[29] On the other hand, Bmp4 has been shown to have significant effects on proper lung formation. *Bmp4* expression is similar to *shh* expression in that it is found primarily at the developing tips in the distal endoderm, but its expression has also been established in the adjacent mesenchyme.[4] A *bmp4* null mutant has been created, however the embryo does not live long enough to see potential effects on lung development as it dies between 6.5 and 9.5 dpc.[30,31] Conversely, an over-expression model for *bmp4* has been created in which *bmp4* is mis-expressed by the SP-C promoter. These lungs are smaller, show a reduction in the amount of branching, have distended terminal buds and also show defects in differentiation with a lower proportion of type II alveolar cells later in gestation.[4] Closer inspection revealed no differences in *shh* expression suggestive that BMP4 has no direct regulatory effects on *shh*. The Fgf family and its receptors have also been shown to be major players in lung branching morphogenesis. Both Fgf1 and Fgf7 have been shown to have effects on lung development in lung culture systems.[32,33] While both induce lung growth, Fgf7 appears to have greater proliferative effects. Fgf1 was also shown to have some minor effects on lung differentiation, but Fgf7 again proved to be the more potent inducer of differentiation by inducing both surfactant proteins A and B, as well as the appearance of clusters of lamellar bodies.[34] This was also shown to be true in the in vivo system, as mice containing a construct over-expressing *Fgf7* in the lung by the SP-C promoter, produce cyst-like structures and show differentiation markers.[35] However, an *Fgf7* null mouse has been created and no lung phenotype was evident, suggesting it is not essential to lung development or can be compensated by other growth factors.[36] Fgf10 on the other hand, is a crucial growth factor pertaining to lung development. Expressed as early as 9.75 dpc, *fgf10* is localized to the distal mesenchyme surrounding the developing lung buds. Expression is quite dynamic, appearing to precede lung bud growth in that it is expressed in areas of the mesenchyme where the next lung bud will form, suggesting interactions between the developing epithelium and adjacent mesenchyme regulate its expression.[37] Recent studies have implicated a couple T-box genes in regulating the expression of *Fgf10*. Several T-box genes have been found to be expressed in the lung, with *tbx1* restricted to the epithelium and *tbx2-5* in the surrounding mesenchyme.[38] By using antisense oligonucleotides to hinder gene expression, it was found that inhibition of *tbx4* and *tbx5* resulted in a dramatic reduction in branching of early embryonic lung cultures, whereas inhibition of *tbx2* and *tbx3* failed to show any effect on branching morphogenesis.[39] Further inspection revealed that there was a loss of *shh* expression in the lung epithelium and that mesenchymal *fgf10* expression was severely reduced in the lung cultures. Reintroduction of exogenous Fgf10 into the culture restored most of the branching defects suggesting that inhibition of *tbx4* and *tbx5* disrupts branching morphogenesis through Fgf10.[39] Removal of *fgf10* from the developing lung results in severe complications. *Fgf10* null mice survive to birth but will quickly die due to lack of proper lung formation. The trachea forms in these mice, but ends in a mass of disorganized mesenchymal cells in which no primary lung buds are visible.[40] An interesting connection between Fgf10 and Shh has been uncovered. Expression levels of *fgf10* increase as development proceeds towards birth.[6] This follows the opposite trend of *shh* expression, although recent studies do not agree with a reduction in *shh* expression at later lung gestation.[7,8] Interestingly, over-expression of *shh* in the distal epithelium causes a reduction in *fgf10* expression.[37] This pattern of interaction suggests that Shh is a potential negative regulator of *fgf10*. However, the *fgf10* null mouse shows no lung formation, with *shh* expression only in the rudimentary trachea, and not in the distal lung endoderm, indicating Fgf10 is upstream of *shh*.[40] When experiments testing the effect of Fgf10 on *shh* expression using beads soaked with Fgf10 implanted in wild-type 11.5 dpc lung explants were performed, no changes in *shh* expression were detected.[41] This is in contrast to more a recent finding in murine palate formation in which exogenous Fgf10 was shown to induce *shh* expression in wild-type palatal epithelium.[42] While the *shh* null lung phenotype develops further than the *fgf10* null lung, as demonstrated by its ability to form two small lung lobes, the trachea and esophagus fail to separate signifying failure in other areas of pulmonary development. This contrasts the *fgf10* null phenotype, in which the trachea and esophagus do manage to separate into individual tubes, but no further lung formation is evident. This suggests that any potential interaction between Shh and Fgf10 would be regulated by intermediate proteins. Furthermore, it has been suggested that the primary developmental functions of these two proteins most likely act in separate parallel pathways during lung development.[43]

If we now take a closer look at the Fgf receptors (Fgfr), further conclusions into a possible relation with Shh signalling can be drawn. While all four *fgfrs* are expressed in the postnatal lung, it is difficult to elucidate their functions through the generation of knockouts. Both

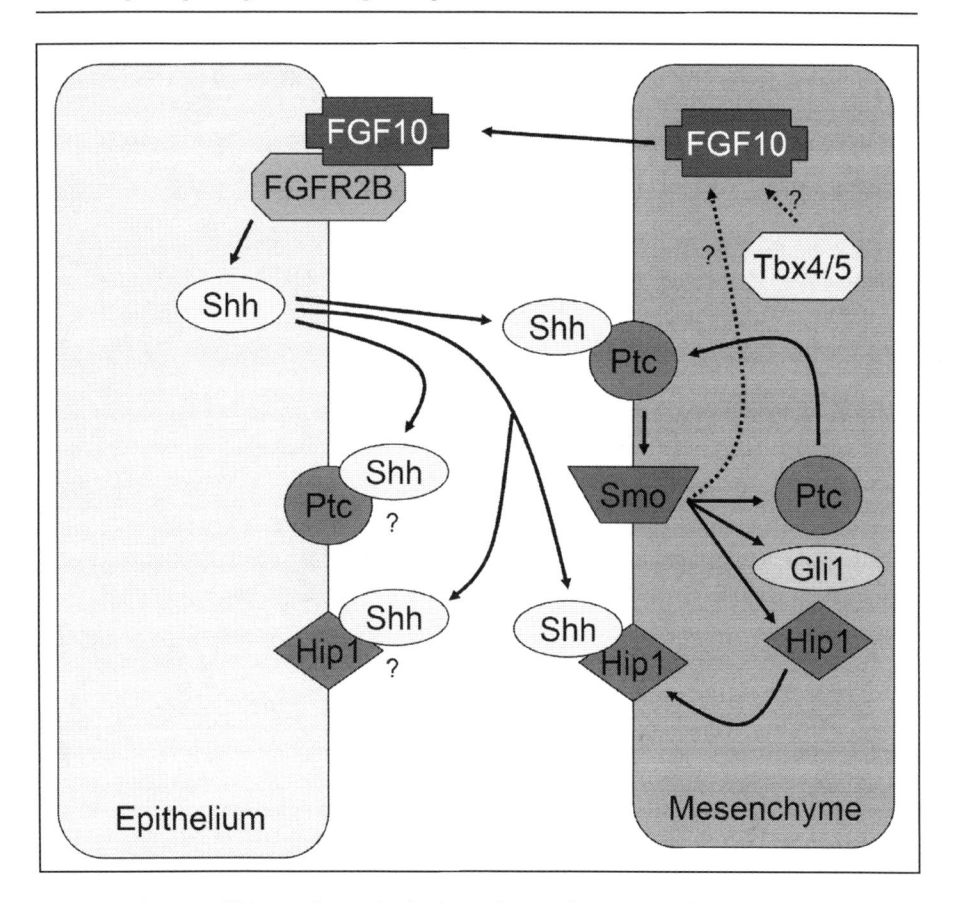

Figure 2. A diagram of Shh signalling in the developing lung. Fgf10 is expressed in the mesenchyme, which will signal to the epithelium to increase *shh* expression through the Fgfr2b receptor. Shh will then signal back to the mesenchyme to the membrane bound Ptc receptor. Shh may also be sequestered by Ptc and/or Hip1 in the epithelium as well as the mesenchyme. Shh binding to Ptc will de-repress Smo which will then signal for the up regulation of *ptc, gli1* and *hip1* expression. Both Ptc and Hip1 will then travel to the cell surface where they will act in a negative feedback mechanism by sequestering the Shh signalling molecule. Smo may also indirectly signal back to the controlling mechanism of Fgf10 production in a negative feedback regulatory loop. Tbx4 and Tbx5 may also be involved in the regulation of Fgf10 expression.

fgfr-1 and *fgfr-2* null mice die very early in development so effects on lung development can not be observed.[44-46] *Fgfr-3* null mice do exhibit some skeletal and inner ear developmental defects, however, no lung phenotype has been reported to date.[47] The *fgfr-4* null phenotype is by far the most mild, as these animals show no gross developmental abnormalities of any kind.[48] However, the combined *fgfr-3/fgfr-4* double homozygous null mutant mouse suffers from dwarfism, and failure to complete alveogenesis postnatally.[48] Most interesting though, is a special *fgfr-2* null mutant, the *fgfr2b$^{-/-}$* (IIIb isoform), generated through fusion chimeras in which mutant embryonic stem cells were combined with wild-type tetraploid embryos to allow survival until birth. These mice fail to develop limbs and lungs, quite similar to the *fgf10* null mouse.[49] When murine *fgfr2b$^{-/-}$* palate explants were given exogenous Fgf10, no induction of *shh* was observed contrary to effects observed in wild-type explants.[42] This indicates that Fgf10 signalling from the mesenchyme, through the epithelial located Fgfr2b receptor regulates *shh* expression in the epithelium. If this holds true in the lung, this could explain previous observations suggesting a relationship between Fgf10 and Shh in the developing lung. While the previously discussed *Fgf10* null mutations resulting in a reduction in *shh* expression fits this model, a possible explanation for *shh* over-expression causing a reduction in *fgf10* now exists. The observed over-expression of *shh* in the distal epithelium causing a reduction in *fgf10* expression could possibly be a form of negative feedback. If mesenchymal *fgf10* expression causes an epithelial increase in *shh* expression signalled through the Fgfr2b receptor, so that Shh can now signal back to the developing mesenchyme to help direct morphogenesis, the prospect of a negative feedback loop to Fgf10 signalling to attenuate the *shh* induction cue could exist.

In summary, development of the lung is a complex process as it not only entails growth and differentiation processes like many other organs, but also the creation of a complicated series of branched airways with their associated vasculature to create the interface for gas exchange. Members of the Shh pathway are found at the epithelial/mesenchymal border and removal of these proteins can have devastating effects on growth and branching morphogenesis. Cross-talk between epithelium and mesenchyme is essential for cordinating growth and branching signals so that the two tissues will successfully grow to form one cohesive functioning unit (see Fig. 2). Feedback mechanisms play an integral part in regulation of developmental signalling mechanisms and the Shh pathway has shown evidence of self-regulation through Ptc and Hip1. While clearly required for pulmonary development, our lack in comprehension of the Hedgehog signalling network and its interactions with other regulatory molecules, clearly demonstrates the need for further investigation.

References

1. Alescio T, Cassini A. Induction in vitro of tracheal buds by pulmonary mesenchyme grafted on tracheal epithelium. J Exp Zool 1962; 150:83-94.
2. Spooner BS, Wessells NK. Mammalian lung development: Interactions in primordium formation and bronchial morphogenesis. J Exp Zool 1970; 175(4):445-454.
3. Shannon JM, Hyatt BA. Epithelial-mesenchymal interactions in the developing lung. Annu Rev Physiol 2004; 66:625-645.
4. Bellusci S, Henderson R, Winnier G et al. Evidence from normal expression and targeted misexpression that bone morphogenetic protein (Bmp-4) plays a role in mouse embryonic lung morphogenesis. Development 1996; 122(6):1693-1702.
5. Litingtung Y, Lei L, Westphal H et al. Sonic hedgehog is essential to foregut development. Nat Genet 1998; 20(1):58-61.
6. Bellusci S, Furuta Y, Rush MG et al. Involvement of Sonic hedgehog (Shh) in mouse embryonic lung growth and morphogenesis. Development 1997; 124(1):53-63.
7. Miller LA, Wert SE, Whitsett JA. Immunolocalization of sonic hedgehog (Shh) in developing mouse lung. J Histochem Cytochem 2001; 49(12):1593-1604.
8. Unger S, Copland I, Tibboel D et al. Down-regulation of sonic hedgehog expression in pulmonary hypoplasia is associated with congenital diaphragmatic hernia. Am J Pathol 2003; 162(2):547-555.
9. Grindley JC, Bellusci S, Perkins D et al. Evidence for the involvement of the Gli gene family in embryonic mouse lung development. Dev Biol 1997; 188(2):337-348.
10. Chiang C, Litingtung Y, Lee E et al. Cyclopia and defective axial patterning in mice lacking Sonic hedgehog gene function. Nature 1996; 383(6599):407-413.
11. Pepicelli CV, Lewis PM, McMahon AP. Sonic hedgehog regulates branching morphogenesis in the mammalian lung. Curr Biol 1998; 8(19):1083-1086.
12. Miller LA, Wert SE, Clark JC et al. Role of Sonic hedgehog in patterning of tracheal-bronchial cartilage and the peripheral lung. Dev Dyn 2004; 231(1):57-71.
13. Watsuji T, Okamoto Y, Emi N et al. Controlled gene expression with a reverse tetracycline-regulated retroviral vector (RTRV) system. Biochem Biophys Res Commun 1997; 234(3):769-773.
14. Goodrich LV, Milenkovic L, Higgins KM et al. Altered neural cell fates and medulloblastoma in mouse patched mutants. Science 1997; 277(5329):1109-1113.
15. Bergstein I, Leopold PL, Sato N et al. In vivo enhanced expression of patched dampens the sonic hedgehog pathway. Mol Ther 2002; 6(2):258-264.
16. Chen Y, Struhl G. Dual roles for patched in sequestering and transducing Hedgehog. Cell 1996; 87(3):553-563.

17. Akiyama H, Shigeno C, Hiraki Y et al. Cloning of a mouse smoothened cDNA and expression patterns of hedgehog signalling molecules during chondrogenesis and cartilage differentiation in clonal mouse EC cells, ATDC5. Biochem Biophys Res Commun 1997; 235(1):142-147.
18. Shannon JM, Srivastava K, Shangguan X et al. [C59] [Poster: G49] disruption of sonic hedgehog signaling alters patterning and gene expression in cultured embryonic lungs. Orlando, Florida, USA: Paper presented at: American Thoracic Society, 2004.
19. Goodrich LV, Johnson RL, Milenkovic L et al. Conservation of the hedgehog/patched signaling pathway from flies to mice: Induction of a mouse patched gene by Hedgehog. Genes Dev 1996; 10(3):301-312.
20. Chuang PT, McMahon AP. Vertebrate Hedgehog signalling modulated by induction of a Hedgehog-binding protein. Nature 1999; 397(6720):617-621.
21. Chuang PT, Kawcak T, McMahon AP. Feedback control of mammalian Hedgehog signaling by the Hedgehog-binding protein, Hip1, modulates Fgf signaling during branching morphogenesis of the lung. Genes Dev 2003; 17(3):342-347.
22. Hui CC, Slusarski D, Platt KA et al. Expression of three mouse homologs of the Drosophila segment polarity gene cubitus interruptus, Gli, Gli-2, and Gli-3, in ectoderm- and mesoderm-derived tissues suggests multiple roles during postimplantation development. Dev Biol 1994; 162(2):402-413.
23. Park HL, Bai C, Platt KA et al. Mouse Gli1 mutants are viable but have defects in SHH signaling in combination with a Gli2 mutation. Development 2000; 127(8):1593-1605.
24. Mo R, Freer AM, Zinyk DL et al. Specific and redundant functions of Gli2 and Gli3 zinc finger genes in skeletal patterning and development. Development 1997; 124(1):113-123.
25. Motoyama J, Liu J, Mo R et al. Essential function of Gli2 and Gli3 in the formation of lung, trachea and oesophagus. Nat Genet 1998; 20(1):54-57.
26. Schimmang T, Lemaistre M, Vortkamp A et al. Expression of the zinc finger gene Gli3 is affected in the morphogenetic mouse mutant extra-toes (Xt). Development 1992; 116(3):799-804.
27. Li Y, Zhang H, Choi SC et al. Sonic hedgehog signaling regulates Gli3 processing, mesenchymal proliferation, and differentiation during mouse lung organogenesis. Dev Biol 2004; 270(1):214-231.
28. King JA, Marker PC, Seung KJ et al. BMP5 and the molecular, skeletal, and soft-tissue alterations in short ear mice. Dev Biol 1994; 166(1):112-122.
29. Jena N, Martin-Seisdedos C, McCue P et al. BMP7 null mutation in mice: Developmental defects in skeleton, kidney, and eye. Exp Cell Res 1997; 230(1):28-37.
30. Winnier G, Blessing M, Labosky PA et al. Bone morphogenetic protein-4 is required for mesoderm formation and patterning in the mouse. Genes Dev 1995; 9(17):2105-2116.
31. Hogan BL, Yingling JM. Epithelial/mesenchymal interactions and branching morphogenesis of the lung. Curr Opin Genet Dev 1998; 8(4):481-486.
32. Nogawa H, Ito T. Branching morphogenesis of embryonic mouse lung epithelium in mesenchyme-free culture. Development 1995; 121(4):1015-1022.
33. Post M, Souza P, Liu J et al. Keratinocyte growth factor and its receptor are involved in regulating early lung branching. Development 1996; 122(10):3107-3115.
34. Cardoso WV, Itoh A, Nogawa H et al. FGF-1 and FGF-7 induce distinct patterns of growth and differentiation in embryonic lung epithelium. Dev Dyn 1997; 208(3):398-405.
35. Simonet WS, DeRose ML, Bucay N et al. Pulmonary malformation in transgenic mice expressing human keratinocyte growth factor in the lung. Proc Natl Acad Sci USA 1995; 92(26):12461-12465.
36. Guo L, Degenstein L, Fuchs E. Keratinocyte growth factor is required for hair development but not for wound healing. Genes Dev 1996; 10(2):165-175.
37. Bellusci S, Grindley J, Emoto H et al. Fibroblast growth factor 10 (FGF10) and branching morphogenesis in the embryonic mouse lung. Development 1997; 124(23):4867-4878.
38. Chapman DL, Garvey N, Hancock S et al. Expression of the T-box family genes, Tbx1-Tbx5, during early mouse development. Dev Dyn 1996; 206(4):379-390.
39. Cebra-Thomas JA, Bromer J, Gardner R et al. T-box gene products are required for mesenchymal induction of epithelial branching in the embryonic mouse lung. Dev Dyn 2003; 226(1):82-90.
40. Sekine K, Ohuchi H, Fujiwara M et al. Fgf10 is essential for limb and lung formation. Nat Genet 1999; 21(1):138-141.
41. Lebeche D, Malpel S, Cardoso WV. Fibroblast growth factor interactions in the developing lung. Mech Dev 1999; 86(1-2):125-136.
42. Rice R, Spencer-Dene B, Connor EC et al. Disruption of Fgf10/Fgfr2b-coordinated epithelial-mesenchymal interactions causes cleft palate. J Clin Invest 2004; 113(12):1692-1700.
43. van Tuyl M, Post M. From fruitflies to mammals: Mechanisms of signalling via the Sonic hedgehog pathway in lung development. Respir Res 2000; 1(1):30-35.

44. Arman E, Haffner-Krausz R, Chen Y et al. Targeted disruption of fibroblast growth factor (FGF) receptor 2 suggests a role for FGF signaling in pregastrulation mammalian development. Proc Natl Acad Sci USA 1998; 95(9):5082-5087.
45. Deng CX, Wynshaw-Boris A, Shen MM et al. Murine FGFR-1 is required for early postimplantation growth and axial organization. Genes Dev 1994; 8(24):3045-3057.
46. Yamaguchi TP, Harpal K, Henkemeyer M et al. fgfr-1 is required for embryonic growth and mesodermal patterning during mouse gastrulation. Genes Dev 1994; 8(24):3032-3044.
47. Colvin JS, Bohne BA, Harding GW et al. Skeletal overgrowth and deafness in mice lacking fibroblast growth factor receptor 3. Nat Genet 1996; 12(4):390-397.
48. Weinstein M, Xu X, Ohyama K et al. FGFR-3 and FGFR-4 function cooperatively to direct alveogenesis in the murine lung. Development 1998; 125(18):3615-3623.
49. Arman E, Haffner-Krausz R, Gorivodsky M et al. Fgfr2 is required for limb outgrowth and lung-branching morphogenesis. Proc Natl Acad Sci USA 1999; 96(21):11895-11899.

CHAPTER 14

New Perspectives in Shh Signalling?

Carolyn E. Fisher*

The Shh-Ptc signalling pathway and its components have been the subject of much research. Previous chapters of this book have illustrated the importance of this pathway and its activation in many aspects of development, regeneration of adult organs, and pathology. The role of Ptc as the primary receptor for Shh and its analogues has been emphasised in these chapters. However, there is mounting evidence that megalin [also known as glycoprotein 330 (gp330) or low-density lipoprotein receptor-related protein 2 (LPR2)], a 600kDa transmembrane glycoprotein belonging to the low density lipoprotein (LDL) receptor family, may be a second receptor for Shh. In this chapter I shall review this evidence; but I shall preface this discussion with an outline of the known properties and biological functions of megalin.

Megalin

Megalin (gp330) was first discovered in 1982 as the pathogenic antigen of Heymann nephritis.[1] The megalin gene has been sequenced in rat and human[2,3] and mapped to chromosome 2.[4] The protein contains a single transmembrane domain,[4] is known to act as an endocytic receptor and is expressed primarily in polarised epithelial cells, and strictly on the apical surfaces of such cells.[5] Both the protein and its mRNA have been identified in human parathyroid cells, placental cytotrophoblasts and epididymal epithelial cells. The protein is also expressed in mammary epithelia, thyroid follicular cells, yolk sacs, the ciliary body of the eye,[6] the intestinal brush border,[7] the male reproductive tract,[8] uterus and oviduct,[9] and gallbladder epithelium.[10] In addition, immortalised foetal rat alveolar pretype II cells,[11] adult rat type II pneumocytes[12,13] and human type II cells[14] have all been reported to express megalin. It has been proposed that in adult lung megalin may be important in supplying vitamin E to type II pneumocytes.[15]

The functions of megalin have been studied in greatest detail in the renal proximal tubule, where it is expressed in the luminal aspects of epithelial cells, and is associated with the sodium-potassium exchanger.[16] It is also involved in renal uptake of angiotensin[17] and the reabsorption of various molecules including calcium[18] and vitamin D.[19]

Cubilin

Interactions between megalin and scaffold proteins, usually mediated through cytosolic adaptors, imply a diversity of functions in cellular communication and signal transduction as well as endocytosis.[20] The 460 kDa receptor protein cubilin (gp 280), which has been sequenced in human[21] and other species, is required for megalin-dependent endocytosis of many ligands in kidney tubules and other epithelial types.[22-27] Cubilin is a peripheral protein, attached to the extracellular face of the epithelial cell membrane by its 110-residue N-terminal sequence,[28] which contains numerous EGF-like, complement (C1r/C1s)-like and bone morphogenic protein-like repeats.[29] In neonate and adult mice, there is significant overlap between

---

*Carolyn E. Fisher—Immunobiology Group, MRC/UoE Centre for Inflammation Research, The Queen's Medical Research Institute, Little France Crescent, Edinburgh EH16 4TJ, Scotland, U.K. Email: carolyn.fisher@ed.ac.uk

Shh and Gli Signalling and Development, edited by Carolyn E. Fisher and Sarah E.M. Howie.
©2006 Landes Bioscience and Springer Science+Business Media.

expression of the cubilin and megalin transcripts.[23] However, cubilin is more restricted in distribution than megalin.[30]

Endocytosis: RAP and Other Adaptor Molecules

The polarised distribution of megalin in epithelial cells results from interactions involving its cytoplasmic (C-terminal) moiety.[31] Several adaptor molecules that bind the cytoplasmic tail of megalin to intracellular proteins have been identified including autosomal recessive hypercholesterolemia (ARH), which facilitates megalin-related endocytosis and might serve as a chaperone during internalization.[32] ARH colocalizes with megalin in clathrin coated pits and in recycling endosomes in the Golgi. Internalised megalin is first seen together with ARH in clathrin coated pits. Then, in sequence, it is seen in early endosomes, pericentriolar tubular recycling endosomes, and finally the cell surface again.

Trafficking through early endosomes might be characteristic of ligand-bound megalin, but in the absence of ligands this process involves the receptor-associated protein, RAP, in humans and rats;[33,34] the mouse homologue is heparin binding protein-44.[35] Megalin-RAP complexes appear to cycle through the late endosomes. From these, megalin is returned to the cell surface while RAP is degraded in the lysosomes.[36] RAP has a chaperone-like function necessary for normal processing and subcellular distribution of megalin. Without RAP, megalin levels in the cell fall significantly; in kidney proximal tubule cells, less is detected on the brush-border membrane and relatively more on the rough endoplasmic reticulum.[37]

Megalin-RAP Binding

In humans, RAP is a 39 kDa protein that copurifies with megalin,[38,39] binds to it with high affinity (K_d = 8nM), and colocalises with it on the apical surfaces of renal tubular epithelial cells.[38] In rats, RAP is a 44 kDa molecule.[39] Binding of RAP to megalin is calcium-dependant.[40,41] Direct binding studies show that there are two primary megalin binding sites within RAP; one between amino acids 85 and 148, and the other between amino acids 178 and 248.[42]

Orlando et al[43] demonstrated that amino acids 1111-1210 represent a binding site on megalin for various ligands including RAP and concluded that there was one common binding site for several ligands. A specific anti-megalin Ab causes only partial inhibition of RAP-binding to megalin[43] and a more recent study suggests that megalin has more than one binding site for RAP;[44] however, the stoichiometry is unknown.[38] Multiple RAP-binding sites on megalin would be consistent with the fact that the most closely-related member of the LDLR family, LDL receptor-related protein (LRP), which is very similar to megalin in overall structural organisation, function and size,[5,45] has multiple RAP-binding sites.[46]

Megalin-Shh-RAP Interactions

Megalin is known to interact with a multitude of molecules, most notably cubilin. However, there is now in vitro evidence that Shh and megalin interact. A radiolabelled ligand binding assay and ELISA showed that N-Shh binds megalin with high affinity. In addition, surface plasmon resonance (SPR) measurements showed that a recombinant fusion protein of N-Shh with glutathione-S-transferase (GST-N-Shh) was able to bind megalin with an affinity constant (K_D) of 21nM in the presence of calcium; in control experiments recombinant GST did not bind to megalin.[44]

Possible interactions among RAP, megalin and Shh were investigated in BN (rat yolk sac cell-line) cells, where it was established that megalin was the only RAP-binding member of the LDLR family present.[44] When these cells were cultured in the presence of GST, intracellular punctate staining consistent with vesicular localisation of GST-N-Shh was evident; addition of RAP, or anti-megalin antibodies, blocked GST-N-Shh uptake. This implies communication among these molecules at some level, and confirms that RAP is a specific inhibitor of megalin in this system. As mentioned earlier, megalin endocytoses its ligands. It has been argued that this leads to lysosomal degradation, evidenced by the presence of TCA-soluble proteolytic fragments of ligands in media after in vitro culture.[11] In contrast, megalin-N-Shh appears resistant to dissociation at pH 4.5, implying stability in the acidic environment of endosomes; moreover, chloroquine (a lysosomal proteinase inhibitor) does not inhibit[32] P-labelled GST-N-Shh degradation in BN cells.[44] How some megalin ligands appear to bypass lysosomal degradation is not clear at present.

Megalin in Development: More Links to Shh

Megalin is expressed on the outer cells of the preimplantation mouse embryo during epithelial differentiation, suggesting a role in early embryo development.[47] Not surprisingly, it participates in the development of the renal proximal tubule,[48] and it is important in development of the forebrain.[49] Megalin-cubilin complexes might be important in placental transport, since application of antibodies to pregnant females has teratogenic effects.[50] However, megalin and cubilin have different expression patterns in the mouse embryo.[51]

Many developmental abnormalities in mice that lack components of the Shh pathway are strikingly similar to those found in megalin -/- mice, suggesting that megalin is a regulatory component of the Shh signalling pathway.[52] Shh -/- and smo -/- mice, smo -/- zebrafish embryos, mice lacking dispatched (Disp), which is critical for the secretion/long-range signalling of N-Shh, and partially rescued Ptc -/- embryos, all display neurodevelopmental abnormalities.[53] Shh and megalin are coexpressed early in the development of the nervous system, and megalin-containing cells internalise the active fragment N-Shh by a mechanism sensitive to anti-megalin antibodies. N-Shh uptake may also be dependent on heparan-sulphate-containing-proteoglycans.[44] McCarthy and Argraves[53] have proposed possible models for the role of megalin in the neurodevelopmental biology of Shh and retinol. N-Shh might signal directly via megalin; it might be internalised by megalin-dependent endocytosis to regulate its availability to Ptc, or in order to deliver it to vesicular pools of Ptc;[54] or it might undergo transcytosis while megalin internalises Ptc and smo.

Proteoglycans and Megalin Interactions

Although heparan sulphate proteoglycans (HSPGs) have been implicated in megalin function,[44] there is no direct interaction; megalin does not specifically bind heparin.[55] RAP possesses binding sites for both megalin and heparin; the heparin-binding site on RAP is between amino acids 261 and 323.[42] The megalin and heparin binding sites within RAP are noncontiguous, consistent with the view that the glycosaminoglycan site is physiologically exposed when RAP is bound to megalin.[42] Therefore it is conceivable that megalin requires the aid of a ligand, such as cell-surface expressed RAP, in order to interact with HSPGs for signalling purposes; alternatively HSPGs might be required for megalin to bind to certain ligands, as evidenced in the thyroid. The Transcytosis of thyroglobulin (Tg) via megalin within the thyroid gland involves HSPGs, and it has been demonstrated that HSPGs bind to the heparin-binding sequence on Tg (between amino acids 2489 and 2503 in rat), facilitating the binding of this prohormone to megalin, and ultimately its transcytosis.[56]

Lung Development and the Role of Megalin

Pulmonary development begins in the mouse at embryonic day (E) 9-9.5 as an endodermal budding from the foregut; two endodermal buds (primary buds) give rise to the left and right lobes of the distal lung. Initially the primary bronchial buds divide asymmetrically, growing ventrally and caudally. Not until E10.5 does lateral branching begin, leading to one left and four right secondary bronchi. As morphogenesis continues, dichotomous branching ensues. Four morphological stages of lung morphogenesis in mammals have been described.[57,58] In mice, the pseudoglandular phase, characterised by dichotomous branching and the establishment of the basic branching pattern of the lungs, begins at E11.5 and ends at approximately E16. The canalicular phase then ensues with centrifugal branching (radially outward from the centre) forming the bronchial airways. E19 signifies the end of the saccular phase, which like

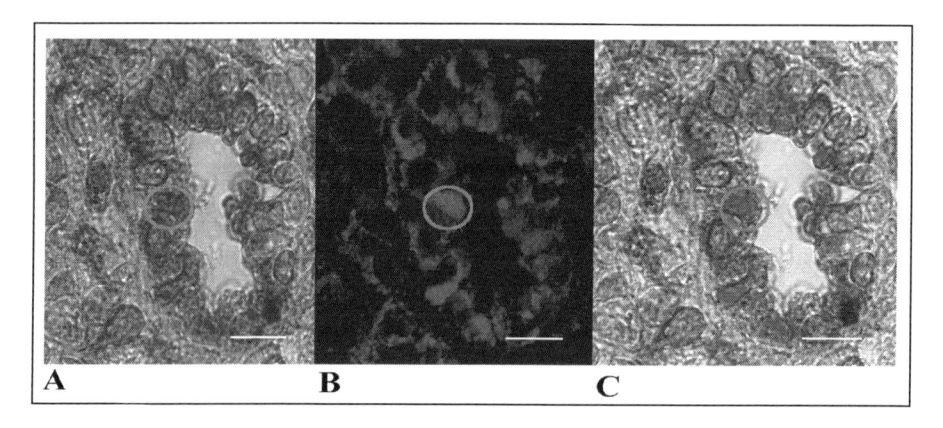

Figure 1. Coexpression of Shh and megalin during murin pulmonary development. The epithelia of embryonic airways express sonic hedgehog protein (A), demonstrated here with chromogenic DAB-based staining; megalin protein is also expressed in some epithelial airway cells (B), as shown here using fluorescent immunohistochemistry. Overlaying the transmission micrograph of Shh staining with the fluorescent image demonstrating megalin expression, shows that these proteins are coexpressed in some airway cells of the developing murine pulmonary system (C). The red circle highlights one such cell. The scale bar represents 10 μm in all of the above images.

the canalicular phase involves centrifugal branching, and results in the formation of distal branches linked to alveolar sacs. The final stage of lung morphogenesis is termed the alveolar phase; mature alveoli are formed by outpouching of alveolar sacs. In mice (and rats) alveolar maturation is an entirely postnatal event.[59]

Whilst the most noticeable abnormalities in megalin-deficient embryos involve the CNS, Willnow et al[49] reported developmental abnormalities in both kidney and lung in megalin-deficient mice. Megalin knockout mice die perinatally of respiratory insufficiency and show abnormalities in epithelia that normally express the protein. In particular, immunohistochemical analysis of the lungs reveals emphysematous areas characterised by enlarged alveoli, and atelectic regions defined by collapsed alveoli and thickened alveolar walls.[49]

Although it has long been known that Shh is expressed in lung during mammalian pulmonary development, the only published evidence that megalin is also expressed during pulmonary development was reported by Kounnas et al[5] who found megalin in the bronchial epithelia of E12.5 murine lungs. Its possible role in pulmonary development has not been investigated until now. Just as Shh expression becomes restricted as pulmonary development progresses, so does megalin expression. Moreover, not only do the expression patterns of the two molecules appear similar but also megalin and Shh are coexpressed in the same cells during pulmonary development (Fig. 1). This probably represents colocalisation, though as yet the evidence is circumstantial.

Although mRNA transcripts for megalin are found in adult mouse lung, there is no coexpression of cubilin.[23] Therefore if megalin is involved in the transport/endocytosis of Shh in this organ, it either does this in isolation or an alternative ligand must be involved. One possible candidate is RAP. RAP is most abundant in the lumen of the ER, but immunohistochemistry and cell surface radioiodination have been used to demonstrate its presence on the apical surface of renal proximal tubule cells,[60] gingival fibroblasts[61] and two carcinoma cell-lines.[55,62] Biochemical studies have shown that RAP present on the surface of cells, or exogenous RAP added to culture, is an effective inhibitor of ligand binding to megalin and LRP. HBP-44 mRNA is present during murine pulmonary morphogenesis[63] and although a study of the immunolocalisation of megalin and RAP proteins during murine embryogenesis, did not determine whether RAP was present in lung,[5] there is no evidence to the contrary.

Indeed, the greatest problem facing researchers intent on studying RAP protein expression in the mouse is the nonavailability, to date, of any specific antibody (personal communications). However, it is likely that RAP protein (HBP-44) is expressed in murine lung, as evidenced by the presence of mRNA.

All seven types of glycosaminoglycans (GAGs) are found within the lungs of postnatal rats,[64] sulphated GAGs are present during chick lung development,[65] and heparin is thought to modulate the kinetics of heparan sulphate binding ligands that drive lung development.[66] Coupled with the observations that megalin and Shh are expressed within the same cells during development of the mouse pulmonary system, and that there is a conserved sequence within Shh for binding heparan containing PGs that is distinct from the binding site for Ptc,[67] this invites conjecture about how all these molecules interact.

At least three models of possible Shh-megalin interactions during neurodevelopment have been proposed by McCarthy and Argraves,[53] though whether megalin constitutes a component of the Shh pathway is not clear.

1. N-Shh signals directly via megalin. In the thyroid, HSPGs aid the binding of megalin to its ligands.[56] By analogy, it could be argued that in those lung cells that coexpress Shh and megalin, HSPGs are bound by Shh, which then facilitates binding to megalin, allowing signalling to ensue. A similar model for Shh binding to Ptc has been proposed[54] on the evidence that HSPGs synthesised by the enzymatic action of *tout velu* regulate Shh movement.[68]
2. The observation that Shh can be internalised by megalin-containing cells during development of the CNS, and the work of McCarthy et al[44] on BN cells, give credence to the possibility that N-Shh might be internalised by megalin-dependent endocytosis, a process reliant on RAP. Although there is no conclusive evidence that RAP protein is expressed in lung, the presence of HBP-44 mRNA suggests that it is likely.
3. N-Shh might undergo transcytosis while megalin internalises Ptc and smo. This would imply that megalin is either a component of the Shh pathway or that it interacts with various elements of it.

The Good, the Bad and the Ugly

As with all relatively new discoveries and models, the interaction(s) between megalin and Shh, the cross-talk between megalin and other components of the Shh signalling network such as Ptc and smo, and the hypothesis that megalin constitutes a regulatory component of the Shh pathway, are by no means fully detailed or articulated.

During mouse embryogenesis, many sites of megalin expression are either identical to those expressing Shh or located in adjacent tissues regulated by Shh signalling,[44] suggesting interaction at some level. Most of the work aimed at elucidating interactions between Shh and megalin has concentrated on development of the CNS, where coexpression of the molecules has been demonstrated. Such studies, and the ex-vivo study of McCarthy et al[44] investigating megalin-Shh interactions in BN cells, strengthens the suggestion made by Herz and Bock[52] that megalin is a component of the Shh signalling pathway. Indeed, when comparisons are made among the abnormalities prevalent in the CNS of Shh -/- mice, megalin KO mice and embryos lacking other components of the Shh signalling cascade, significant crossover is apparent. Since megalin and Shh are also coexpressed during development of the murine pulmonary system, it is tempting to speculate that the interactions demonstrated during neurodevelopment are conserved across different organs, and possibly different species.

However, while there is evidence that Shh and megalin are coexpressed during pulmonary development, just as they are during development of the CNS, the data concerning lung organogenesis from KO studies do not directly implicate megalin in the Shh pathway. Shh -/- mice essentially have no lungs; it is an embryonic lethal phenotype. Megalin -/- mice die perinatally due to respiratory insufficiency, lungs do form but there appear to be problems with differentiation/specialisation. This suggests that even if megalin acts as a second receptor for Shh during early lung development, in conjunction with the primary receptor Ptc, it is neither sufficient nor indispensable for lung branching morphogenesis. In addition, the phenotypic evidence from studies on KO mice clearly indicate that megalin cannot be substituted for Ptc or Shh i.e., this is not an example of redundancy in nature. The most plausible conclusion from this evidence is that the function(s) of megalin during pulmonary morphogenesis is (are) unrelated to the Shh pathway. This of course does not preclude interactions between these molecules, or the possibility that Shh is a ligand for megalin, but it challenges the view that megalin is a component of the Shh pathway.

There are data suggesting that Shh and megalin directly interact, and phenotypic data strongly indicating that megalin constitutes a member of the Shh signalling pathway (the 'good'). But there are also data inconsistent with this conclusion (the 'bad'), leaving us with conflicting information and a need for clarification (the 'ugly'). Such confusion is inevitable at this stage; the Shh pathway/network has not been fully elucidated. It will not be an easy task to integrate our understanding of this pathway with megalin, a very versatile protein that interacts with a vast number of ligands and serves many different functions that remain incompletely understood. It is clear that the importance of Shh-megalin interactions varies among different organs, as does the relative importance of the individual proteins at different stages of development. Further research will be required to elucidate these interactions.

References

1. Kerjaschki D, Farquhar MG. The pathogenic antigen of Heymann nephritis is a membrane glycoprotein of the renal proximal tubule brush border. Proc Natl Acad Sci USA 1982; 79(18):5557-5561.
2. Saito A, Pietromonaco S, Loo AK et al. Complete cloning and sequencing of rat gp330/'megalin,' a distinctive member of the low density lipoprotein receptor gene family. Proc Natl Acad Sci USA 1994; 91:9725-9729.
3. Hjälm G, Murray E, Crumley G et al. Cloning and sequencing of human gp330, a Ca^{2+}-binding receptor with potential intracellular signaling properties. Eur J Biochem 1996; 239:132-137.
4. Xia YR, Bachinsky DR, Smith JA et al. Mapping of the glycoprotein 330 (Gp330) gene to mouse chromosome 2. Genomics 1993; 17:780-781.
5. Kounnas MZ, Haudenschild CC, Strickland DK et al. Immunological localization of glycoprotein 330, low density lipoprotein receptor related protein and 39 kDa receptor associated protein in embryonic mouse tissues. In Vivo 1994; 8:343-352.
6. Lundgren S, Carling T, Hjälm G et al. Tissue distribution of human gp330/megalin, a putative Ca(2+)-sensing protein. J Histochem Cytochem 1997; 45:383-392.
7. Yammani RR, Seetharam S, Seetharam B. Cubilin and megalin expression and their interaction in the rat intestine: Effect of thyroidectomy. Am J Physiol Endocrinol Metab 2001; 281:E900-E907.
8. Van Praet O, Argraves WS, Morales CR. Coexpression and interaction of cubilin and megalin in the adult male rat reproductive system. Mol Reprod Dev 2003; 64:129-135.
9. Argraves WS, Morales CR. Immunolocalization of cubilin, megalin, apolipoprotein J, and apolipoprotein A-I in the uterus and oviduct. Mol Reprod Dev 2004; 69:419-427.
10. Erranz B, Miquel JF, Argraves WS et al. Megalin and cubilin expression in gallbladder epithelium and regulation by bile acids. J Lipid Res 2004; 45:2185-2198.
11. Stefansson S, Kounnas MZ, Henkin J et al. Gp330 on type II pneumocytes mediates endocytosis leading to degradation of pro-urokinase, plasminogen activator inhibitor-1 and urokinase-plasminogen activatior inhibitor-1 complex. J Cell Sci 1995; 108:2361-2368.
12. Zheng G, Bachinsky DR, Stamenkovic I et al. Organ distribution in rats of two members of the low-density lipoprotein receptor gene family, gp330 and LRP/alpha 2MR, and the receptor associated protein (RAP). J Histochem Cytochem 1994; 42:531-542.
13. Chatelet F, Brianti E, Ronco P et al. Ultrastructural localization by monoclonal antibodies of brush border antigens expressed by glomeruli. II. Extrarenal distribution. Am J Pathol 1986; 122:512-519.
14. Lundgren S, Carling T, Hjälm G et al. Tissue distribution of human gp330/megalin, a putative Ca^{2+}-sensing protein. J Histochem Cytochem 1997; 45:383-392.
15. Kolleck I, Sinha P, Rustow B. Vitamin E as an antioxidant of the lung: mechanisms of vitamin E delivery to alveolar type II cells. Am J Respir Crit Care Med 2002; 166:S62-S66.
16. Biemesderfer D, Nagy T, DeGray B et al. Specific association of megalin and the Na^+/H^+ exchanger isoform NHE3 in the proximal tubule. J Biol Chem 1999; 274:17518-17524.
17. Gonzalez-Villalobos R, Klassen RB, Allen PL et al. Megalin binds and internalizes Angiotensin II. Am J Physiol Renal Physiol 2005; 288(2):F420-427.

18. Frick KK, Bushinsky DA. Molecular mechanisms of primary hypercalciuria. J Am Soc Nephrol 2003; 14:1082-1095.
19. Hilpert J, Wogensen L, Thykjaer T et al. Expression profiling confirms the role of endocytic receptor megalin in renal vitamin D3 metabolism. Kidney Int 2002; 62:1672-1681.
20. Gotthardt M, Trommsdorff M, Nevitt MF et al. Interactions of the low density lipoprotein receptor gene family with cytosolic adaptor and scaffold proteins suggest diverse biological functions in cellular communication and signal transduction. J Biol Chem 2000; 275:25616-25624.
21. Kozyraki R, Kristiansen M, Silahtaroglu A et al. The human intrinsic factor-vitamin B12 receptor, cubilin: Molecular characterization and chromosomal mapping of the gene to 10p within the autosomal recessive megaloblastic anemia (MGA1) region. Blood 1998; 91:3593-3600.
22. Moestrup SK, Verroust PJ. Megalin- and cubilin-mediated endocytosis of protein-bound vitamins, lipids, and hormones in polarized epithelia. Ann Rev Nutr 2001; 21:407-428.
23. Hammad SM, Barth JL, Knaak C et al. Megalin acts in concert with cubilin to mediate endocytosis of HDL. J Biol Chem 2000; 275:12003-12008.
24. Argraves WS, Barth JL. Cubilin and megalin: Partners in lipoprotein and vitamin metabolism. Trends in Cardiovascular Medicine 2001; 11:26-31.
25. Birn H, Fyfe JC, Jacobsen C et al. Cubilin is an albumin binding protein important for renal tubular albumin reabsorption. J Clin Invest 2000a; 105:1353-1361.
26. Nykjaer A, Fyfe JC, Kozyraki R et al. Cubilin dysfunction causes abnormal metabolism of the steroid hormone 25(OH) vitamin D(3). Proc Natl Acad Sci USA 2001; 98:13895-13900.
27. Christensen EI, Devuyst O, Dom G et al. Loss of chloride channel ClC-5 impairs endocytosis by defective trafficking of megalin and cubilin in kidney proximal tubules. Proc Natl Acad Sci USA 2003; 100:8472-8477.
28. Bork P, Beckmann G. The CUB domain. A widespread module in developmentally regulated proteins. J Mol Biol 1993; 231:539-545.
29. Kristiansen M, Kozyraki R, Jacobsen C et al. Molecular dissection of the intrinsic factor-vitamin B12 receptor, cubilin, discloses regions important for membrane association and ligand binding. J Biol Chem 1999; 274:20540-20544.
30. Christensen EI, Birn H, Verroust P et al. Membrane receptors for endocytosis in the renal proximal tubule. Int Rev Cytol 1998; 180:237-284.
31. Marzolo MP, Yuseff MI, Retamal C et al. Differential distribution of low-density lipoprotein-receptor-related protein (LRP) and megalin in polarized epithelial cells is determined by their cytoplasmic domains. Traffic 2003; 4:273-288.
32. Nagai M, Meerloo T, Takeda T et al. The adaptor protein ARH escorts megalin to and through endosomes. Mol Biol Cell 2003; 14:4984-4996.
33. Bu G, Geuze HJ, Strous GJ et al. 39 kDa receptor-associated protein is an ER resident protein and molecular chaperone for LDL receptor-related protein. EMBO J 1995; 14:2269-2280.
34. Willnow TE, Armstrong SA, Hammer RE et al. Functional expression of low density lipoprotein receptor-related protein is controlled by receptor-associated protein in vivo. Proc Natl Acad Sci USA 1995; 92:4537-4541.
35. Furukawa T, Ozawa M, Huang R-P et al. A heparin binding protein whose expression increases during differentiation of embryonal carcinoma cells to parietal endoderm cells: cDNA cloning and sequence analysis. J Biochem (Tokyo) 1990; 108(2):297-302.
36. Czekay RP, Orlando RA, Woodward L et al. Endocytic trafficking of megalin/RAP complexes: Dissociation of the complexes in late endosomes. Mol Biol Cell 1997; 8:517-532.
37. Birn H, Vorum H, Verroust PJ et al. Receptor-associated protein is important for normal processing of megalin in kidney proximal tubules. J Am Soc Nephrol 2000b; 11:191-202.
38. Kounnas MZ, Argraves WS, Strickland DK. The 39-kDa receptor-associated protein interacts with two members of the low density lipoprotein receptor family, α2-macroglobulin receptor and glycoprotein 330. J Biol Chem 1992; 267:21162-21166.
39. Orlando RA, Kerjaschki D, Kurihara H et al. Gp330 associates with a 44-kDa protein in the rat kidney to form the Heymann nephritis antigenic complex. Proc Natl Acad Sci USA 1992; 89:6698-6702.
40. Christensen EI, Gliemann J, Moestrup SK. Renal tubule gp330 is a calcium binding receptor for endocytic uptake of protein. J Histochem Cytochem 1992; 40(10):1481-1490.
41. Biemesderfer D, Dekan G, Aronson PS et al. Biosynthesis of the gp330/44-kDa Heymann nephritis antigenic complex: Assembly takes place in the ER. Am J Physiol 1993; 264(6 pt 2):F1011-F1020.
42. Orlando RA, Farquhar MG. Functional domains of the receptor-associated protein (RAP). Proc Natl Acad Sci USA 1994; 91:3161-3165.
43. Orlando RA, Exner M, Czekay R-P et al. Identification of the second cluster of ligand-binding repeats in megalin as a site for receptor-ligand interactions. Proc Natl Acad Sci USA 1997; 94:2368-2373.

44. McCarthy RA, Barth JL, Chintalapudi MR et al. Megalin functions as an endocytic sonic hedgehog receptor. J Biol Chem 2002; 277:25660-25667.
45. Herz J. The LDL receptor gene family: (Un)expected signal transducers in the brain. Neuron 2001; 29:571-581.
46. Williams SE, Ashcom JD, Argraves WS et al. A novel mechsanism for controlling the activity of alpha 2-macroglobulin receptor/low density lipoprotein receptor-related protein. Multiple regulatory sites for 39-kDa receptor-associated protein. J Biol Chem 1992; 267:9035-9040.
47. Gueth-Hallonet C, Santa-Maria A, Verroust P et al. Gp330 is specifically expressed in outer cells during epithelial differentiation in the preimplantation mouse embryo. Development 1994; 120:3289-3299.
48. Christiensen EI, Verroust PJ. Megalin and cubilin, role in proximal tubule function and during development. Pediatr Nephrol 2002; 17:993-999.
49. Willnow TE, Hilpert J, Armstrong SA et al. Defective forebrain development in mice lacking gp330/megalin. Proc Natl Acad Sci USA 1996; 93:8460-8464.
50. Sahali D, Mulliez N, Chatelet F et al. Characterization of a 280-kD protein restricted to the coated pits of the renal brush border and the epithelial cells of the yolk sac. Teratogenic effect of the specific monoclonal antibodies. J Exp Med 1988; 167:213-218.
51. Drake CJ, Fleming PA, Larue AC et al. Differential distribution of cubilin and megalin expression in the mouse embryo. Anat Rec A Discov Mol Cell Evol Biol 2004; 277:163-170.
52. Herz J, Bock HH. Lipoprotein receptors in the nervous system. Ann Rev Biochem 2002; 71:405-434.
53. McCarthy RA, Argraves WS. Megalin and the neurodevelopmental biology of sonic hedgehog and retinol. J Cell Sci 2003; 116:955-960.
54. Cohen MM. The hedgehog signaling network. Am J Med Genet A 2003; 123:5-28.
55. Orlando RA, Farquhar, MG. Identification of a cell line that expresses a cell surface and a soluble form of the gp330/receptor-associated protein (RAP) Heymann nephritis antigenic complex. Proc Natl Acad Sci USA 1993; 90(9):4082-4086.
56. Lisi S, Pinchera A, McClusky RT et al. Preferential megalin-mediated transcytosis of low hormonogenic thyroglobulin: A control mechanism for thyroid hormone release. Proc Natl Acad Sci USA 2003; 100(25):14858-14863.
57. Hislop AA. Airway and blood vessel interaction during lung development. J Anat 2002; 201:325-334.
58. McMurtry IF. Introduction: Pre and postnatal lung development, maturation, and plasticity. Am J Physiol Lung Cell Mol Physiol 2002; 282(3):L341-L344.
59. Burri PH. The postnatal growth of the rat lung III Morphology. Anat Rec 1974; 180:77-98.
60. Pietromonaco S, Kerjaschki D, Binder R et al. Molecular cloning of a cDNA encoding a major pathogenic domain of the Heymann nephritis antigen gp330. Proc Natl Acad Sci USA 1990; 87(5):1811-1815.
61. Strickland DK, Ashcom JD, Williams S et al. Sequence identity between the alpha 2-macroglobulin receptor and low density lipoprotein receptor-related protein suggests that this molecule is a multifunctional receptor. J Biol Chem 1990; 265(29):17401-4.
62. Czekay RP, Orlando RA, Woodward L et al. The expression of megalin (gp330) and LRP diverges during F9 cell differentiation. J Cell Sci 1995; 108(4):1433-41.
63. Lorent K, Overbergh L, Delabie J. The distribution of mRNA coding for 2 alpha macroglobulin, the murino globulins, the alpha2macroglobulin receptor and the alpha 2 macroglobulin receptor associated protein during mouse embryogenesis and in adult tissues. Differentiation 1994; 55:213-223.
64. Vaccaro CA, Brody JS. Ultrastructural localisation and characterisation of proteoglycans in pulmonary alveolus. Am Review Respir Disease 1979; 120:901-910.
65. Calvitti M, Baroni T, Calastrini C et al. Bronchial branching correlates with specific glycosidase activity, extracellular glycosaminoglycan accumulation, TCFβ_2, and IL-1 localization during chick embryo lund development. J Histochem Cytochem 2004; 52:325-334.
66. Jesudason EC, Connell MG, fernig DG et al. Heparin and in vitro experimental lung hypoplasia. Ped Surg Int 2000; 16:247-251.
67. Rubin JB, Choi Y, Segal RA. Cerebellar proteoglycans regulate sonic hedgehog responses during development. Development 2002; 129:223-2232.
68. Bellaiche Y, The I, Perrimon N. Tout-velu is a Drosophila homologue of the putative tumour suppressor EXT-1 and is needed for Hh diffusion. Nature 1998; 394:85-88.

Index

A

Activator 5, 6, 12, 23, 24, 40, 75, 83, 85-88, 120, 125, 129
Alveolar 75, 117, 127, 142, 147, 150
Antagonism 83, 84, 88
Antibody 18, 51, 74, 110-113, 120, 121, 126, 148, 149, 151
Apical ectodermal ridge (AER) 80-84, 87-89, 94, 96
Apoptosis 18, 19, 44, 48, 50, 76, 81, 84, 88, 112, 140

B

β-selection 109, 111
Basal epidermal layer (BEL) 96, 99-102, 104
Bax 113
Bcl-2 24, 30, 113
Blastema 93, 99-104
Bone morphogenetic protein (Bmp) 2, 16, 17, 24, 28, 30, 31, 39, 44-46, 52, 66, 74, 87, 88, 102, 103, 114, 116, 121-123, 129, 132-134, 137, 138, 140-142, 147
Brain 4, 6, 8, 13, 18, 19, 23-25, 28, 37, 44, 46-49, 51, 58, 72, 134
Branching morphogenesis 7, 121-123, 127-129, 133, 138, 140, 142, 144, 152
Bronchi (Br) 128, 137-139, 149
Bronchial epithelia 127, 150

C

Cell cycle 18, 19, 41, 59, 62, 110
Cell fate 5, 6, 14, 19, 28, 44, 45, 81, 111, 131
Central nervous system (CNS) 6, 8, 12, 16, 18, 19, 24, 26, 36, 37, 46, 54, 58, 150, 151
Cerebral hemisphere 48
Chicken 13, 16, 83
Cholesterol 4, 5, 12, 52-54, 72, 85, 86
Conditional mutant 75, 76, 134
Craniofacial 6, 44-54, 66, 72, 76
Craniofacial development 6, 44, 48, 50, 51, 54
Cubilin 147-150
Cyclopamine 4, 18, 39, 48, 50-52, 59-62, 64, 66, 102-104, 120-122, 127, 129, 131, 132

D

Desert (Dhh) hedgehog 4, 7, 13, 23, 38, 93, 110, 112, 117, 139
Development 1, 2, 4-8, 12-19, 23-25, 27-31, 36-41, 44-52, 54, 58-61, 63-66, 69-76, 79-84, 86, 87, 89, 93-99, 104, 107-114, 116-123, 125-127, 129-134, 137-144, 147, 149-152
Differentiation 4-8, 13, 14, 18, 19, 37, 40, 41, 44, 46, 58-64, 69, 70, 75, 80, 89, 102, 104, 107, 109-111, 113, 116-118, 120-123, 125-127, 129, 131, 132, 134, 138, 142, 144, 149, 151
Dispatched (Disp) 4, 12, 86, 97, 149
Dorsal midline 12, 15, 16, 24, 25, 28, 30
Dorsoventral patterning 23, 25, 28, 30, 50
Drosophila 2-5, 7, 8, 12, 23, 24, 39, 58, 62, 72, 82, 85, 86, 117, 140
Ductal branching 116-118, 120, 121, 123
Ductal differentiation 117, 118

E

Embryogenesis 1, 2, 5, 7, 8, 12, 24, 46, 52, 76, 108, 110, 111, 131, 132, 150, 151
Endocytosis 4, 147-151
Endoskeleton 94-96, 99, 104
Epimorphic 99
Epithelial bud 116-118, 132, 138
Epithelial-mesenchymal signalling 69, 117
Epithelium 7, 16, 17, 44-46, 48, 51, 52, 58, 59, 69, 70, 72-76, 102, 112, 116-120, 127, 129-133, 137-144, 147, 150
Etiology 54, 66
Exoskeleton 94
Eye 4, 6-8, 15, 25, 48, 49, 58-60, 63-66, 96, 147

F

Facial development 46, 50
Feedback loop 82, 83, 86-89, 102, 139, 144
Fibroblast growth factor (Fgf) 2, 30, 39, 44-46, 48, 50-52, 65, 66, 74, 81-84, 87-89, 98, 99, 102, 104, 116, 121-123, 127, 129, 131, 137-140, 142-144
Fin bud 93-99, 101, 104

Fin ray 93-95, 99-101, 103
Floor plate 5, 6, 12, 13, 37, 39, 40, 95, 98

G

Ganglion cell 58, 61-65
Gastrointestinal tract 125, 129
Gli 4-8, 12, 18, 19, 23, 24, 27, 29, 31, 36, 39, 41, 71, 76, 79, 85-87, 89, 107, 110, 112, 117, 120-122, 125, 127, 137, 139-141
Gli gene 5-8, 19, 39, 71, 76, 120, 121, 140
Gli protein 4, 6, 24, 29, 31, 85, 89, 141
Gli3 5-8, 12, 17, 23, 24, 26, 29-31, 36, 39-41, 48, 52, 54, 72, 76, 79, 83-85, 87, 88, 110, 112, 120, 125, 127, 129, 133, 134, 138, 140, 141
Gut 7, 17, 125, 126, 129, 131, 132, 134, 140

H

Hedgehog (Hh) 2-8, 12-14, 17, 18, 23, 24, 26-31, 36-39, 44-51, 58-66, 69, 71-74, 76, 79, 82, 84-87, 93, 94, 96, 98, 101-104, 107, 109-114, 116, 117, 120-123, 125, 127, 129, 131, 132, 134, 137, 139, 141, 144, 150
Hedgehog-interacting protein 5, 72, 86, 125, 139
Heparan sulphate 4, 149, 151
Hip1 5, 71-74, 127, 139, 143, 144
Homologue 2, 4, 5, 8, 12, 23, 26, 39, 82, 87, 117, 139, 148
Human 5, 8, 25, 28, 44, 46, 48, 49, 51-54, 66, 70, 74, 82, 84, 87, 94, 107, 110-114, 116-118, 125, 127, 131-133, 147, 148

I

Indian hedgehog (Ihh) 4, 6, 7, 13, 17, 23, 26, 38, 93, 100, 101, 110, 112, 117, 123, 127, 132, 133, 139
Intestine 125, 130, 132, 134
Invertebrate 5, 8, 69

K

Kidney 75, 128, 133, 134, 137, 147, 148, 150

L

Ligand 3, 4, 12, 15, 28, 31, 48, 51, 73, 74, 85, 86, 88, 102, 104, 117, 119, 120, 123, 147-152
Limb induction 79, 83
Limb prepattern 83, 84, 89
Lineage 15, 16, 28, 36, 37, 79, 110, 111
Long-range 2, 12, 27, 72, 86, 88, 89, 149
Lung 5-8, 125-129, 134, 137-144, 147, 149-152
Lung branching 127, 142, 152
Lung development 7, 127, 129, 137-143, 149, 151, 152
Lung growth 127, 142
Lung lobe 137, 138, 140-142

M

Mammalian 1, 2, 4, 5, 27, 44, 63, 66, 72, 101, 116, 117, 125, 137, 139, 140, 149, 150
Megalin 147-152
Mesenchyme 6-8, 18, 44-46, 48, 51, 52, 69-76, 79-85, 87-89, 94-96, 99, 101-103, 107, 116, 117, 119-122, 127-129, 131-134, 137-144
Midline 5, 12-19, 25, 28, 30, 36, 37, 46, 47, 51, 53, 58, 66, 76, 133
Mitogen 5, 6, 8, 14, 19, 29, 75, 127, 131, 134
Morphogen 2, 4-6, 8, 12, 13, 17, 23, 37, 44, 79, 81, 82, 85, 88, 114, 137, 139, 147
Morphogenesis 1, 2, 5, 7, 8, 39, 45, 46, 48, 51, 54, 69, 73-76, 81, 84, 86, 88, 89, 95, 96, 99, 102, 114, 116-118, 120-123, 125, 127-129, 131, 133, 134, 138-140, 142, 144, 149, 150, 152
Mouse 5-8, 13, 15-19, 23-31, 38, 46, 48, 49, 51, 63, 69, 70, 72, 74-76, 80-84, 86-89, 104, 107, 109-111, 113, 114, 116-121, 125, 127, 129, 131-134, 137-143, 147-152
Mouse mutant 8, 29, 87, 131, 133, 140

N

Neonate 121, 147
Nephron 133
Neurodevelopment 149, 151
Neurogenesis 18, 19, 40, 41, 58, 63-66
Notch 2, 14, 15, 17, 19, 111
Notochord 2, 5, 6, 13-19, 25, 37, 58, 95, 131

Index

O

Odontogenesis 51, 69-76
Odontogenic initiation 75, 76
Oligodendrocyte 13, 27-29, 36-41

P

Palate 45, 48, 49, 51-53, 142, 143
Palmitoylation 85, 86
Pancreas 8, 125, 126, 128, 131, 132
Patched (Ptc) 2-4, 12, 13, 18, 19, 23, 28, 29, 31, 54, 59, 63, 74, 84-86, 107, 110, 112, 120-122, 125, 138-140, 143, 144, 147, 149, 151, 152
Patterning zone (PZ) 100-102
Photoreceptor 7, 58-63
Polarization 69, 73, 76, 82, 83
PreTCR 109, 110
Progenitor 6, 13, 17, 19, 24, 25, 27, 28, 36-41, 58, 62, 63, 79, 99, 101, 107-109, 111-113, 127, 132, 134
Proliferation 4, 6-8, 13, 18, 19, 24, 28, 29, 31, 40, 44, 46, 48, 51, 58, 62, 63, 70, 73, 75, 76, 79, 81, 93, 94, 97, 99, 101, 102, 107, 109, 111, 113, 117, 120-122, 125, 127, 129, 131, 132, 134, 138, 140
Prostate 7, 8, 116-123
Prostatic bud 117-122
Proteoglycan 4, 12, 86, 149
Proximo-distal patterning 65, 66
Ptc1 3, 4, 30, 70, 71, 73-75, 100-102, 119, 120

R

Rat 7, 19, 63, 116, 121, 127, 147-151
Receptor 12, 23, 31, 39, 51, 59, 70, 71, 75, 83-86, 96, 102, 112, 117, 120, 125, 127, 133, 139, 143, 144, 147, 148, 151, 152
Receptor-associated protein (RAP) 148-151
Recombinant 37, 48, 51, 75, 110, 148
Regeneration 1, 8, 93, 96, 99-102, 104, 147
Repressor 2, 3, 5, 6, 12, 15, 23, 24, 30, 36, 40, 41, 84-87, 125, 129, 141
Retina 7, 58-66
Retinal pigmented epithelium (RPE) 58-64
Retinoic acid (RA) 46, 48, 49, 54, 66, 83, 85, 96, 97, 101, 102, 104, 127, 133
Rodent 62, 63, 70, 116-118, 120, 121

S

Scleroblast 95, 99-102, 104
Short-range 2, 12, 13, 17, 72, 85
Signalling 2-8, 12-19, 23, 26-31, 36-41, 44-46, 48-54, 58-66, 69-76, 79-89, 93, 94, 96, 97, 101, 102, 104, 107, 109-114, 116, 117, 119-123, 125, 127, 129, 131-134, 137, 139-144, 147, 149, 151, 152
Signalling network 85, 141, 144, 151
Skeleton 98, 139
Skinny hedgehog (Skn) 85, 86
Smo 2-4, 13, 14, 17, 19, 23, 26-29, 31, 48, 51, 76, 84, 85, 87, 107, 110, 112, 138, 139, 143, 149, 151
Smoothened 13, 23, 39, 48, 59, 64, 84, 97, 107, 110, 138, 139
Sonic hedgehog (Shh) 2, 4-8, 12-19, 23-31, 36-41, 44-54, 59-63, 66, 69-76, 79, 81-89, 93, 95-104, 107, 110-113, 116-123, 125-134, 137-144, 147-152
Spatial gradient 88
Spatiotemporal gradient 62
Spinal cord 5, 13, 17, 18, 23, 24, 26, 30, 31, 36-41, 61, 72
Stomach 7, 125, 129-132, 134, 141

T

T cell 107, 109-111, 113, 114
Telencephalon 19, 23-31, 37, 46, 48
Thymocyte 8, 107-113
Thymus 107-112
Tiggywinkle hedgehog (Twhh) 59, 61, 62, 93, 95, 97
Tissue interaction 44, 70, 73
Tongue 7, 51, 52
Tooth development 7, 51, 69, 71-76
Transcription factor 2, 3, 5, 7, 8, 12, 13, 16-18, 23, 25, 28, 36, 39, 61-64, 66, 74, 76, 83, 86, 98, 104, 107, 110, 120, 125, 129, 133, 139, 140
Transforming growth factor-β (TGF-β) 2, 102, 116, 121-123, 141
Transgenic 19, 25, 61, 62, 69, 82, 83, 116, 120, 127, 131-133, 141

U

Urogenital sinus (UGS) 116-121

V

Vertebrate 2, 4-8, 12-14, 16, 23, 24, 29, 39, 58, 59, 61, 63, 66, 69, 71, 72, 79, 80, 82, 84-87, 89, 93, 117

X

Xenopus 7, 8, 14, 15, 17, 18, 61, 66, 93

Z

Zebrafish (Zf) 2, 4, 7, 13, 14, 16, 17, 58-62, 64, 66, 83, 93-99, 101, 103, 104, 149

Zinc finger 3, 5, 86, 110, 140

Zone of polarizing activity (ZPA) 2, 6, 79-85, 87-89, 94-96